# Matrix Iterative
# Analysis

Prentice-Hall
Series in Automatic Computation
*George Forsythe, editor*

# Matrix Iterative Analysis

### RICHARD S. VARGA
*Professor of Mathematics*
*Case Institute of Technology*

PRENTICE-HALL, INC.
*Englewood Cliffs, New Jersey*

PRENTICE-HALL INTERNATIONAL INC., *London*
PRENTICE-HALL OF AUSTRALIA PTY., LTD., *Sydney*
PRENTICE-HALL OF CANADA, LTD., *Toronto*
PRENTICE-HALL OF JAPAN, INC., *Tokyo*

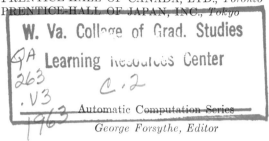

Automatic Computation Series

*George Forsythe, Editor*

Introduction to Algol; Bauer et al
Computers and their Uses; Desmonde
Matrix Iterative Analysis; Varga
Digital Processing; Schultz
Rounding Errors in Algebraic Processes; Wilkinson
Iterative Methods for the Solution of Equations; Traub

© 1962 by
Prentice-Hall, Inc.
Englewood Cliffs, N. J.

KANAWHA VALLEY GRADUATE CENTER

Current printing (last digit):
10   9   8   7   6   5   4

Library of Congress Catalog Number 62-21277
Printed in the United States of America
56550C

To Esther

# PREFACE

With the growth in speed and complexity of modern digital computers has come an increase in the use of computers by those who wish to find or approximate the solutions of partial differential equations in several varibles. This increasing use of computers for such problems has correspondingly interested many mathematicians in the underlying theory for that smaller branch of numerical analysis concerned with the efficient solution of matrix problems arising from discrete approximations to elliptic partial differential equations. This current interest has generated sufficient important mathematical contributions to warrant a survey of this mathematical theory. Accordingly, our first major aim is to survey the basic results pertaining to this topic.

The basic material for this book is closely aligned with modern computing methods. The author was fortunate to have been associated with the Mathematics Group of the Bettis Atomic Power Laboratory where very large matrix problems (of order 20,000 in two dimensions!) are solved on fast computers in the design of nuclear reactors. This valuable experience, gratefully acknowledged by the author, showed that present usage of computers to solve large scale elliptic partial differential equations is almost exclusively confined to *cyclic* iterative methods. In contrast, *non-cyclic* methods—such as Southwell's relaxation method, which has been widely used for many years on desk calculators—have received far less use on computers. Accordingly, we shall look only into the mathematical theory of cyclic iterative methods. Interestingly enough, the basis for the analysis of such modern cyclic iterative methods can be traced back to fundamental research by Perron and Frobenius on non-negative matrices, and our first aim is more nearly to survey the basic results on cyclic iterative methods, using the Perron-Frobenius theory as a basis.

The material given here is intended as a text for first year graduate students in mathematics. This material, an outgrowth of courses given at the University of Pittsburgh (1957–58) and Case Institute of Technology (1960–61), assumes familiarity with basic knowledge in matrix and linear algebra. For the most part, the material makes unstinting use of familiar matrix results and notations. But the author has not hesitated to introduce nonalgebraic items. For example, the useful notion of a *directed graph* is introduced early in Chapter 1 to help clarify the concept of irreducibility of a matrix. Later, it plays a useful role in deriving matrix properties of discrete approximations to elliptic partial differential equations. Similarly, the classical notion of *Padé rational approximations* of functions is used in Chapter 8 as a basis for generating numerical methods for parabolic partial differential equations.

To serve as an aid to the instructor using this material in the classroom, exercises are included after each section of a chapter. These often theoretically extend the material in the section. Occasionally, the exercises are numerical in nature; and even limited numerical experience will be of value to the reader.

A brief summary of the contents follows: Chapter 1 introduces vector and matrix norms, as well as directed graph theory and diagonally dominant matrices. Chapter 2 discusses the Perron-Frobenius theory of nonnegative matrices. The next three chapters are basically concerned with the analysis of variants of the successive overrelaxation (SOR) iterative method. Chapter 6 presents several viewpoints on the derivation of difference approximations to elliptic differential equations, including the Ritz form of the variational method. Chapter 7 is devoted to variants of the alternating direction implicit (ADI) methods. Chapter 8 investigates parabolic partial differential equations and obtains an association between the nature of basic iterative methods and parabolic partial differential equations. Chapter 9 treats theoretically the practical problem of the estimation of optimum iteration parameters. Finally, the two appendices contain numerical results.

While writing this manuscript, I have received valuable suggestions from many unselfish friends, colleagues, and students. To all, I give my sincere thanks. I especially want to thank Professors Garrett Birkhoff, David Young, George Forsythe, and Alston Householder and Raymond Nelson for their encouragement and helpful comments on early manuscripts. I also wish to thank R. Laurence Johnston and Louis A. Hageman, who carried out the numerical calculations; Harvey S. Price, who diligently checked all the exercises; and Martin Levy and William Roudebush, who carefully read the manuscript. Finally, sincere thanks are due to Mrs. Sarolta Petro, who, with great patience and fortitude, typed all the versions of the manuscript.

R.S.V.

# TABLE OF CONTENTS

# DEFINITIONS OF
# BASIC MATRIX PROPERTIES

# GLOSSARY OF SYMBOLS

$\mathbf{x}^T$    transpose of $\mathbf{x}$, 7

$\mathbf{x}^*$    conjugate transpose of $\mathbf{x}$, 7

$\|\mathbf{x}\|$    Euclidean norm of $\mathbf{x}$, 8

$\rho(A)$    spectral radius of $A$, 9

$\|A\|$    spectral norm of $A$, 9

$A^T$    transpose of $A$, 11

$A^*$    conjugate transpose of $A$, 11

$\|\mathbf{x}\|_1$    $l_1$-norm of $\mathbf{x}$, 15

$\|\mathbf{x}\|_\infty$    $l_\infty$-norm of $\mathbf{x}$, 15

$\|A\|_1$    $l_1$-norm of $A$, 15

$\|A\|_\infty$    $l_\infty$-norm of $A$, 15

$\det B$    determinant of $B$, 26

$O$    null matrix, 26

$\phi(t)$    characteristic polynomial, 31

$\gamma(A)$    index of primitivity of $A$, 42

$\mathrm{tr}(A)$    trace of $A$, 44

$R(A^m)$    average rate of convergence, 62

$R_\infty(A)$    asymptotic rate of convergence, 67

$\exp(A)$    exponential of $A$, 87

# CHAPTER 1

# MATRIX PROPERTIES AND CONCEPTS

## 1.1. INTRODUCTION

The title of this book, *Matrix Iterative Analysis*, suggests that we might consider here all matrix numerical methods which are iterative in nature. However, such an ambitious goal is in fact replaced by the more practical one where we seek to consider in some detail that smaller branch of numerical analysis concerned with the efficient solution, by means of iteration, of matrix equations arising from discrete approximations to partial differential equations. These matrix equations are generally characterized by the property that the associated square matrices are *sparse*, i.e., a large percentage of the entries of these matrices are zero. Furthermore, the nonzero entries of these matrices occur in some natural pattern, which, relative to a digital computer, permits even very large-order matrices to be efficiently stored. Cyclic iterative methods are ideally suited for such matrix equations, since each step requires relatively little digital computer storage or arithmetic computation. As an example of the magnitude of problems that have been successfully solved on digital computers by cyclic iterative methods, the Bettis Atomic Power Laboratory of the Westinghouse Electric Corporation had in daily use in 1960 a two-dimensional program which would treat as a special case, Laplacian-type matrix equations of order 20,000.†

The idea of solving large systems of linear equations by iterative methods is certainly not new, dating back at least to Gauss (1823). Later, Southwell (1946) and his school gave real impetus to the use of iterative methods when they systematically considered the numerical solution of

---

† This program, called "PDQ-4," was specifically written for the Philco-2000 computer with 32,000 words of core storage. Even more staggering is Bettis' use of a three-dimensional program, "TNT-1," which treats coupled matrix equations of order 108,000.

practical physics and engineering problems. The iterative method of *relaxation* advocated by Southwell, a *noncyclic* iterative method, was successfully used for many years by those who used either pencil and paper or desk calculators to carry out the necessary arithmetical steps, and this method was especially effective when human insight guided the entire course of the computations. With the advent of large-scale digital computers, this human insight was generally difficult to incorporate efficiently into computer programs. Accordingly, mathematicians began to look for ways of accelerating the convergence of basic *cyclic* or systematic iterative methods, methods which when initially prescribed are *not* to be altered in the course of solving matrix equations—in direct contrast with the noncyclic methods. We will concern ourselves here only with cyclic iterative methods (which for brevity we call *iterative methods*); the theory and applications of noncyclic iterative methods have been quite adequately covered elsewhere,† and these latter iterative methods generally are not used on large digital computers.

The basis for much of the present activity in this area of numerical analysis concerned with cyclic iterative methods is a series of papers by Frankel (1950), Geiringer (1949), Reich (1949), Stein and Rosenberg (1948), and Young (1950), all of which appeared when digital computers were emerging with revolutionary force. Because of the great impact of these papers on the stream of current research in this area, we have found it convenient to define *modern* matrix iterative analysis as having begun in about 1948 with the work of the above-mentioned authors. Starting at this point, our *first aim* is to describe the basic results of modern matrix iterative analysis from its beginning to the present.

We have presupposed here a basic knowledge of matrix and linear algebra theory, material which is thoroughly covered in the outstanding books by Birkhoff and MacLane (1953), Faddeeva (1959) and Bellman (1960). Thus, the reader is assumed to know, for example, what the Jordan normal form of a square complex matrix is.

Except for several isolated topics, which can be read independently, our *second aim* is to have the material here reasonably self-contained and complete. As we shall see, our development of matrix iterative analysis depends fundamentally on the early research of Perron (1907) and Frobenius (1908–12) on matrices with non-negative entries; thus, our first aim is not only to describe the basic results in this field, but also to use the Perron-Frobenius theory of non-negative matrices as a foundation for the exposition of these results. With the goal of having the material self-contained, we have devoted Chapter 2 to the Perron-Frobenius theory, although recently an excellent book by Gantmakher (1959) has also devoted a chapter to this topic.

† References are given in the Bibliography and Discussion at the end of this chapter.

Our *third aim* is to present sufficient numerical detail for those who are ultimately interested in the practical applications of the theory to the numerical solution of partial differential equations. To this end, included in Appendices A and B are illustrative examples which show the transition through the stages from problem formulation, derivation of matrix equations, application of various iterative methods, to the final examination of numerical results typical of digital computer output. Those interested in actual numerical applications are strongly urged to carry through in detail the examples presented in these Appendices. We have also included exercises for the reader in each chapter; these not only test the mastery of the material of the chapter, but in many cases allow us to indicate interesting theoretical results and extensions which have not been covered in the text. Starred exercises may require more effort on the part of the reader.

The material in this book is so organized that the general derivation of matrix equations (Chapter 6) from self-adjoint elliptic partial differential equations is not discussed until a large body of theory has been presented. The unsuspecting reader may feel he has been purposely burdened with a great number of "unessential" (from the numerical point of view) theorems and lemmas before any applications have appeared. In order to ease this burden, and to give motivation to this theory, in the next section we shall consider an especially simple example arising from the numerical solution of the Dirichlet problem showing how non-negative matrices occur naturally. Finally, the remainder of Chapter 1 deals with some fundamental concepts and results of matrix numerical analysis.

There are several important associated topics which for reasons of space are only briefly mentioned. The analysis of the effect of rounding errors and the question of convergence of the discrete solution of a system of linear equations to the continuous solution of the related partial differential equation as the mesh size tends to zero in general require mathematical tools which are quite different from those used in the matrix analysis of iterative methods. We have listed important references for these topics in the Bibliography and Discussion for this chapter.

## 1.2. A SIMPLE EXAMPLE

We now consider the numerical solution of the Dirichlet problem for the unit square, i.e., we seek approximations to the function $u(x, y)$ defined in the closed unit square which satisfies Laplace's equation

$$(1.1) \quad \frac{\partial^2 u(x, y)}{\partial x^2} + \frac{\partial^2 u(x, y)}{\partial y^2} = u_{xx}(x, y) + u_{yy}(x, y) = 0,$$

$$0 < x, y < 1,$$

in the interior of the unit square. If $\Gamma$ denotes the boundary of the square, then in addition to the differential equation of (1.1), $u(x, y)$ is to satisfy the Dirichlet boundary condition

$$(1.2) \qquad\qquad u(x, y) = g(x, y), \qquad (x, y) \in \Gamma,$$

where $g(x, y)$ is some specified function defined on $\Gamma$. We now impose a uniform square mesh of side $h = \frac{1}{3}$ on this unit square, and we number the interior and boundary intersections (mesh points) of the horizontal and vertical line segments by means of appropriate subscripts, as shown in Figure 1. Instead of attempting to find the function $u(x, y)$ satisfying (1.1) for *all* $0 < x, y < 1$ and the boundary condition of (1.2), we seek only approximations to this function $u(x, y)$ at just the interior mesh points of the unit square. Although there are a number of different ways (Chapter 6) of finding such approximations of $u(x, y)$, one simple procedure begins by expanding the function $u(x, y)$ in a Taylor's series in two variables. Assuming that $u(x, y)$ is sufficiently differentiable, then

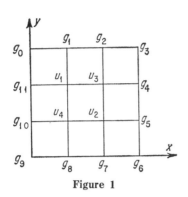

**Figure 1**

$$(1.3) \qquad u(x_0 \pm h, y_0) = u(x_0, y_0) \pm h u_x(x_0, y_0) + \frac{h^2}{2} u_{xx}(x_0, y_0)$$

$$\pm \frac{h^3}{3!} u_{xxx}(x_0, y_0) + \frac{h^4}{4!} u_{xxxx}(x_0, y_0) \pm \cdots,$$

$$(1.3') \qquad u(x_0, y_0 \pm h) = u(x_0, y_0) \pm h u_y(x_0, y_0) + \frac{h^2}{2} u_{yy}(x_0, y_0)$$

$$\pm \frac{h^3}{3!} u_{yyy}(x_0, y_0) + \frac{h^4}{4!} u_{yyyy}(x_0, y_0) \pm \cdots,$$

where the point $(x_0, y_0)$ and its four neighboring points $(x_0 \pm h, y_0)$, $(x_0, y_0 \pm h)$ are points of the closed unit square. We find then that

$$(1.4) \qquad \frac{1}{h^2} \{ u(x_0 + h, y_0) + u(x_0 - h, y_0) + u(x_0, y_0 + h)$$

$$+ u(x_0, y_0 - h) - 4u(x_0, y_0) \}$$

$$= \{ u_{xx}(x_0, y_0) + u_{yy}(x_0, y_0) \}$$

$$+ \frac{h^2}{12} \{ u_{xxxx}(x_0, y_0) + u_{yyyy}(x_0, y_0) \} + \cdots.$$

From (1.1), the first term of the right side of (1.4) is zero, and if we neglect terms with coefficients $h^2$ or higher, we have approximately

(1.4') $\qquad u(x_0, y_0) \doteqdot \frac{1}{4}\{u(x_0 + h, y_0) + u(x_0 - h, y_0) + u(x_0, y_0 + h)$

$$+ u(x_0, y_0 - h)\}.$$

If we let

$$u_1 \equiv u(\tfrac{1}{3}, \tfrac{2}{3}), \quad u_2 \equiv u(\tfrac{2}{3}, \tfrac{1}{3}), \quad u_3 \equiv u(\tfrac{2}{3}, \tfrac{2}{3}), \quad \text{and} \quad u_4 \equiv u(\tfrac{1}{3}, \tfrac{1}{3}),$$

and similarly $g_9$ is the value of the specified function $g(x, y)$ at the origin $x = y = 0$, etc., we now define respectively approximations $w_i$ for the values $u_i$, $1 \le i \le 4$, by means of (1.4'):

(1.5)
$$w_1 = \tfrac{1}{4}(w_3 + w_4 + g_1 + g_{11}),$$
$$w_2 = \tfrac{1}{4}(w_3 + w_4 + g_5 + g_7),$$
$$w_3 = \tfrac{1}{4}(w_1 + w_2 + g_2 + g_4),$$
$$w_4 = \tfrac{1}{4}(w_1 + w_2 + g_8 + g_{10}),$$

which are then four linear equations in the four unknowns $w_i$, each equation representing the approximate value of the unknown function $u(x, y)$ at an interior mesh point as an average of the approximate values of $u(x, y)$ at neighboring mesh points. In matrix notation, (1.5) can be written as

(1.5') $\qquad\qquad\qquad A\mathbf{w} = \mathbf{k},$

where

(1.6) $\quad A = \begin{bmatrix} 1 & 0 & -\tfrac{1}{4} & -\tfrac{1}{4} \\ 0 & 1 & -\tfrac{1}{4} & -\tfrac{1}{4} \\ -\tfrac{1}{4} & -\tfrac{1}{4} & 1 & 0 \\ -\tfrac{1}{4} & -\tfrac{1}{4} & 0 & 1 \end{bmatrix}, \quad \mathbf{w} = \begin{bmatrix} w_1 \\ w_2 \\ w_3 \\ w_4 \end{bmatrix}, \quad \text{and} \quad \mathbf{k} = \frac{1}{4}\begin{bmatrix} g_1 + g_{11} \\ g_5 + g_7 \\ g_2 + g_4 \\ g_8 + g_{10} \end{bmatrix}.$

Here, $\mathbf{k}$ is a vector whose components can be calculated from the known boundary values $g_i$. Now, it is obvious that the matrix $A$ can be written as $I - B$, where

(1.7) $\qquad\qquad\qquad B = \frac{1}{4}\begin{bmatrix} 0 & 0 & 1 & 1 \\ 0 & 0 & 1 & 1 \\ 1 & 1 & 0 & 0 \\ 1 & 1 & 0 & 0 \end{bmatrix}.$

Evidently, both the matrices $A$ and $B$ are real and symmetric, and it is clear that the entries of the matrix $B$ are all non-negative real numbers. The characteristic polynomial of the matrix $B$ turns out to be simply

$$(1.8) \qquad \phi(\mu) = \det (\mu I - B) = \mu^2(\mu^2 - \tfrac{1}{4}),$$

so that the eigenvalues of $B$ are $\mu_1 = -\tfrac{1}{2}$, $\mu_2 = 0 = \mu_3$, and $\mu_4 = \tfrac{1}{2}$, and thus

$$\max_{1 \leq i \leq 4} | \mu_i | = \frac{1}{2}.$$

Since the eigenvalues $\nu_i$ of $A$ are of the form $1 - \mu_i$, the eigenvalues of $A$ are evidently positive real numbers, and it follows that $A$ is a real, symmetric, and positive definite matrix. As the matrix $A$ is nonsingular, its inverse matrix $A^{-1}$ is uniquely defined and is given explicitly by

$$(1.9) \qquad A^{-1} = \frac{1}{6} \begin{bmatrix} 7 & 1 & 2 & 2 \\ 1 & 7 & 2 & 2 \\ 2 & 2 & 7 & 1 \\ 2 & 2 & 1 & 7 \end{bmatrix},$$

and thus the entries of the matrix $A^{-1}$ are all positive real numbers. We shall see later (in Chapter 6) that these simple conclusions, such as the matrix $B$ having its eigenvalues in modulus less than unity and the matrix $A^{-1}$ having only positive real entries, hold quite generally for matrix equations derived from self-adjoint second-order elliptic partial differential equations.

Since we can write the matrix equation (1.5′) equivalently as

$$(1.10) \qquad \mathbf{w} = B\mathbf{w} + \mathbf{k},$$

we can now generate for this simple problem our first (cyclic) iterative method, called the *point Jacobi* or *point total-step method*.† If $\mathbf{w}^{(0)}$ is an arbitrary real or complex vector approximation of the unique (since $A$ is nonsingular) solution vector $\mathbf{w}$ of (1.5′), then we successively define a sequence of vector iterates $\mathbf{w}^{(m)}$ from

$$(1.11) \qquad \mathbf{w}^{(m+1)} = B\mathbf{w}^{(m)} + \mathbf{k}, \qquad m \geq 0.$$

The first questions we would ask concern the convergence of (1.11), i.e., does each $\lim_{m \to \infty} w_j^{(m)}$ exist, and assuming these limits exist, does each limit

---

† Other names are also associated with this iterative method. See Sec. 3.1.

equal $w_j$ for every component $j$? To begin to answer this, let

$$\varepsilon^{(m)} \equiv \mathbf{w}^{(m)} - \mathbf{w}, \qquad m \geq 0,$$

where $\varepsilon^{(m)}$ is the *error vector* associated with the vector iterate $\mathbf{w}^{(m)}$. Subtracting (1.10) from (1.11), we obtain

$$\varepsilon^{(m+1)} = B\varepsilon^{(m)},$$

from which it follows inductively that

(1.12) $$\varepsilon^{(m)} = B^m \varepsilon^{(0)}, \qquad m \geq 0.$$

For any component $j$, it is clear that $\lim_{m \to \infty} \epsilon_j^{(m)}$ exists if and only if $\lim_{m \to \infty} w_j^{(m)}$ exists, and if these limits both exist then $\lim_{m \to \infty} w_j^{(m)} = w_j$ if and only if $\lim_{m \to \infty} \epsilon_j^{(m)} = 0$. Therefore, with (1.12), if we wish each component of the error vector to vanish in the limit, we seek conditions which insure that

(1.13) $$\lim_{m \to \infty} B^m \varepsilon^{(0)} = \mathbf{0},$$

for *all* vectors $\varepsilon^{(0)}$. But seeking conditions to insure (1.13) is equivalent to determining when

(1.14) $$\lim_{m \to \infty} B^m = O,$$

where $O$ is the null $n \times n$ matrix. This will be discussed in the next section.

## 1.3 NORMS AND SPECTRAL RADII

The concepts of vector norms, matrix norms, and the spectral radii of matrices play an important role in iterative numerical analysis. Just as it is convenient to compare two vectors in terms of their lengths, it will be similarly convenient to compare two matrices by some measure or norm. As we shall see, this will be the basis for deciding which of two iterative methods is more rapidly convergent, in some precise sense.

To begin with, let $V_n(C)$ be the $n$-dimensional vector space over the field of complex numbers $C$ of column vectors $\mathbf{x}$, where the vector $\mathbf{x}$, its transpose $\mathbf{x}^T$, and its conjugate transpose $\mathbf{x}^*$ are denoted by

$$\mathbf{x} = \begin{bmatrix} x_1 \\ x_2 \\ \cdot \\ \cdot \\ \cdot \\ x_n \end{bmatrix}, \qquad \mathbf{x}^T = [x_1 \ x_2 \ \cdots \ x_n], \qquad \mathbf{x}^* = [\bar{x}_1 \ \bar{x}_2 \ \cdots \ \bar{x}_n],$$

where $x_1, x_2, \cdots, x_n$ are complex numbers, and $\bar{x}_i$ is the complex conjugate of $x_i$.

DEFINITION 1.1. Let $\mathbf{x}$ be a (column) vector of $V_n(C)$. Then,

$$
(1.15) \qquad \| \mathbf{x} \| \equiv (\mathbf{x}^*\mathbf{x})^{1/2} = \left( \sum_{i=1}^{n} | x_i |^2 \right)^{1/2}
$$

is the *Euclidean norm* (or length) of $\mathbf{x}$.

With this definition, the following results are well known.

**Theorem 1.1.**   *If* $\mathbf{x}$ *and* $\mathbf{y}$ *are vectors of* $V_n(C)$, *then*

$$
\| \mathbf{x} \| > 0, \quad \text{unless } \mathbf{x} = 0;
$$

$$
(1.16) \qquad \text{if } \alpha \text{ is a scalar, then } \| \alpha\mathbf{x} \| = | \alpha | \cdot \| \mathbf{x} \|;
$$

$$
\| \mathbf{x} + \mathbf{y} \| \leq \| \mathbf{x} \| + \| \mathbf{y} \|.
$$

If we have an infinite sequence $\mathbf{x}^{(0)}, \mathbf{x}^{(1)}, \mathbf{x}^{(2)}, \cdots$ of vectors of $V_n(C)$, we say that this sequence *converges* to a vector $\mathbf{x}$ of $V_n(C)$ if

$$
\lim_{m \to \infty} x_j^{(m)} = x_j, \qquad \text{for all } 1 \leq j \leq n,
$$

where $x_j^{(m)}$ and $x_j$ are respectively the $j$th components of the vectors $\mathbf{x}^{(m)}$ and $\mathbf{x}$. Similarly, by the convergence of an infinite series $\sum_{m=0}^{\infty} \mathbf{y}^{(m)}$ of vectors of $V_n(C)$ to a vector $\mathbf{y}$ of $V_n(C)$, we mean that

$$
\lim_{N \to \infty} \sum_{m=0}^{N} y_j^{(m)} = y_j, \qquad \text{for all } 1 \leq j \leq n.
$$

In terms of Euclidean norms, it then follows from Definition 1.1 that

$$
\| \mathbf{x}^{(m)} - \mathbf{x} \| \to 0, \qquad m \to \infty,
$$

if and only if the sequence $\mathbf{x}^{(0)}, \mathbf{x}^{(1)}, \cdots$ of vectors converges to the vector $\mathbf{x}$, and similarly

$$
\left\| \sum_{m=0}^{N} \mathbf{y}^{(m)} - \mathbf{y} \right\| \to 0, \qquad N \to \infty,
$$

if and only if the infinite series $\sum_{m=0}^{\infty} \mathbf{y}^{(m)}$ converges to the vector $\mathbf{y}$.

Our next basic definition, which will be repeatedly used in subsequent developments, is

DEFINITION 1.2. Let $A = (a_{i,j})$ be an $n \times n$ complex matrix with eigenvalues $\lambda_i$, $1 \leq i \leq n$. Then

$$(1.17) \qquad \rho(A) \equiv \max_{1 \leq i \leq n} |\lambda_i|$$

is the *spectral radius* of the matrix $A$.

Geometrically, if all the eigenvalues $\lambda_i$ of $A$ are plotted in the complex $z$-plane, then $\rho(A)$ is the radius of the smallest disk† $|z| \leq R$, with center at the origin, which includes all the eigenvalues of the matrix $A$.

Now, we shall assign to each $n \times n$ matrix $A$ with complex entries a non-negative real number which, like the vector norm $\| \mathbf{x} \|$, has properties of length similar to those of (1.16).

DEFINITION 1.3. If $A = (a_{i,j})$ is an $n \times n$ complex matrix, then

$$(1.18) \qquad \| A \| = \sup_{\mathbf{x} \neq 0} \frac{\| A\mathbf{x} \|}{\| \mathbf{x} \|}.$$

is the *spectral norm* of the matrix $A$.

Basic properties of the spectral norm of a matrix, analogous to those obtained for the Euclidean norm of the vector $\mathbf{x}$, are given in

**Theorem 1.2.** *If $A$ and $B$ are two $n \times n$ matrices, then*

$$\| A \| > 0, \text{ unless } A \equiv O, \text{ the null matrix;}$$

$$\text{if } \alpha \text{ is a scalar, } \| \alpha A \| = | \alpha | \cdot \| A \|;$$

$$(1.19)$$

$$\| A + B \| \leq \| A \| + \| B \|;$$

$$\| A \cdot B \| \leq \| A \| \cdot \| B \|.$$

*Moreover,*

$$(1.20) \qquad \| A\mathbf{x} \| \leq \| A \| \cdot \| \mathbf{x} \|$$

*for all vectors $\mathbf{x}$, and there exists a nonzero vector $\mathbf{y}$ in $V_n(C)$ for which*

$$(1.21) \qquad \| A\mathbf{y} \| = \| A \| \cdot \| \mathbf{y} \|.$$

*Proof.* The results of (1.19) and (1.20) follow directly from Theorem 1.1 and Definition 1.3. To establish (1.21), observe that the ratio $\| A\mathbf{x} \| / \| \mathbf{x} \|$ is unchanged if $\mathbf{x}$ is replaced by $\alpha\mathbf{x}$, where $\alpha$ is a scalar. Hence, we can write that

$$\| A \| = \sup_{\|\mathbf{x}\|=1} \| A\mathbf{x} \|.$$

† To be precise, the set of points for which $|z - a| \leq R$ is called a *disk*, whereas its subset, defined by $|z - a| = R$, is called a *circle*.

But as the set of all vectors $\mathbf{x}$ with $\| \mathbf{x} \| = 1$ is compact, and $A\mathbf{x}$ is a continuous function defined on this set, then there exists a vector $\mathbf{y}$ with $\| \mathbf{y} \| = 1$ such that

$$\| A \| = \sup_{\|\mathbf{x}\|=1} \| A\mathbf{x} \| = \| A\mathbf{y} \|,$$

which completes the proof.

To connect the spectral norm and spectral radius of Definitions 1.2 and 1.3, we have the

**Corollary.** *For an arbitrary $n \times n$ complex matrix $A$,*

$$(1.22) \qquad\qquad \| A \| \geq \rho(A).$$

*Proof.* If $\lambda$ is any eigenvalue of $A$, and $\mathbf{x}$ is an eigenvector associated with the eigenvalue $\lambda$, then $A\mathbf{x} = \lambda\mathbf{x}$. Thus, from Theorems 1.1 and 1.2,

$$| \lambda | \cdot \| \mathbf{x} \| = \| \lambda\mathbf{x} \| = \| A\mathbf{x} \| \leq \| A \| \cdot \| \mathbf{x} \| ,$$

from which we conclude that $\| A \| \geq | \lambda |$ for *all* eigenvalues of $A$, which proves (1.22).

In the terminology of Faddeeva (1959) and Householder (1958), (1.20) states that the spectral norm of a matrix is *consistent* with the Euclidean vector norm, and (1.21) states that the spectral norm of a matrix is *subordinate* to the Euclidean vector norm. There are many other ways of defining vector and matrix norms that satisfy the properties of Theorems 1.1 and 1.2. Although some other definitions of norms are given in the exercises for this section, we have concentrated only on the Euclidean vector norm and the matrix spectral norm, as these will generally be adequate for our future purposes.

Matrix spectral norms can also be expressed in terms of the spectral radii. If

$$A = \begin{bmatrix} a_{1,1} & a_{1,2} & \cdots & a_{1,n} \\ a_{2,1} & a_{2,2} & \cdots & a_{2,n} \\ \cdot & & & \cdot \\ \cdot & & & \cdot \\ \cdot & & & \cdot \\ a_{n,1} & a_{n,2} & \cdots & a_{n,n} \end{bmatrix}, \quad A^T = \begin{bmatrix} a_{1,1} & a_{2,1} & \cdots & a_{n,1} \\ a_{1,2} & a_{2,2} & \cdots & a_{n,2} \\ \cdot & & & \cdot \\ \cdot & & & \cdot \\ \cdot & & & \cdot \\ a_{1,n} & a_{2,n} & \cdots & a_{n,n} \end{bmatrix},$$

$$A^* = \begin{bmatrix} \bar{a}_{1,1} & \bar{a}_{2,1} & \cdots & \bar{a}_{n,1} \\ \bar{a}_{1,2} & \bar{a}_{2,2} & \cdots & \bar{a}_{n,2} \\ \cdot & & & \cdot \\ \cdot & & & \cdot \\ \cdot & & & \cdot \\ \bar{a}_{1,n} & \bar{a}_{2,n} & \cdots & \bar{a}_{n,n} \end{bmatrix}$$

denote, respectively, the $n \times n$ matrix $A$ with complex entries $a_{i,j}$, its *transpose* $A^T$, and its *conjugate transpose* $A^*$, then the matrix product $A^*A$ is also an $n \times n$ matrix.

**Theorem 1.3.** *If* $A = (a_{i,j})$ *is an* $n \times n$ *complex matrix, then*

(1.23) $$\| A \| = [\rho(A^*A)]^{1/2}.$$

*Proof.* The matrix $A^*A$ is a Hermitian and non-negative definite matrix, i.e.,

$$(A^*A)^* = A^*A \quad \text{and} \quad \mathbf{x}^*A^*A\mathbf{x} = \| A\mathbf{x} \|^2 \geq 0$$

for any vector $\mathbf{x}$. As $A^*A$ is Hermitian, let $\{ \alpha_i \}_{i=1}^n$ be an orthogonal set of eigenvectors of $A^*A$, i.e., $A^*A\,\alpha_i = \nu_i\alpha_i$ where $0 \leq \nu_1 \leq \nu_2 \leq \cdots \leq \nu_n$, and $\alpha_i^*\alpha_j = 0$ for $i \neq j$, and $\alpha_i^*\alpha_i = 1$ for all $1 \leq i, j \leq n$. If

$$\mathbf{x} = \sum_{i=1}^n c_i\alpha_i$$

is any nonzero vector, then by direct computation,

$$\left(\frac{\| A\mathbf{x} \|}{\| \mathbf{x} \|}\right)^2 = \frac{\mathbf{x}^*A^*A\mathbf{x}}{\mathbf{x}^*\mathbf{x}} = \frac{\displaystyle\sum_{i=1}^n | c_i |^2\nu_i}{\displaystyle\sum_{j=1}^n | c_j |^2},$$

so that

$$0 \leq \nu_1 \leq \left(\frac{\| A\mathbf{x} \|}{\| \mathbf{x} \|}\right)^2 \leq \nu_n.$$

Moreover, choosing $\mathbf{x} = \alpha_n$ shows that equality is possible on the right. Thus, from Definition 1.3,

$$\| A \|^2 = \sup_{\mathbf{x} \neq 0} \left(\frac{\| A\mathbf{x} \|}{\| \mathbf{x} \|}\right)^2 = \nu_n = \rho(A^*A),$$

which completes the proof.

**Corollary.** *If* $A$ *is an* $n \times n$ *Hermitian matrix, then*

(1.24) $$\| A \| = \rho(A).$$

*Moreover, if* $g_m(x)$ *is any real polynomial of degree* $m$ *in* $x$, *then*

(1.24′) $$\| g_m(A) \| = \rho(g_m(A)).$$

*Proof.* If $A$ is Hermitian, then by definition $A^* = A$. Thus,

$$\| A \|^2 = \rho(A^*A) = \rho(A^2) = \rho^2(A),$$

proving that $\| A \| = \rho(A)$. But the general case now follows quickly, since, if $g_m(x)$ is a real polynomial in the variable $x$, it is easy to verify that the matrix $g_m(A)$ is also Hermitian, which proves (1.24′).

In particular, we now know that the spectral norm and the spectral radius of a Hermitian matrix are necessarily equal. This is not true for general matrices. To show that equality is *not* in general valid in (1.22), it suffices to consider the simple matrix

$$A = \begin{pmatrix} \alpha & 1 \\ 0 & \alpha \end{pmatrix},$$

where $\alpha$ is any complex number. As $A$ is a triangular matrix with diagonal entries equal to $\alpha$, its eigenvalues are then equal to $\alpha$ so that $\rho(A) = |\alpha|$. On the other hand, direct computation shows that

$$A^*A = \begin{bmatrix} |\alpha|^2 & \bar{\alpha} \\ \alpha & |\alpha|^2 + 1 \end{bmatrix},$$

which has eigenvalues $|\alpha|^2 + \frac{1}{2} \pm \sqrt{|\alpha|^2 + \frac{1}{4}}$, so that

$$\| A \| = \{|\alpha|^2 + \tfrac{1}{2} + \sqrt{|\alpha|^2 + \tfrac{1}{4}}\}^{1/2} > |\alpha| = \rho(A).$$

Analogous to our definitions of convergence of vector sequences and series, we can similarly define what we mean by the convergence of matrix sequences and series. If $A^{(0)} = (a_{i,j}^{(0)})$, $A^{(1)} = (a_{i,j}^{(1)})$, $\cdots$ is an infinite sequence of $n \times n$ complex matrices, we say that this sequence *converges* to an $n \times n$ complex matrix $A = (a_{i,j})$ if

$$\lim_{m \to \infty} a_{i,j}^{(m)} = a_{i,j} \qquad \text{for all } 1 \leq i, j \leq n.$$

Similarly, by the convergence of an infinite series $\sum\limits_{m=0}^{\infty} B^{(m)}$ of $n \times n$ complex matrices $B^{(m)} = (b_{i,j}^{(m)})$ to the matrix $B = (b_{i,j})$, we mean

$$\lim_{N \to \infty} \sum_{m=0}^{N} b_{i,j}^{(m)} = b_{i,j} \qquad \text{for all } 1 \leq i, j \leq n.$$

In terms of spectral norms, it follows from Theorem 1.2 that

$$\| A^{(m)} - A \| \to 0 \quad \text{as} \quad m \to \infty$$

if and only if the sequence $A^{(0)}$, $A^{(1)}$, $\cdots$ of matrices converges to the matrix $A$, and similarly,

$$\left\| \sum_{m=0}^{N} B^{(m)} - B \right\| \to 0 \quad \text{as} \quad N \to \infty$$

if and only if the infinite series $\sum_{m=0}^{\infty} B^{(m)}$ converges to the matrix $B$.

We now return to the problem posed in the previous section. If $A$ is an $n \times n$ complex matrix, when does the sequence $A$, $A^2$, $A^3$, $\cdots$ of powers of the matrix $A$ converge to the null matrix $O$? As sequences of this form are basic to our development, we now make the

DEFINITION 1.4. Let $A$ be $n \times n$ complex matrix. Then, $A$ is *convergent* (to zero) if the sequence of matrices $A$, $A^2$, $A^3$, $\cdots$ converges to the null matrix $O$, and is *divergent* otherwise.

**Theorem 1.4.** *If $A$ is an $n \times n$ complex matrix, then $A$ is convergent if and only if $\rho(A) < 1$.*

*Proof.* For the given matrix $A$, there exists a nonsingular $n \times n$ matrix $S$ which reduces the matrix $A$ to its Jordan normal form,† i.e.,

$$(1.25) \qquad SAS^{-1} \equiv \tilde{A} \equiv \begin{bmatrix} J_1 & & & & \\ & J_2 & & & O \\ & & \ddots & & \\ & & & \ddots & \\ O & & & & J_r \end{bmatrix},$$

where each of the $n_l \times n_l$ submatrices $J_l$ has the form

$$(1.26) \qquad J_l = \begin{bmatrix} \lambda_l & 1 & & & \\ & \lambda_l & 1 & & O \\ & & \lambda_l & 1 & \\ & & & \ddots & \ddots \\ & O & & & 1 \\ & & & & \lambda_l \end{bmatrix}, \qquad 1 \le l \le r.$$

† See, for example, Birkhoff and MacLane (1953, p. 334) for a proof of this basic result.

Since each submatrix $J_l$ is upper triangular, so is $\tilde{A}$, and thus the set $\{\lambda_l\}_{l=1}^r$ includes all the distinct eigenvalues of the matrices $A$ and $\tilde{A}$, which are similar matrices from (1.25). By direct computation with (1.25),

$$
(1.27) \quad (\tilde{A})^m =
\begin{bmatrix}
J_1^m & & & & & \\
 & J_2^m & & & & \\
 & & \cdot & & & \\
 & & & \cdot & & \\
 & & & & \cdot & \\
 & & & & & J_r^m
\end{bmatrix}, \quad m \geq 1.
$$

Because of the special form of the matrices $J_l$ of (1.26), it is easy to determine the entries of the powers of the matrix $J_l$. Indeed,

$$
J_l^2 =
\begin{bmatrix}
\lambda_l^2 & 2\lambda_l & 1 & & \\
 & \lambda_l^2 & 2\lambda_l & 1 & \\
 & & & & \\
 & & & 2\lambda_l & \\
 & & & & \lambda_l^2
\end{bmatrix}
\quad \text{if } n_l \geq 3,
$$

and in general, if $J_l^m \equiv (d_{i,j}^{(m)}(l))$, $1 \leq i, j \leq n_l$, then

$$
(1.28) \quad d_{i,j}^{(m)}(l) =
\begin{cases}
0, & j < i, \\[2mm]
\dbinom{m}{j-i} \lambda_l^{m-i+i} & \text{for } i \leq j \leq \min(n_l, m+i), \\[2mm]
0, & m+i < j \leq n_l,
\end{cases}
$$

where

$$
\binom{m}{k} = \frac{m!}{k!(m-k)!}
$$

is the familiar binomial coefficient. If $A$ is convergent, then $A^m \to O$ as $m \to \infty$. But as $(\tilde{A})^m = SA^m S^{-1}$, it follows that $(\tilde{A})^m \to O$ as $m \to \infty$. Consequently, each $(J_l)^m \to O$ as $m \to \infty$, so that the diagonal entries

$\lambda_l$ of $J_l$ must satisfy $|\lambda_l| < 1$ for all $1 \le l \le r$. Clearly,

$$\rho(A) = \rho(\tilde{A}) = \max_{1 \le l \le r} |\lambda_l| < 1,$$

which proves the first part. On the other hand, if $\rho(A) = \rho(\tilde{A}) < 1$, then $|\lambda_l| < 1$ for all $1 \le l \le r$. By making direct use of (1.28) and the fact that $|\lambda_l| < 1$, it follows that

$$\lim_{m \to \infty} d_{i,j}^{(m)}(l) = 0, \qquad \text{for all } 1 \le i, j \le n_l,$$

so that each $J_l$ is convergent. Thus, from (1.27), $\tilde{A}$ is also convergent, and finally, as $A^m = S^{-1}(\tilde{A})^m S$, we have that the matrix $A$ is convergent, completing the proof.

If $A$ is an $n \times n$ convergent matrix, we thus far know only that $\| A^m \| \to 0$ as $m \to \infty$. More precise results on the behavior of $\| A^m \|$ as a function of $m$ will be obtained in Sec. 3.2.

## EXERCISES

**1.** If $\mathbf{x}$ is any column vector with complex components $x_1, x_2, \cdots, x_n$, let

$$\| \mathbf{x} \|_1 \equiv \sum_{i=1}^{n} |x_i| \quad \text{and} \quad \| \mathbf{x} \|_\infty \equiv \max_{1 \le i \le n} |x_i|.$$

Show that the results (1.16) of Theorem 1.1 are satisfied for these definitions of $\| \mathbf{x} \|$.

**2.** If $A = (a_{i,j})$ is any $n \times n$ complex matrix, let

$$\| A \|_1 \equiv \max_{1 \le i \le n} \sum_{i=1}^{n} |a_{i,j}| \quad \text{and} \quad \| A \|_\infty \equiv \max_{1 \le i \le n} \sum_{j=1}^{n} |a_{i,j}|.$$

Show that these definitions of matrix norms satisfy all the results of Theorem 1.2 relative to the vector norms $\| \mathbf{x} \|_1$ and $\| \mathbf{x} \|_\infty$ of Exercise 1 above, respectively. Moreover, show that the result (1.22) of the Corollary is equally valid for the matrix norms $\| A \|_1$ and $\| A \|_\infty$.

**3.** Let $A$ be an $n \times n$ *normal* matrix, i.e., $AA^* = A^*A$. If $\| A \|$ is the spectral norm of $A$, show that $\| A \| = \rho(A)$. (*Hint:* Use the fact† that $A = U^*DU$, where $U$ is unitary and $D$ is diagonal.)

**4.** A matrix norm is *unitarily invariant* if $\| A \| = \| AU \| = \| UA \|$ for every complex $n \times n$ matrix $A$ and every unitary $n \times n$ matrix $U$, i.e., $U^* = U^{-1}$. Show that the spectral norm is unitarily invariant. Is this true for the matrix norms of Exercise 2?

† See, for example, Bellman (1960, p. 197).

**5.** If $A$ is any nonsingular $n \times n$ complex matrix, show that the eigenvalues $\lambda$ of $A$ satisfy

$$\frac{1}{\| A^{-1} \|} \leq |\lambda| \leq \| A \|$$

(Browne (1928)).

**\*6.** It is known (Stein (1952)) that if $B$ is any $n \times n$ complex matrix with $\rho(B) < 1$, there exists a positive definite Hermitian matrix $A$ such that $A - B^*AB$ is positive definite. Conversely, if $A$ and $A - B^*AB$ are both positive definite Hermitian $n \times n$ matrices, then $\rho(B) < 1$. Prove this necessary and sufficient condition that $B$ is convergent with the simplifying assumption that $B$ is similar to a diagonal matrix.

**7.** Following the proof of Theorem 1.4, let $A$ be an $n \times n$ complex matrix, and suppose that $\lim_{m \to \infty} A^m$ exists. What can be said about the Jordan normal form of $A$? In particular, prove that

$$\lim_{m \to \infty} A^m = I$$

if and only if $A = I$.

**8.** Let $S$ be a positive definite $n \times n$ Hermitian matrix, and let

$$\| \mathbf{x} \| \equiv (\mathbf{x}^* S \mathbf{x})^{1/2}.$$

Show that this quantity is a norm, i.e., it satisfies the properties of (1.16).

## 1.4. BOUNDS FOR THE SPECTRAL RADIUS OF A MATRIX AND DIRECTED GRAPHS

It is generally difficult to determine precisely the spectral radius of a given matrix. Nevertheless, upper bounds can easily be found from the following theorem of Gerschgorin (1931).

**Theorem 1.5.** *Let $A = (a_{i,j})$ be an arbitrary $n \times n$ complex matrix, and let*

$$(1.29) \qquad \Lambda_i \equiv \sum_{\substack{j=1 \\ j \neq i}}^{n} | a_{i,j} |, \qquad 1 \leq i \leq n.$$

*Then, all the eigenvalues $\lambda$ of $A$ lie in the union of the disks*

$$| z - a_{i,i} | \leq \Lambda_i, \qquad 1 \leq i \leq n.$$

*Proof.* Let $\lambda$ be any eigenvalue of the matrix $A$, and let $\mathbf{x}$ be an eigenvector of $A$ corresponding to $\lambda$. We normalize the vector $\mathbf{x}$ so that its largest component in modulus is unity. By definition,

$$(1.30) \qquad (\lambda - a_{i,i})x_i = \sum_{\substack{j=1 \\ j \neq i}}^{n} a_{i,j}x_j, \qquad 1 \leq i \leq n.$$

In particular, if $|x_r| = 1$, then

$$|\lambda - a_{r,r}| \leq \sum_{\substack{j=1 \\ j \neq r}}^{n} |a_{r,j}| \cdot |x_j| \leq \sum_{\substack{j=1 \\ j \neq r}}^{n} |a_{r,j}| = \Lambda_r.$$

Thus, the eigenvalue $\lambda$ lies in the disk $|z - a_{r,r}| \leq \Lambda_r$. But since $\lambda$ was an arbitrary eigenvalue of $A$, it follows that all the eigenvalues of the matrix $A$ lie in the *union* of the disks $|z - a_{i,i}| \leq \Lambda_i$, $1 \leq i \leq n$, completing the proof.

Since the disk $|z - a_{i,i}| \leq \Lambda_i$ is a subset of the disk $|z| \leq |a_{i,i}| + \Lambda_i$, we have the immediate result of

**Corollary 1.** *If $A = (a_{i,j})$ is an arbitrary $n \times n$ complex matrix, and*

$$(1.31) \qquad \nu \equiv \max_{1 \leq i \leq n} \sum_{j=1}^{n} |a_{i,j}|,$$

*then $\rho(A) \leq \nu$.*

Thus, the maximum of the row sums of the moduli of the entries of the matrix $A$ gives us a simple upper bound for the spectral radius $\rho(A)$ of the matrix $A$. But, as $A$ and $A^T$ have the same eigenvalues, the application of Corollary 1 to $A^T$ similarly gives us

**Corollary 2.** *If $A = (a_{i,j})$ is an arbitrary $n \times n$ complex matrix, and*

$$(1.31') \qquad \nu' \equiv \max_{1 \leq j \leq n} \sum_{i=1}^{n} |a_{i,j}|,$$

*then $\rho(A) \leq \nu'$.*

Both the row and column sums of the moduli of the entries of the matrix $A$ are readily calculated, and a better upper bound for $\rho(A)$ is clearly the minimum of $\nu$ and $\nu'$. To improve further on these results, we use now the fact that similarity transformations on a matrix leave its eigenvalues invariant.

**Corollary 3.** *If* $A = (a_{i,j})$ *is an arbitrary* $n \times n$ *complex matrix, and* $x_1, x_2, \cdots, x_n$ *are any* $n$ *positive real numbers, let*

$$(1.32) \qquad \nu \equiv \max_{1 \le i \le n} \left\{ \frac{\left| \sum_{j=1}^{n} |a_{i,j}| x_j \right|}{x_i} \right\}; \qquad \nu' \equiv \max_{1 \le j \le n} \left\{ x_j \sum_{i=1}^{n} \frac{|a_{i,j}|}{x_i} \right\}.$$

*Then,* $\rho(A) \le \min (\nu, \nu')$.

*Proof.* Let $D = \text{diag } (x_1, x_2, \cdots, x_n)$ be the $n \times n$ diagonal matrix whose diagonal entries are $x_1, x_2, \cdots, x_n$, and apply Corollaries 1 and 2 to the matrix $D^{-1}AD$, whose spectral radius necessarily coincides with that of $A$.

To illustrate this last result, let $x_1 = 1 = x_4$ and $x_2 = 2 = x_3$, and consider in particular the matrix given in (1.7). Then,

$$\nu = \nu' = \tfrac{3}{4},$$

and thus $\rho(B) \le \tfrac{3}{4}$. However, if $x_1 = x_2 = x_3 = x_4 = 1$, then

$$\nu = \nu' = \tfrac{1}{2},$$

which is the exact value of $\rho(B)$ for this matrix from (1.8). This shows that equality is possible in Corollaries 1, 2, and 3.

In an attempt to improve upon, say, Corollary 1, suppose that the row sums of the moduli of the entries of the matrix $A$ were not all equal to $\nu$ of (1.31). Could we hope to conclude that $\rho(A) < \nu$? The counterexample given by the matrix

$$(1.33) \qquad A = \begin{pmatrix} 1 & 1 \\ 0 & 3 \end{pmatrix}$$

shows this to be false, since $\nu$ of (1.31) is 3, but $\rho(A) = 3$ also. This leads us to the following important definition.

DEFINITION 1.5. For $n \ge 2$, an $n \times n$ complex matrix $A$ is *reducible* if there exists an $n \times n$ permutation matrix† $P$ such that

$$(1.34) \qquad PAP^{T} = \begin{bmatrix} A_{1,1} & A_{1,2} \\ O & A_{2,2} \end{bmatrix},$$

where $A_{1,1}$ is an $r \times r$ submatrix and $A_{2,2}$ is an $(n - r) \times (n - r)$ sub-matrix, where $1 \le r < n$. If no such permutation matrix exists, then $A$ is

---

† A permutation matrix is a square matrix which in each row and each column has some one entry unity, all others zero.

*irreducible.* If $A$ is a $1 \times 1$ complex matrix, then $A$ is irreducible if its single entry is nonzero, and reducible otherwise.

It is evident from Definition 1.5 that the matrix of (1.33) is reducible.

The term *irreducible* (*unzerlegbar*) was introduced by Frobenius (1912); it is also called *unreduced* and *indecomposable* in the literature.† The motivation for calling matrices such as those in (1.34) *reducible* is quite clear, for if we seek to solve the matrix equation $\tilde{A}\mathbf{x} = \mathbf{k}$, where $\tilde{A} = PAP^T$ is the partitioned matrix of (1.34), we can partition the vectors $\mathbf{x}$ and $\mathbf{k}$ similarly so that the matrix equation $\tilde{A}\mathbf{x} = \mathbf{k}$ can be written as

$$A_{1,1}\mathbf{x}_1 + A_{1,2}\mathbf{x}_2 = \mathbf{k}_1,$$

$$A_{2,2}\mathbf{x}_2 = \mathbf{k}_2.$$

Thus, by solving the second equation for $\mathbf{x}_2$ and with this known solution for $\mathbf{x}_2$ solving the first equation for $\mathbf{x}_1$, we have *reduced* the solution of the original matrix equation to the solution of two *lower*-order matrix equations.

The geometrical interpretation of the concept of irreducibility by means of graph theory is quite useful, and we now consider some elementary notions from the theory of graphs (König (1950)). Let $A = (a_{i,j})$ be any $n \times n$ complex matrix, and consider any $n$ distinct points $P_1, P_2, \cdots, P_n$ in the plane, which we shall call *nodes*. For every nonzero entry $a_{i,j}$ of the matrix, we connect the node $P_i$ to the node $P_j$ by means of a path‡ $\overrightarrow{P_iP_j}$, directed from $P_i$ to $P_j$, as shown in Figure 2. In this way, with every $n \times n$ matrix $A$ can be associated a *finite directed* graph $G(A)$. As an example, the matrix $B$ of (1.7) has a directed graph

**Figure 2**

(1.35)     $G(B)$:

Similarly, the matrix $A$ of (1.33) has the directed graph

(1.36)     $G(A)$:

$P_1$     $P_2$

---

† See also Romanovsky (1936), Debreu and Herstein (1953), Wielandt (1950), and Geiringer (1949).

‡ For a diagonal entry $a_{i,i} \neq 0$, the path joining the node $P_i$ to itself is called a *loop*. For an illustration, see $G(A)$ in (1.36).

DEFINITION 1.6. A directed graph is *strongly connected* if, for any ordered pair of nodes $P_i$ and $P_j$, there exists a directed path

$$\overrightarrow{P_i P_{l_1}}, \quad \overrightarrow{P_{l_1} P_{l_2}}, \quad \cdots, \quad \overrightarrow{P_{l_{r-1}} P_{l_r=j}}$$

connecting $P_i$ to $P_j$.

We shall say that such a path has *length* $r$. By inspection, we see that the directed graph $G(B)$ of (1.35) is strongly connected. On the other hand, for the directed graph $G(A)$ of (1.36), there exists no path from $P_2$ to $P_1$, so that $G(A)$ is *not* strongly connected.

The indicated equivalence of the matrix property of irreducibility of Definition 1.5 with the concept of the strongly connected directed graph of a matrix is brought out precisely in the following theorem, whose proof is left as an exercise.

**Theorem 1.6.** *An $n \times n$ complex matrix $A$ is irreducible if and only if its directed graph $G(A)$ is strongly connected.*

It is obvious, from either equivalent formulation of irreducibility, that all the off-diagonal entries of any row or column of a matrix cannot vanish if the matrix is irreducible.

With the concept of irreducibility, we can sharpen the result of Theorem 1.5 as follows:

**Theorem 1.7.** *Let $A = (a_{i,j})$ be an irreducible $n \times n$ complex matrix, and assume that $\lambda$, an eigenvalue of $A$, is a boundary point of the union of the disks $|z - a_{i,i}| \leq \Lambda_i$. Then, all the $n$ circles $|z - a_{i,i}| = \Lambda_i$ pass through the point $\lambda$.*

*Proof.* If $\lambda$ is an eigenvalue of the matrix $A$ and $\mathbf{x}$ is an eigenvector of $A$ corresponding to $\lambda$, where $|x_r| = 1 \geq |x_i|$ for all $1 \leq i \leq n$, then as in the proof of Theorem 1.5 we know that

$$|\lambda - a_{r,r}| \leq \sum_{\substack{j=1 \\ j \neq r}}^{n} |a_{r,j}| \cdot |x_j| \leq \sum_{\substack{j=1 \\ j \neq r}}^{n} |a_{r,j}| = \Lambda_r.$$

But as $\lambda$ is assumed to be a boundary point of the union of all the disks $|z - a_{i,i}| \leq \Lambda_i$, we necessarily have that $|\lambda - a_{r,r}| = \Lambda_r$, and therefore

$$|\lambda - a_{r,r}| = \sum_{\substack{j=1 \\ j \neq r}}^{n} |a_{r,j}| \cdot |x_j| = \Lambda_r.$$

Consequently, for all $a_{r,l} \neq 0$ we have that $|x_l| = 1$. Since $A$ is irreducible, there exists at least one $p \neq r$ for which $a_{r,p} \neq 0$, and we conclude that

$|x_p|$ is also unity. Repeating the above argument with $r = p$, we now have that

$$|\lambda - a_{p,p}| = \sum_{\substack{j=1 \\ j \neq p}}^{n} |a_{p,j}| \cdot |x_j| = \Lambda_p,$$

making use once again of the hypothesis that $\lambda$ is a boundary point of the union of all the disks $|z - a_{i,i}| \leq \Lambda_i$, and therefore $|x_j| = 1$ for all $a_{p,j} \neq 0$. From Theorem 1.6, we can find a sequence (path) of nonzero entries of the matrix $A$ of the form $a_{r,p}, a_{p,r_2}, \cdots, a_{r_{m-1},j}$ for *any* integer $j$, $1 \leq j \leq n$, which permits us to continue the argument; thus we conclude that

$$|\lambda - a_{j,j}| = \Lambda_j, \qquad \text{for all } 1 \leq j \leq n,$$

proving that all the $n$ circles $|z - a_{i,i}| = \Lambda_i$ pass through the point $\lambda$, which completes the proof.

This sharpening of Theorem 1.5 immediately gives rise to the sharpened form of Corollary 3 of Theorem 1.5.

**Corollary.** *Let $A = (a_{i,j})$ be an irreducible $n \times n$ complex matrix, and $x_1, x_2, \cdots, x_n$ be any $n$ positive real numbers. If*

(1.37)
$$\frac{\sum_{j=1}^{n} |a_{i,j}| x_j}{x_i} \leq \nu$$

*for all $1 \leq i \leq n$, with strict inequality for at least one $i$, then $\rho(A) < \nu$. Similarly, if*

(1.37′)
$$x_j \sum_{i=1}^{n} \frac{|a_{i,j}|}{x_i} \leq \nu'$$

*for all $1 \leq j \leq n$, with strict inequality for at least one $j$, then $\rho(A) < \nu'$.*

*Proof.* For simplicity, we shall prove only the first assertion under the assumption that $x_1 = x_2 = \cdots = x_n = 1$. The complete proof follows easily in the manner of the proof of Corollary 3 of Theorem 1.5. If

$$\sum_{j=1}^{n} |a_{i,j}| \leq \nu$$

for all $1 \leq i \leq n$, with strict inequality for at least one $i$, it is clear that all the circles

$$|z - a_{i,i}| = \Lambda_i = \sum_{\substack{j=1 \\ j \neq i}}^{n} |a_{i,j}|$$

do not pass through a common point $\sigma$ where $|\sigma| = \nu$. Moreover, all the disks $|z - a_{i,i}| \leq \Lambda_i$ are subsets of the disk $|z| \leq \nu$. Thus, by Theorem 1.7, no point of the circle $|z| = \nu$ is an eigenvalue of $A$, proving that $\rho(A) < \nu$.

## EXERCISES

*1. Using the notation of Theorem 1.5, if the union of $m$ particular disks $|z - a_{i,i}| \leq \Lambda_i$ is disjoint from the union of the remaining disks, show that this union of $m$ disks contains precisely $m$ eigenvalues of the matrix $A$ (Gerschgorin (1931)).

2. Using the vector and matrix norms $||\mathbf{x}||_1$ and $||\mathbf{x}||_\infty$, show that the results of Exercise 2, Sec. 1.3, directly imply the results of Corollaries 1 and 2 of Theorem 1.5.

3. Prove Theorem 1.6.

4. Let $A = (a_{i,j})$ be an $n \times n$ complex matrix, $n \geq 2$, and let $W = \{1, 2, \cdots, n\}$ be the set of the first $n$ positive integers. Prove that $A$ is irreducible if and only if, for every two disjoint nonempty subsets $S$ and $T$ of $W$ with $S \cup T = W$, there exists an element $a_{i,j} \neq 0$ of $A$ with $i \in S, j \in T$.

5. Using Exercise 1 above, what are the strongest statements you can make about the eigenvalues of the matrix

$$
A = \begin{bmatrix}
1 & -1/2 & -1/2 & 0 \\
-1/2 & 3/2 & i & 0 \\
0 & -i/2 & 5 & i/2 \\
-1 & 0 & 0 & 5i
\end{bmatrix}?
$$

*6. Prove the following generalization of Theorem 1.5 (Brauer (1947)). Let $A = (a_{i,j})$ be an arbitrary $n \times n$ complex matrix, and let $\Lambda_i$, $1 \leq i \leq n$, be defined as in (1.29). Then, all the eigenvalues $\lambda$ of $A$ lie in the union of the $n(n - 1)/2$ ovals of Cassini

$$
|z - a_{i,i}| \cdot |z - a_{j,j}| \leq \Lambda_i \cdot \Lambda_j, \quad 1 \leq i, j \leq n, \quad i \neq j
$$

What stronger result, like that of Theorem 1.7, is valid when $A$ is assumed to be irreducible?

## 1.5. DIAGONALLY DOMINANT MATRICES

As we shall see in later sections, diagonally dominant matrices play an important role in the numerical solution of certain partial differential equations.

DEFINITION 1.7. An $n \times n$ complex matrix $A = (a_{i,j})$ is *diagonally dominant* if

$$(1.38) \qquad |a_{i,i}| \geq \sum_{\substack{j=1 \\ j \neq i}}^{n} |a_{i,j}| = \Lambda_i$$

for all $1 \leq i \leq n$. An $n \times n$ matrix $A$ is *strictly diagonally dominant* if strict inequality in (1.38) is valid for all $1 \leq i \leq n$. Similarly, $A$ is *irreducibly diagonally dominant*† if $A$ is irreducible and diagonally dominant, with strict inequality in (1.38) for at least one $i$.

**Theorem 1.8.** *Let* $A = (a_{i,j})$ *be an* $n \times n$ *strictly or irreducibly diagonally dominant complex matrix. Then, the matrix* $A$ *is nonsingular. If all the diagonal entries of* $A$ *are in addition positive real numbers, then the eigenvalues* $\lambda_i$ *of* $A$ *satisfy*

$$(1.39) \qquad \operatorname{Re} \lambda_i > 0, \qquad 1 \leq i \leq n.$$

*Proof.* We shall consider only the case when the matrix $A$ is strictly diagonally dominant, since the proof for the case when $A$ is irreducibly diagonally dominant is similar. With $A$ strictly diagonally dominant, the union of the disks $|z - a_{i,i}| \leq \Lambda_i$ does not include the origin $z = 0$ of the complex plane, and thus from Theorem 1.5, $\lambda = 0$ is not an eigenvalue of $A$, proving that $A$ is nonsingular. If the diagonal entries of the matrix $A$ are all positive real numbers, it is clear that the union of the disks $|z - a_{i,i}| \leq \Lambda_i$ in this case contains only points in the complex plane having their real parts positive, which completes the proof.

Since a Hermitian matrix has real eigenvalues, we have as an immediate consequence of Theorem 1.8 the

**Corollary.** *If* $A = (a_{i,j})$ *is a Hermitian* $n \times n$ *strictly diagonally dominant or irreducibly diagonally dominant matrix with positive real diagonal entries, then* $A$ *is positive definite.*

## EXERCISES

**1.** Complete the proof of Theorem 1.8 for the case where the $n \times n$ matrix $A$ is irreducibly diagonally dominant.

**2.** Given the matrix

$$A = \begin{bmatrix} 3 & -\frac{1}{2} & 0 \\ -\frac{1}{2} & 1 & -\frac{1}{2} \\ 0 & -\frac{1}{2} & 1 \end{bmatrix},$$

† Note that a matrix can be irreducible and diagonally dominant *without* being irreducibly diagonally dominant.

what are the strongest results you can state about

(a) all the eigenvalues of $A$,
(b) the spectral radius of $A$,
(c) the spectral norm of $A$, using the results of the last two sections?

**\*3.** Let $A = (a_{i,j})$ be an $n \times n$ irreducible matrix which is diagonally dominant in the sense of Definition 1.7. If $a_{i,i} \geq 0$, $1 \leq i \leq n$, and $a_{i,j} \leq 0$ for all $i \neq j$, prove that $\lambda = 0$ is an eigenvalue of $A$ if and only if

$$\sum_{k=1}^{n} a_{i,k} = 0 \qquad \text{for all } 1 \leq i \leq n$$

(Taussky (1949)).

**4.** Let $A = (a_{i,j})$ be an $n \times n$ diagonally dominant matrix with non-negative real diagonal entries. Show that the eigenvalues $\lambda_i$ of $A$ satisfy

$$\text{Re } \lambda_i \geq 0, \qquad 1 \leq i \leq n.$$

Moreover, if $A$ is Hermitian, conclude that $A$ is non-negative definite.

**5.** From Theorem 1.8, if $A$ is any strictly diagonally dominant matrix, then $\det (A) \neq 0$. Using this fact, prove directly Gerschgorin's Theorem 1.5.

## BIBLIOGRAPHY AND DISCUSSION

**1.1.** Probably the earliest mention of iterative methods dates back to Gauss (1823) who, in a letter to his pupil Gerling, suggested the use of iterative methods for solving systems of linear equations by iteration. For other informative historical remarks, see Bodewig (1959, p. 145), and Ostrowski (1956).

The theoretical convergence properties of noncyclic iterative methods such as the method of relaxation and more generally free-steering methods are considered by Temple (1939), Ostrowski (1954 and 1956), and Schechter (1959). The practical application of certain of these noncyclic iterative methods to the solution of systems of linear equations arising from physical and engineering problems is given in Southwell (1946), Shaw (1953), and Allen (1954).

For a general discussion and bibliographical coverage of the effect of rounding errors in practical computations and the convergence of discrete solutions of matrix equations to the continuous solution of an associated partial differential equation, refer to the recent excellent books by Forsythe and Wasow (1960) and Richtmyer (1957).

**1.3.** An excellent introductory treatment of matrix and vector norms is given in Faddeeva (1959) and Ostrowski (1960). For a thorough discussion of matrix and vector norms, showing the elegant geometrical relationship between matrix and vector norms and convex bodies, see Householder (1958); for norms and convex bodies in more general spaces, see Kolmogoroff (1934); for a more general treatment of matrix norms, see Ostrowski (1955) and Werner Gautschi (1953a), (1953b).

The origin of Theorem 1.4, which gives a necessary and sufficient condition that a matrix be convergent, goes back to Oldenburger (1940). It is interesting that there are several proofs of this fundamental result which make no use of the Jordan normal form of a matrix. Householder (1958), for example, proves this result using only matrix norm concepts, whereas Arms and Gates (1956) prove this result by means of complex variable theory. Another necessary and sufficient condition for a matrix to be convergent was given by Stein (1952); see Exercise 6 of Sec. 1.3.

The topic of unitarily invariant matrix norms (see Exercise 3 of Sec. 1.3) is considered by von Neumann (1937) and Fan and Hoffman (1955).

**1.4.** From the basic Theorem 1.5 of Gerschgorin (1931), rediscovered by A. Brauer (1946), there has followed a succession of papers which generalize this result. A. Brauer (1947) proves a similar result in which ovals of Cassini replace the disks of Theorem 1.5; see Exercise 6. For other definitions of the radii $\Lambda_i$ of (1.29), see Fan (1958) and references given therein.

The definition of the matrix property of irreducibility is due to Frobenius (1912). The equivalent formulation of irreducibility as given in Exercise 4 is due to Geiringer (1949). Its use in Theorem 1.7 to sharpen Gerschgorin's Theorem (1.4) is due to Taussky (1948). See also Taussky (1962).

For introductory material of finite directed graphs, see König (1950) and Berge (1958). There also arise some new computational problems of determining when a given large square matrix $A$ is irreducible. For recent applications of graph theoretic arguments to matrix theory, see Harary (1959 and 1960). Of course, all these concepts of irreducible and reducible non-negative matrices, directed graphs etc., have connections with the theory of finite Markov chains. See Kemeny and Snell (1960).

**1.5.** That strictly or irreducibly diagonally dominant matrices are necessarily nonsingular is a result which many have discovered independently. Taussky (1949) has collected a long list of contributors to this result and is an excellent reference for related material. See also Bodewig (1949, pp. 68–70). The second part of Theorem 1.8, which states that the real parts of the eigenvalues of such matrices are positive, is due to Taussky (1949).

# CHAPTER 2

# NON-NEGATIVE MATRICES

## 2.1. SPECTRAL RADII OF NON-NEGATIVE MATRICES

In investigations of the rapidity of convergence of various iterative methods, the spectral radius $\rho(A)$ of the corresponding iteration matrix $A$ plays a major role. For the practical estimation of this constant $\rho(A)$, we have thus far only upper bounds for $\rho(A)$ of Sec. 1.4 provided by the extensions of Gerschgorin's Theorem 1.5. In this section, we shall look closely into the Perron-Frobenius theory of square matrices having non-negative real numbers as entries. Not only will this theory provide us with both nontrivial upper and lower bounds for the spectral radius for this class of matrices, but the structure of this theory will be decidedly germane to our subsequent development of iterative methods. We begin with

DEFINITION 2.1. Let $A = (a_{i,j})$ and $B = (b_{i,j})$ be two $n \times r$ matrices. Then, $A \geq B$ ($>B$) if $a_{i,j} \geq b_{i,j}$ ($>b_{i,j}$) for all $1 \leq i \leq n$, $1 \leq j \leq r$. If $O$ is the null matrix and $A \geq O$ ($>O$), we say that $A$ is a *non-negative* (*positive*) *matrix*. Finally, if $B = (b_{i,j})$ is an arbitrary complex $n \times r$ matrix, then $|B|$ denotes† the matrix with entries $|b_{i,j}|$.

Since column vectors are $n \times 1$ matrices, we shall use the terms *non-negative* and *positive* vector throughout. In developing the Perron-Frobenius theory, we shall first establish a series of lemmas on non-negative irreducible square matrices.

**Lemma 2.1.** *If* $A \geq O$ *is an irreducible* $n \times n$ *matrix, then*

$$(I + A)^{n-1} > O.$$

† This is not to be confused with the determinant of a square matrix $B$, which we denote by *det B*.

*Proof.* It is sufficient to show that for any nonzero vector $\mathbf{x} \geq \mathbf{0}$, $(I + A)^{n-1}\mathbf{x} > \mathbf{0}$. Defining the sequence of non-negative vectors $\mathbf{x}_{k+1} = (I + A)\mathbf{x}_k$, $0 \leq k \leq n - 2$, where $\mathbf{x}_0 = \mathbf{x}$, the proof will follow by showing that $\mathbf{x}_{k+1}$ has *fewer* zero components† than does $\mathbf{x}_k$ for every $0 \leq k \leq n - 2$. Since $\mathbf{x}_{k+1} = \mathbf{x}_k + A\mathbf{x}_k$, it is clear that $\mathbf{x}_{k+1}$ has no more zero components than does $\mathbf{x}_k$. If $\mathbf{x}_{k+1}$ and $\mathbf{x}_k$ have exactly the same number of zero components, then for a suitable $n \times n$ permutation matrix $P$, we may write

$$(2.1) \qquad P\mathbf{x}_{k+1} = \begin{pmatrix} \alpha \\ 0 \end{pmatrix}, \qquad P\mathbf{x}_k = \begin{pmatrix} \beta \\ 0 \end{pmatrix}, \qquad \alpha > \mathbf{0}, \beta > \mathbf{0},$$

where the vectors $\alpha$ and $\beta$ both, have say, $m$ components, $1 \leq m < n$. Partitioning the corresponding new matrix $PAP^T$ relative to the partitioning in (2.1), we can write

$$(2.2) \qquad \begin{pmatrix} \alpha \\ 0 \end{pmatrix} = \begin{pmatrix} \beta \\ 0 \end{pmatrix} + \begin{bmatrix} A_{1,1} & A_{1,2} \\ A_{2,1} & A_{2,2} \end{bmatrix} \cdot \begin{pmatrix} \beta \\ 0 \end{pmatrix},$$

where $A_{1,1}$ and $A_{2,2}$ are square, and $A_{1,1}$ is of order $m$. This implies that $A_{2,1}\beta = 0$, but as $A_{2,1} \geq O$ and $\beta > \mathbf{0}$, this can only occur when $A_{2,1} = O$, which contradicts the hypothesis that $A$ is irreducible. Thus, $\mathbf{x}_{k+1}$ has fewer zero components than does $\mathbf{x}_k$, and since $\mathbf{x}_0$ has at most $(n - 1)$ zero components, then $\mathbf{x}_k$ has at most $(n - k - 1)$ zero components. Thus,

$$\mathbf{x}_{n-1} = (I + A)^{n-1}\mathbf{x}_0$$

is a positive vector, completing the proof.

If $A = (a_{i,j}) \geq O$ is an irreducible $n \times n$ matrix and $\mathbf{x} \geq \mathbf{0}$ is any nonzero vector, let

$$(2.3) \qquad r_\mathbf{x} \equiv \min \left\{ \frac{\displaystyle\sum_{j=1}^{n} a_{i,j}x_j}{x_i} \right\},$$

where the minimum is taken over all $i$ for which $x_i > 0$. Clearly, $r_\mathbf{x}$ is a non-negative real number and is the supremum of all numbers $\rho \geq 0$ for which

$$(2.4) \qquad\qquad\qquad A\mathbf{x} \geq \rho\mathbf{x}.$$

We now consider the non-negative quantity $r$ defined by

$$(2.5) \qquad\qquad\qquad r = \sup_{\substack{\mathbf{x} \geq \mathbf{0} \\ \mathbf{x} \neq \mathbf{0}}} \{r_\mathbf{x}\}.$$

† Provided, of course, that $\mathbf{x}_k$ has at least one zero component.

As $r_{\mathbf{x}}$ and $r_{\alpha\mathbf{x}}$ have the same value for any scalar $\alpha > 0$, we need consider only the set $P$ of vectors $\mathbf{x} \geq \mathbf{0}$ with $\|\mathbf{x}\| = 1$, and we correspondingly let $Q$ be the set of all vectors $\mathbf{y} = (I + A)^{n-1}\mathbf{x}$ where $\mathbf{x} \in P$. From Lemma 2.1, $Q$ consists only of positive vectors. Multiplying both sides of the inequality $A\mathbf{x} \geq r_{\mathbf{x}}\mathbf{x}$ by $(I + A)^{n-1}$, we obtain

$$A\mathbf{y} \geq r_{\mathbf{x}}\mathbf{y},$$

and we conclude from (2.4) that $r_{\mathbf{y}} \geq r_{\mathbf{x}}$. Therefore, the quantity $r$ of (2.5) can be defined equivalently as

$$(2.5') \qquad\qquad r = \sup_{\mathbf{y} \in Q} \{r_{\mathbf{y}}\}.$$

As $P$ is a compact set of vectors, so is $Q$, and as $r_{\mathbf{y}}$ is a continuous function on $Q$, there necessarily exists a positive vector $\mathbf{z}$ for which

$$(2.6) \qquad\qquad A\mathbf{z} \geq r\mathbf{z},$$

and no vector $\mathbf{w} \geq \mathbf{0}$ exists for which $A\mathbf{w} > r\mathbf{w}$. We shall call *all* non-negative nonzero vectors $\mathbf{z}$ satisfying (2.6) *extremal vectors* of the matrix $A$.

**Lemma 2.2.** *If $A \geq O$ is an irreducible $n \times n$ matrix, the quantity $r$ of (2.5) is positive. Moreover, each extremal vector $\mathbf{z}$ is a positive eigenvector of the matrix $A$ with corresponding eigenvalue $r$, i.e., $A\mathbf{z} = r\mathbf{z}$, and $\mathbf{z} > \mathbf{0}$.*

*Proof.* If $\xi$ is the positive vector whose components are all unity, then since the matrix $A$ is irreducible, no row of $A$ can vanish, and consequently no component of $A\xi$ can vanish. Thus, $r_{\xi} > 0$, proving that $r > 0$.

For the second part of this lemma, let $\mathbf{z}$ be an extremal vector with $A\mathbf{z} - r\mathbf{z} = \eta$, where $\eta \geq \mathbf{0}$. If $\eta \neq \mathbf{0}$, then some component of $\eta$ is positive; multiplying through by the matrix $(I + A)^{n-1}$, we have

$$A\mathbf{w} - r\mathbf{w} > \mathbf{0}$$

where $\mathbf{w} = (I + A)^{n-1}\mathbf{z} > \mathbf{0}$. It would then follow that $r_{\mathbf{w}} > r$, contradicting the definition of $r$ in (2.5'). Thus, $A\mathbf{z} = r\mathbf{z}$, and since $\mathbf{w} > \mathbf{0}$ and $\mathbf{w} = (1 + r)^{n-1}\mathbf{z}$, then $\mathbf{z} > \mathbf{0}$, completing the proof.

**Lemma 2.3.** *Let $A = (a_{i,j}) \geq O$ be an irreducible $n \times n$ matrix, and let $B = (b_{i,j})$ be an $n \times n$ complex matrix with $|B| \leq A$. If $\beta$ is any eigenvalue of $B$, then*

$$(2.7) \qquad\qquad |\beta| \leq r,$$

*where r is the positive constant of (2.5). Moreover, equality is valid in (2.7), i.e., $\beta = re^{i\phi}$, if and only if $|B| = A$, and where B has the form*

$$(2.8) \qquad\qquad B = e^{i\phi}DAD^{-1},$$

*and D is a diagonal matrix whose diagonal entries have modulus unity.*

*Proof.* If $\beta \mathbf{y} = B\mathbf{y}$ where $\mathbf{y} \neq \mathbf{0}$, then

$$\beta y_i = \sum_{j=1}^{n} b_{i,j} y_j, \qquad 1 \leq i \leq n.$$

Using the hypotheses of the lemma and the notation of Definition 2.1, it follows that

$$(2.9) \qquad\qquad |\beta| \, |\mathbf{y}| \leq |B| \, |\mathbf{y}| \leq A \, |\mathbf{y}|,$$

which implies that $|\beta| \leq r_{|\mathbf{y}|} \leq r$, proving (2.7). If $|\beta| = r$, then $|\mathbf{y}|$ is an extremal vector of $A$. Therefore, from Lemma 2.2, $|\mathbf{y}|$ is a positive eigenvector of $A$ corresponding to the positive eigenvalue $r$. Thus,

$$(2.10) \qquad\qquad r\,|\mathbf{y}| = |B| \, |\mathbf{y}| = A\,|\mathbf{y}|,$$

and since $|\mathbf{y}| > \mathbf{0}$, we conclude from (2.10) and the hypothesis $|B| \leq A$ that

$$(2.11) \qquad\qquad |B| = A.$$

For the vector $\mathbf{y}$, where $|\mathbf{y}| > \mathbf{0}$, let

$$D = \operatorname{diag}\left\{\frac{y_1}{|y_1|}, \cdots, \frac{y_n}{|y_n|}\right\}.$$

It is clear that the diagonal entries of $D$ have modulus unity, and

$$(2.12) \qquad\qquad \mathbf{y} = D\,|\mathbf{y}|.$$

Setting $\beta = re^{i\phi}$, then $B\mathbf{y} = \beta\mathbf{y}$ can be written as

$$(2.13) \qquad\qquad C\,|\mathbf{y}| = r\,|\mathbf{y}|,$$

where

$$(2.14) \qquad\qquad C \equiv e^{-i\phi}D^{-1}BD.$$

From (2.10) and (2.13), equating terms equal to $r\,|\mathbf{y}|$ we have

$$(2.15) \qquad\qquad C\,|\mathbf{y}| = |B| \, |\mathbf{y}| = A\,|\mathbf{y}|.$$

From the definition of the matrix $C$ in (2.14), $|C| = |B|$. Combining with (2.11), we have

$$(2.16) \qquad\qquad |C| = |B| = A.$$

Thus, from (2.15) we conclude that $C|\mathbf{y}| = |C||\mathbf{y}|$, and as $|\mathbf{y}| > \mathbf{0}$, it follows that $C = |C|$ and thus $C = A$ from (2.16). Combining this result with (2.14) gives the desired result that $B = e^{i\phi}DAD^{-1}$. Conversely, it is obvious that if $B$ has the form in (2.8), then $|B| = A$, and $B$ has an eigenvalue $\beta$ with $|\beta| = r$, which completes the proof.

Setting $B = A$ in Lemma 2.3 immediately gives us the

**Corollary.** *If $A \geq O$ is an irreducible $n \times n$ matrix, then the positive eigenvalue $r$ of Lemma 2.2 equals the spectral radius $\rho(A)$ of $A$.*

In other words, if $A \geq O$ is an irreducible $n \times n$ matrix, its spectral radius $\rho(A)$ is positive, and the intersection in the complex plane of the circle $|z| = \rho(A)$ with the positive real axis *is* an eigenvalue of $A$.

**Lemma 2.4.** *If $A \geq O$ is an irreducible $n \times n$ matrix, and $B$ is any principal square submatrix† of $A$, then $\rho(B) < \rho(A)$.*

*Proof.* If $B$ is any principal submatrix of $A$, then there is an $n \times n$ permutation matrix $P$ such that $B = A_{1,1}$ where

$$(2.17) \qquad C \equiv \begin{bmatrix} A_{1,1} & O \\ O & O \end{bmatrix}; \qquad PAP^T = \begin{bmatrix} A_{1,1} & A_{1,2} \\ A_{2,1} & A_{2,2} \end{bmatrix}.$$

Here, $A_{1,1}$ and $A_{2,2}$ are, respectively, $m \times m$ and $(n - m) \times (n - m)$ principal square submatrices of $PAP^T$, $1 \leq m < n$. Clearly, $O \leq C \leq PAP^T$, and $\rho(C) = \rho(B) = \rho(A_{1,1})$, but as $C = |C| \neq PAP^T$, the conclusion follows immediately from Lemma 2.3 and its corollary.

We now collect the above results into the following main theorem of Perron (1907) and Frobenius (1912).

**Theorem 2.1.** *Let $A \geq O$ be an irreducible $n \times n$ matrix. Then,*

1. *$A$ has a positive real eigenvalue equal to its spectral radius.*
2. *To $\rho(A)$ there corresponds an eigenvector $\mathbf{x} > \mathbf{0}$.*
3. *$\rho(A)$ increases when any entry of $A$ increases.*
4. *$\rho(A)$ is a simple eigenvalue of $A$.*

† A principal square submatrix of an $n \times n$ matrix $A$ is any matrix obtained by crossing out any $j$ rows and the corresponding $j$ columns of $A$, where $1 \leq j < n$.

*Proof.* Parts 1 and 2 follow immediately from Lemma 2.2 and the Corollary to Lemma 2.3. To prove part 3, suppose we increase some entry of the matrix $A$, giving us a new irreducible matrix $\tilde{A}$, where $\tilde{A} \geq A$ and $\tilde{A} \neq A$. Applying Lemma 2.3, we conclude that $\rho(\tilde{A}) > \rho(A)$. To prove that $\rho(A)$ is a simple eigenvalue of $A$, i.e., $\rho(A)$ is a zero of multiplicity one of the characteristic polynomial $\phi(t) = \det (tI - A)$, we make use of the fact that $\phi'(t)$ is the sum of the determinants of the principal $(n - 1)$ $\times$ $(n - 1)$ submatrices of $tI - A$.† If $A_i$ is any principal submatrix of $A$, then from Lemma 2.4, $\det [tI - A_i]$ cannot vanish for any $t \geq \rho(A)$. From this it follows that

$$\det [\rho(A)I - A_i] > 0,$$

and thus

$$\phi'(\rho(A)) > 0.$$

Consequently, $\rho(A)$ cannot be a zero of $\phi(t)$ of multiplicity greater than one, and thus $\rho(A)$ is a simple eigenvalue of $A$. This result of Theorem 2.1 incidentally shows us that if $A\mathbf{x} = \rho(A)\mathbf{x}$ where $\mathbf{x} > \mathbf{0}$ and $\| \mathbf{x} \| = 1$, we cannot find another eigenvector $\mathbf{y} \neq \gamma\mathbf{x}$, $\gamma$ a scalar, of $A$ with $A\mathbf{y} = \rho(A)\mathbf{y}$, so that the eigenvector $\mathbf{x}$ so normalized is *uniquely* determined.

Historically, Perron (1907) proved this theorem assuming that $A > O$. Whereas this obviously implies that the matrix $A$ is irreducible, in this case, however, Perron showed that there was but *one* eigenvalue of *modulus* $\rho(A)$. Later, Frobenius (1912) extended most of Perron's results to the class of non-negative irreducible matrices. In the next section, we shall investigate precisely how many eigenvalues of a non-negative irreducible $n \times n$ matrix can have their moduli equal to $\rho(A)$.

We now return to the problem of finding bounds for the spectral radius of a matrix. In Sec. 1.4 we found nontrivial upper-bound estimates of $\rho(A)$. For non-negative matrices, the Perron-Frobenius theory gives us nontrivial *lower-bound* estimates of $\rho(A)$. Coupled with Gerschgorin's Theorem 1.5, we thus obtain inclusion theorems for the spectral radius of a non-negative irreducible square matrix.

**Lemma 2.5.** *If $A = (a_{i,j}) \geq O$ is an irreducible $n \times n$ matrix, then either*

$$(2.18) \qquad \sum_{j=1}^{n} a_{i,j} = \rho(A) \qquad for\ all\ 1 \leq i \leq n,$$

*or*

$$(2.19) \qquad \min_{1 \leq i \leq n} \left( \sum_{j=1}^{n} a_{i,j} \right) < \rho(A) < \max_{1 \leq i \leq n} \left( \sum_{j=1}^{n} a_{i,j} \right).$$

† For example, see Birkhoff and MacLane (1953, p. 288).

*Proof.* First, suppose that all the row sums of $A$ are equal to $\sigma$. If $\xi$ is the vector with all components unity, then obviously $A\xi = \sigma\xi$, and $\sigma \leq \rho(A)$ by Definition 1.2. But, Corollary 1 of Theorem 1.5 shows us that $\rho(A) \leq \sigma$, and we conclude that $\rho(A) = \sigma$, which is the result of (2.18). If all the row sums of $A$ are not equal, we can construct a non-negative irreducible $n \times n$ matrix $B = (b_{i,j})$ by *decreasing* certain positive entries of $A$ so that for all $1 \leq i \leq n$,

$$\sum_{j=1}^{n} b_{i,j} = \alpha = \min_{1 \leq i \leq n} \left( \sum_{j=1}^{n} a_{i,j} \right),$$

where $O \leq B \leq A$ and $B \neq A$. As all the row sums of $B$ are equal to $\alpha$, we can apply the result of (2.18) to the matrix $B$, and thus $\rho(B) = \alpha$. Now, by statement 3 of Theorem 2.1, we must have that $\rho(B) < \rho(A)$, so that

$$\min_{1 \leq i \leq n} \left( \sum_{j=1}^{n} a_{i,j} \right) < \rho(A).$$

On the other hand, we can similarly construct a non-negative irreducible $n \times n$ matrix $C = (c_{i,j})$ by *increasing* certain of the positive entries of $A$ so that all the row sums of the matrix $C$ are equal, where $C \geq A$ and $C \neq A$. It follows that

$$\rho(A) < \rho(C) = \max_{1 \leq i \leq n} \left( \sum_{j=1}^{n} a_{i,j} \right),$$

and the combination of these inequalities gives the desired result of (2.19).

**Theorem 2.2.** *Let* $A = (a_{i,j}) \geq O$ *be an irreducible* $n \times n$ *matrix, and let* $P^*$ *be the hyperoctant of vectors* $\mathbf{x} > \mathbf{0}$. *Then, for any* $\mathbf{x} \in P^*$, *either*

$$(2.20) \qquad \min_{1 \leq i \leq n} \left[ \frac{\sum_{j=1}^{n} a_{i,j}x_j}{x_i} \right] < \rho(A) < \max_{1 \leq i \leq n} \left[ \frac{\sum_{j=1}^{n} a_{i,j}x_j}{x_i} \right],$$

*or*

$$(2.21) \qquad \frac{\sum_{j=1}^{n} a_{i,j}x_j}{x_i} = \rho(A) \qquad for\ all\ 1 \leq i \leq n.$$

*Moreover,*

$$(2.22) \qquad \sup_{\mathbf{x} \in P^*} \left\{ \min_{1 \leq i \leq n} \left[ \frac{\sum_{j=1}^{n} a_{i,j}x_j}{x_i} \right] \right\} = \rho(A) = \inf_{\mathbf{x} \in P^*} \left\{ \max_{1 \leq i \leq n} \left[ \frac{\sum_{j=1}^{n} a_{i,j}x_j}{x_i} \right] \right\}.$$

*Proof.* For any positive vector $\mathbf{x} \in P^*$, let

$$D = \text{diag}\ (x_1, x_2, \cdots, x_n),$$

and consider, as in the proof of Corollary 3 of Theorem 1.5, the matrix $B = D^{-1}AD$. It is clear that $B$ is a non-negative irreducible matrix, and (2.20) and (2.21) now follow directly from Lemma 2.5. Although we must certainly have from (2.20) and (2.21) that

$$(2.23) \qquad \sup_{\mathbf{x} \in P^*} \left\{ \min_{1 \le i \le n} \left[ \frac{\sum_{j=1}^{n} a_{i,j} x_j}{x_i} \right] \right\} \le \rho(A) \le \inf_{\mathbf{x} \in P^*} \left\{ \max_{1 \le i \le n} \left[ \frac{\sum_{j=1}^{n} a_{i,j} x_j}{x_i} \right] \right\},$$

choosing the positive eigenvector $\mathbf{y} \in P^*$ corresponding to the eigenvalue $\rho(A)$ shows that equality is valid throughout (2.23), completing the proof of Theorem 2.2.

As in Corollary 3 of Theorem 1.5, we obviously could have made comparable statements about the column sums of $A$ in both Lemma 2.5 and Theorem 2.2, since $A^T$ is a non-negative irreducible $n \times n$ matrix if $A$ is.

For a simple illustration of the results of Theorem 2.2, let us return to the problem of estimating the spectral radii of the matrices $B$ and $A^{-1}$ defined respectively in (1.7) and (1.9). Both of these matrices are non-negative and irreducible, and as their row (and column) sums are all equal, we conclude by direct observation that

$$\rho(B) = \tfrac{1}{2}, \qquad \rho(A^{-1}) = 2.$$

This method will be of considerable practical value to us in that both upper and lower bounds for the spectral radius of a non-negative irreducible matrix can be obtained by the simple arithmetic methods of Theorem 2.2, without resorting to the determination of the characteristic polynomial of the matrix.

## EXERCISES

**1.** Consider the matrix

$$A(\alpha) = \begin{bmatrix} 1 & 4 & 1 + 0.9\alpha \\ 3 & 0.8\alpha & 3 \\ \alpha & 5 & 1 \end{bmatrix}, \qquad \alpha \ge 0.$$

Show that, as $\alpha$ increases, $\rho[A(\alpha)]$ is strictly increasing for all $\alpha \ge 0$, and determine upper and lower bounds for $\rho[A(\alpha)]$ for all $\alpha \ge 0$.

2. Let $A \geq O$ be an irreducible $n \times n$ matrix, and let $\mathbf{x}^{(0)}$ be an arbitrary column vector with $n$ positive components. Defining

$$\mathbf{x}^{(r)} = A\mathbf{x}^{(r-1)} = \cdots = A^r\mathbf{x}^{(0)}, \qquad r \geq 1,$$

let

$$\bar{\lambda}_r \equiv \max_{1 \leq i \leq n} \left(\frac{x_i^{(r+1)}}{x_i^{(r)}}\right) \text{ and } \underline{\lambda}_r \equiv \min_{1 \leq i \leq n} \left(\frac{x_i^{(r+1)}}{x_i^{(r)}}\right).$$

Show that $\underline{\lambda}_0 \leq \underline{\lambda}_1 \leq \underline{\lambda}_2 \leq \cdots \leq \rho(A) \leq \cdots \leq \bar{\lambda}_2 \leq \bar{\lambda}_1 \leq \bar{\lambda}_0$. Thus, the inequalities for $\rho(A)$ of Theorem 2.2 can be improved by this method.

3. Applying the method of the previous exercise to the matrix $A(\alpha)$ of Exercise 1, find bounds $l(\alpha) \leq \rho[A(\alpha)] \leq u(\alpha)$ such that

$$1 \leq \frac{u(\alpha)}{l(\alpha)} \leq 1.01$$

uniformly in $0 \leq \alpha \leq 1$.

4. Let $H_n$ be the $n \times n$ matrix with entries

$$h_{i,j} = \frac{1}{i + j - 1}, \qquad \text{for } 1 \leq i, j \leq n.$$

$H_n$ is called the $n$th segment of the (infinite) *Hilbert matrix*. Prove that, as $\alpha$ increases, $\rho(H_n)$ is strictly increasing for all $n \geq 1$.

5. If $A \geq O$ is an irreducible $n \times n$ matrix, prove that the matrix $A$ cannot have two linearly independent eigenvectors with non-negative real components (*Hint*: Apply Theorem 2.2 and statement 4 of Theorem 2.1 to linear combinations of these eigenvectors.)

6. Prove the result of the Corollary of Theorem 1.7, using Lemma 2.3 and Theorem 2.2.

7. Let $A \geq O$ be an irreducible $n \times n$ matrix. With the notation of Theorem 2.2, prove that $\rho(A)$ can also be characterized by

$$\max_{\mathbf{x} \epsilon P^*} \left\{ \min_{\mathbf{y} \epsilon P^*} \frac{\mathbf{y}^T A \mathbf{x}}{\mathbf{y}^T \mathbf{x}} \right\} = \rho(A) = \min_{\mathbf{y} \epsilon P^*} \left\{ \max_{\mathbf{x} \epsilon P^*} \frac{\mathbf{y}^T A \mathbf{x}}{\mathbf{y}^T \mathbf{x}} \right\}$$

(Birkhoff and Varga (1958)).

8. Let $B \geq O$ be an irreducible $n \times n$ matrix with $\rho(B) < 1$, and let $C \geq O$ be an $n \times n$ matrix which is not the null matrix. Show that there exists a unique positive value of $\lambda$ such that

$$\rho\left(\frac{C}{\lambda} + B\right) = 1$$

(Birkhoff and Varga (1958)).

## 2.2. CYCLIC AND PRIMITIVE MATRICES

In the discussion following the Perron-Frobenius Theorem 2.1 of the previous section, we stated that in his early work Perron had actually proved that a positive square matrix $A$ has but one eigenvalue of modulus $\rho(A)$. For non-negative irreducible square matrices, investigations into the cases where there exist several eigenvalues of the matrix $A$ of modulus $\rho(A)$ reveal a beautiful structure due to Frobenius (1912). We begin with

DEFINITION 2.2. Let $A \geq O$ be an irreducible $n \times n$ matrix, and let $k$ be the number of eigenvalues of $A$ of modulus $\rho(A)$. If $k = 1$, then $A$ is *primitive*. If $k > 1$, then $A$ is *cyclic of index $k$*.

If $A \geq O$ is an irreducible $n \times n$ cyclic matrix of index $k > 1$, then there are $k$ eigenvalues

$$\alpha_j = \rho(A) \exp{(i\phi_j)}, \quad 0 \leq j \leq k - 1,$$

of the matrix $A$, which we order as

$$0 = \phi_0 \leq \phi_1 \leq \cdots \leq \phi_{k-1} < 2\pi.$$

We can apply Lemma 2.3 with $B \equiv A$, and we conclude from (2.8) that

$$(2.24) \qquad A = \exp{(i\phi_j)} \cdot D_j A D_j^{-1} \qquad \text{for all} \ \ 0 \leq j \leq k - 1,$$

where $D_j$ is a diagonal matrix whose diagonal entries have modulus unity. Since $\rho(A)$ is a simple eigenvalue of $A$ from Theorem 2.1, then (2.24) shows that each eigenvalue $\alpha_j$ is also a simple eigenvalue of $A$, and thus the $\phi_j$'s are *distinct*. By direct computation with (2.24), we also find that

$$(2.25) \qquad A = \exp{[i(\phi_j \pm \phi_p)]}(D_j D_p^{\pm 1}) A (D_j D_p^{\pm 1})^{-1},$$

$$\text{for all} \ \ 0 \leq j, p \leq k - 1.$$

Thus, all the complex numbers $\rho(A) \cdot \exp{[i(\phi_j \pm \phi_p)]}, 0 \leq j, p \leq k - 1$, are eigenvalues of $A$, and we conclude that the numbers $\exp{(i\phi_j)}$ form a *finite multiplicative group*. Since the order of every element in a finite group is a divisor of the order of the group (Birkhoff and MacLane (1953, p. 147)), then

$$\exp{(ik\phi_j)} = 1 \qquad \text{for all} \ \ 0 \leq j \leq k - 1,$$

and thus

$$(2.26) \qquad \phi_j = \frac{2\pi j}{k}, \qquad \text{for all} \ \ 0 \leq j \leq k - 1.$$

In other words, the $k$ eigenvalues of modulus $\rho(A)$ of the cyclic matrix $A$ of index $k$ are precisely the roots of the equation

$$(2.27) \qquad \lambda^k - \rho^k(A) = 0.$$

Continuing with our investigations in the cyclic case, let $\mathbf{y}_1$ be a complex eigenvector of $A$ associated with the eigenvalue

$$\alpha_1 = \rho(A) \cdot \exp\left(\frac{2\pi i}{k}\right).$$

Letting $D_1$ be the diagonal matrix of (2.12) where $\mathbf{y}_1 = D_1 \mid \mathbf{y}_1 \mid$, then since $D_1$ has diagonal entries of modulus unity, we can find an $n \times n$ permutation matrix $P$ (which permutes the rows and columns of $D_1$) such that

$$(2.28) \qquad PD_1P^T =
\begin{bmatrix}
I_1 e^{i\delta_1} & & & \bigcirc \\
& I_2 e^{i\delta_2} & & \\
& & \ddots & \\
\bigcirc & & & I_s e^{i\delta_s}
\end{bmatrix}, \qquad s > 1,\dagger$$

where the square matrices $I_j$ are the identity matrices of not necessarily the same order, and the real numbers $\delta_j$ are *distinct* and satisfy $0 \le \delta_j < 2\pi$. Without loss of generality, we may assume that $\delta_1 = 0$, since $\exp(-i\delta_1)\,\mathbf{y}_1$ is also a (complex) eigenvector of $A$ associated with the eigenvalue $\alpha_1$. With the permutation matrix $P$ of (2.28), we form the matrix $PAP^T$ and partition it, following the partition of (2.28), into $s^2$ submatrices $A_{p,q}$, $1 \le p, q \le s$, where

$$(2.29) \qquad PAP^T =
\begin{bmatrix}
A_{1,1} & A_{1,2} & \cdots & A_{1,s} \\
A_{2,1} & A_{2,2} & \cdots & A_{2,s} \\
\vdots & & & \vdots \\
A_{s,1} & A_{s,2} & \cdots & A_{s,s}
\end{bmatrix},$$

and the diagonal submatrices $A_{i,i}$ are square.

† Clearly, $s$ cannot be unity, for then the eigenvector $\mathbf{y}_1$ is a scalar times the non-negative vector $\mid \mathbf{y}_1 \mid$ and could not correspond to the complex eigenvalue

$$\rho(A) \cdot \exp[i(2\pi/k)].$$

We now show that the number $s$ of distinct $\delta_i$'s of (2.28) is necessarily equal to $k$, the cyclic index of $A$, and that only particular submatrices $A_{i,j}$ of (2.29) are non-null. From (2.24) we deduce that

$$PAP^T = \exp\,(i\phi_1)\,(PD_1P^T)\,(PAP^T)\,(PD_1P^T)^{-1}$$

where $\phi_1 = 2\pi/k$. Therefore, with the explicit representations of $PD_1P^T$ and $PAP^T$ in (2.28) and (2.29), we find that

$$(2.30) \qquad A_{p,q} = \exp\left[i\left(\frac{2\pi}{k} + \delta_p - \delta_q\right)\right]A_{p,q}, \qquad \text{for all } 1 \le p, q \le s.$$

If $\delta_q \not\equiv \{(2\pi/k) + \delta_p\}$ (mod $2\pi$), then $A_{p,q} = O$. Since the $\delta_i$'s are by assumption distinct, then for each fixed $p$ there is *at most* one $q(\ne p)$ for which $A_{p,q} \ne O$. On the other hand, since $A$ is irreducible, no row (or rows) of $A$ can have only zero entries, so that there is *at least* one $q$ for which $A_{p,q} \ne O$; therefore for each $p$ there is *precisely* one $q$, say $q(p)$, for which $A_{p,q(p)} \ne O$, and

$$(2.31) \qquad \frac{2\pi}{k} + \delta_p \equiv \delta_{q(p)} \text{ (mod } 2\pi).$$

With $p = 1$ and $\delta_1 = 0$, we see that $\delta_{q(1)} = 2\pi/k$. By permuting once again the rows and columns (by blocks) of the matrices $PAP^T$ and $PDP^T$, we may assume that $\delta_{q(1)} = \delta_2$. Continuing in this way, we find that

$$(2.32) \qquad \delta_{q(j-1)} = \delta_j = \frac{2\pi(j-1)}{k}, \qquad 2 \le j \le \min\,(k, s).$$

If $1 < s < k$, then (2.32) shows that

$$\frac{2\pi}{k} + \delta_s - \delta_r \ne 0 \text{ (mod } 2\pi) \qquad \text{for any } 1 \le r \le s - 1;$$

thus $A_{s,r} = O$ for all $1 \le r \le s - 1$, which implies that $A$ is reducible. Since this is a contradiction, then $s \ge k$. Similarly, if $s > k$, we have from (2.32) that

$$\frac{2\pi}{k} + \delta_p - \delta_r = \frac{2\pi p}{k} - \delta_r$$

for all $1 \le p \le k$, $k < r \le s$. As the $\delta_j$'s are *distinct* by construction, we conclude that

$$\frac{2\pi}{k} + \delta_p - \delta_r \not\equiv 0 \text{ (mod } 2\pi),$$

so that from (2.30) $A_{p,r} = O$ for all $1 \leq p \leq k$, $k < r \leq s$. As this again implies that the matrix $A$ is reducible, we conclude finally that $s = k$. Now, with the particular choice of $\delta_j$'s of (2.32), the relations of (2.30) show us that the matrix $PAP^T$ simplifies to

(2.33)
$$PAP^T = \begin{bmatrix} O & A_{1,2} & O & \cdots & O \\ O & O & A_{2,3} & \cdots & O \\ \vdots & & & & \vdots \\ \vdots & & & & \vdots \\ O & O & O & & A_{k-1,k} \\ A_{k,1} & O & O & \cdots & O \end{bmatrix},$$

where the null diagonal submatrices are square. This is the basis† for

**Theorem 2.3.** *Let $A \geq O$ be an irreducible $n \times n$ cyclic matrix of index $k$ ($> 1$). Then the $k$ eigenvalues of modulus $\rho(A)$ are of the form*

$$\rho(A) \cdot \exp\left[i\left(\frac{2\pi j}{k}\right)\right], \qquad 0 \leq j \leq k - 1.$$

*Moreover, all the eigenvalues of $A$ have the property that rotations of the complex plane about the origin through angles of $2\pi/k$, but through no smaller angles, carry the set of eigenvalues into itself. Finally, there is an $n \times n$ permutation matrix $P$ such that $PAP^T$ has the form (2.33).*

*Proof.* From the previous discussion, all that needs to be proved is the rotational invariance of the set of eigenvalues of $A$. To prove this rotational invariance, we simply make use of (2.24) and (2.26), which allows us to express the matrix $A$ in the form

(2.34)
$$A = \exp\left(\frac{2\pi i}{k}\right) D_1 A D_1^{-1}.$$

But since the matrix product $D_1 A D_1^{-1}$ is itself similar to $A$, and hence has the same eigenvalues as $A$, we conclude from (2.34) that rotations of the complex plane about the origin through angles of $2\pi/k$ carry the set of eigenvalues of $A$ into itself. The fact that there are *exactly* $k$ eigenvalues of modulus $\rho(A)$, by Definition 2.2, shows that this rotational invariance cannot be valid for smaller angles, which completes the proof.

† Theorem 2.3 is a fundamental result of Frobenius (1912). The method of proof is due to Wielandt (1950).

**Corollary.** *If $A \geq O$ is an irreducible $n \times n$ cyclic matrix of index $k$, then its characteristic polynomial $\phi(t) = \det (tI - A)$ is of the form*

$$(2.35) \qquad \phi(t) = t^m [t^k - \rho^k(A)][t^k - \delta_2 \rho^k(A)] \cdots [t^k - \delta_r \rho^k(A)],$$

*where $|\delta_i| < 1$ for $1 < i \leq r$ if $r > 1$.*

*Proof.* From Theorem 2.3, if $\lambda$ is any nonzero eigenvalue of $A$, then all the roots of $t^k - \lambda^k = 0$ are also eigenvalues of $A$. Hence, we necessarily have

$$\phi(t) = \det (\lambda I - A) = \prod_{i=1}^{n} (t - \lambda_i) = t^m \prod_{i=1}^{r} (t^k - \sigma_i^k),$$

where each $\sigma_i$ is a nonzero eigenvalue of $A$. This, coupled with the fact that the matrix $A$ has precisely $k$ eigenvalues of modulus $\rho(A)$, by Definition 2.2, proves (2.35).

The specific representation (2.33) of a non-negative irreducible cyclic matrix of index $k > 1$ can, of course, be permuted into several different, but equivalent forms. Actually, the particular form that will be more useful in later developments (Chapters 4, 5, and 7) is the following:

$$(2.33') \qquad PAP^T = \begin{bmatrix} O & O & \cdots & O & A_{1,k} \\ A_{2,1} & O & & O & O \\ O & A_{3,2} & & O & O \\ \cdot & & & & \cdot \\ \cdot & & & & \cdot \\ \cdot & & & & \cdot \\ O & O & \cdots & A_{k,k-1} & O \end{bmatrix}.$$

We say that $(2.33')$ is the *normal form* of an irreducible $n \times n$ matrix $A \geq O$, which is cyclic of index $k$ $(>1)$. More generally, we have

DEFINITION 2.3. An $n \times n$ complex matrix $A$ (not necessarily non-negative or irreducible) is *weakly cyclic of index $k$* $(>1)$ if there exists an $n \times n$ permutation matrix $P$ such that $PAP^T$ is of the form $(2.33')$, where the null diagonal submatrices are square.

It is interesting to note that a matrix can be simultaneously weakly cyclic of different indices. See Exercise 8 of this section.

We now remark that the rotational invariance in Theorem 2.3 of the eigenvalues of a cyclic matrix could have been established *directly* from the normal form $(2.33')$ of the matrix. In fact, using this normal form in the general case of weakly cyclic matrices, one similarly establishes the following result of Romanovsky (1936). See Exercise 12.

**Theorem 2.4.** *Let $A = (a_{i,j})$ be an $n \times n$ weakly cyclic matrix of index $k > 1$. Then*

$$(2.36) \qquad \phi(t) = \det(tI - A) = t^m \prod_{i=1}^{r} (t^k - \sigma_i^k).$$

As a simple example of a cyclic matrix, consider the non-negative matrix $B$ of (1.7). As we have already mentioned, $B$ is irreducible and, as defined, is in the normal form of a matrix which is cyclic of index 2. Its characteristic polynomial of (1.8) is, moreover, exactly of the form (2.35) where $k = 2$.

We see from (2.33), that if a matrix $A$ is cyclic of index $k$ ($>1$), it necessarily has zero diagonal entries. However, direct computation of the powers of the matrix $A$ by means of (2.33) shows that each matrix $A^p$, $p > 1$, necessarily contains some zero entries. We shall now contrast this property of cyclic matrices with another fundamental result of Frobenius (1912) on primitive matrices.

**Lemma 2.6.** *If $A > O$ is an $n \times n$ matrix, then $A$ is primitive.*

*Proof.* Since $A > O$, $A$ is obviously non-negative and irreducible. If $A$ were cyclic of some index $k > 1$, then from Theorem 2.3 $A$ would necessarily have some zero entries. This contradicts the hypothesis $A > O$ and thus $A$ is primitive, which completes the proof.

We remark that, by Definition 2.2, a primitive matrix is necessarily both non-negative and irreducible.

**Lemma 2.7.** *If $A$ is primitive, then $A^m$ is also primitive for all positive integers $m$.*

*Proof.* Since $\rho(A)$ is a simple eigenvalue of $A$ and the only eigenvalue of $A$ with modulus $\rho(A)$, then $(\rho(A))^m$ is a simple eigenvalue of $A^m$ and the only eigenvalue of $A^m$ of modulus $(\rho(A))^m$. Thus, as $A^m \geq O$ for $m \geq 1$, it suffices to show that $A^m$ is irreducible for all $m \geq 1$. On the contrary, suppose that for some positive integer $\nu$, $A^\nu$ is reducible and has the form

$$(2.37) \qquad A^\nu = \begin{bmatrix} B & C \\ O & D \end{bmatrix},$$

where $B$ and $D$ are square submatrices. Let $A\mathbf{x} = \rho(A)\mathbf{x}$ where $\mathbf{x} > \mathbf{0}$. Thus,

$$A^{\nu}\mathbf{x} = (\rho(A))^{\nu}\mathbf{x},$$

and if

$$\mathbf{x} = \begin{bmatrix} \mathbf{x}_1 \\ \mathbf{x}_2 \end{bmatrix},$$

then from (2.37),

$$D\mathbf{x}_2 = (\rho(A))^{\nu}\mathbf{x}_2,$$

which gives $(\rho(A))^{\nu}$ as an eigenvalue of $D$. As $A^T$, where $A^T$ is the transpose of $A$, is itself a non-negative and irreducible $n \times n$ matrix, Theorem 2.1 applied to $A^T$ shows that there exists a vector $\mathbf{y} > \mathbf{0}$ such that $A^T\mathbf{y} = \rho(A)\,\mathbf{y}$. From the form of (2.37), we similarly conclude that $(\rho(A))^{\nu}$ is an eigenvalue of $B^T$, and therefore of $B$ also. But as the eigenvalues of $A^{\nu}$ are just the eigenvalues of $B$ and $D$, we necessarily conclude that $(\rho(A))^{\nu}$ is a multiple eigenvalue of $A^{\nu}$, which is a contradiction. Thus, $A^{\nu}$ is irreducible for all $\nu \geq 1$, which completes the proof.

**Lemma 2.8.** *If* $A = (a_{i,j}) \geq O$ *is an irreducible* $n \times n$ *matrix with* $a_{i,i} > 0$ *for all* $1 \leq i \leq n$, *then* $A^{n-1} > O$.

*Proof.* Actually, this result follows almost immediately from Lemma 2.1, since it is obvious that we can construct a non-negative irreducible $n \times n$ matrix $B$ such that $A \geq \gamma(I + B)$ where $\gamma > 0$. But as

$$A^{n-1} \geq \gamma^{n-1}(I + B)^{n-1} > O,$$

we have the desired result.

With the above lemmas,[†] we prove the following important theorem of Frobenius (1912):

**Theorem 2.5.** *Let* $A \geq O$ *be an* $n \times n$ *matrix. Then* $A^m > O$ *for some positive integer* $m$ *if and only if* $A$ *is primitive.*

*Proof.* If an $n \times n$ matrix $B$ is reducible, then there exists by Definition 1.5 an $n \times n$ permutation matrix $P$ for which

$$PBP^T = \begin{bmatrix} B_{1,1} & B_{1,2} \\ O & B_{2,2} \end{bmatrix},$$

† Due to Herstein (1954).

where the diagonal submatrices $B_{1,1}$ and $B_{2,2}$ are square. From this, it follows that the powers of the matrix $PBP^T$ necessarily are of the form

$$PB^m P^T = \begin{bmatrix} B_{1,1}^m & B_{1,2}^{(m)} \\ 0 & B_{2,2}^m \end{bmatrix},$$

so that every power of a reducible matrix is also reducible. Returning to the hypothesis of this lemma that $A^m > 0$, we must surely have then that $A$ is irreducible. If $A$ is not primitive, then $A$ is cyclic of some index $k > 1$. Thus, by Definition 2.2 there are $k$ eigenvalues of $A$ with modulus $\rho(A)$, and thus there are $k$ eigenvalues of $A^m$ with modulus $(\rho(A))^m$. This contradicts Lemma 2.6, which proves that if $A^m > 0$, then $A$ is primitive. Conversely, if $A$ is primitive, then $A$ is by definition irreducible, which implies that there exists a closed path (see the discussion of Sec. 1.4), $a_{1,i_1}, a_{i_1,i_2}, \cdots, a_{i_{r_1-1}, i_{r_1=1}}$ of nonzero entries of $A$. Thus, $A^{r_1} \equiv (a_{i,j}^{(1)})$, which is primitive by Lemma 2.7, has $a_{1,1}^{(1)} > 0$. Since $A^{r_1}$ is irreducible, there again exists a closed path $a_{2,i_1}^{(1)}, a_{i_1,i_2}^{(1)}, \cdots, a_{i_{r_2-1}, i_{r_2=2}}^{(1)}$ of nonzero entries of $A^{r_1}$. Thus $(A^{r_1})^{r_2} \equiv (a_{i,j}^{(2)})$ has $a_{1,1}^{(2)} > 0$, as well as $a_{2,2}^{(2)} > 0$. Continuing in this way, we see that $A^{r_1 r_2 \cdots r_n}$ is irreducible, non-negative, and has all its diagonal entries positive. Now, with Lemma 2.8, some power of $A$ is positive, which completes the proof.

If $A$ is an $n \times n$ primitive matrix, then by Theorem 2.4 some positive integer power of $A$ is positive, and it is clear from the irreducibility of $A$ that all subsequent powers of $A$ are also positive. Thus, there exists a *least* positive integer $\gamma(A)$, called the *index of primitivity* of $A$, for which $A^{\gamma(A)} > 0$. As an example, let $A = (a_{i,j}) \geq 0$ be an irreducible $n \times n$ matrix with $a_{i,i} > 0$ for all $1 \leq i \leq n$. From Lemma 2.8 and Theorem 2.5, $A$ is primitive and $\gamma(A) \leq n - 1$. If only $a_{1,1} > 0$, then again $A$ must be primitive since cyclic matrices have only zero diagonal entries and it can be shown in this case that $\gamma(A) \leq 2n - 2$.[†] In general, for an arbitrary $n \times n$ primitive matrix, it is known[‡] that

$$(2.38) \qquad\qquad \gamma(A) \leq n^2 - 2n + 2.$$

We conclude this section with a result that emphasizes the difference between the behavior of powers of cyclic and primitive matrices.[§] This result will have important applications in Chapter 5.

[†] See Exercise 1 of this section.
[‡] Wielandt (1950), Holladay and Varga (1958), Rosenblatt (1957), and Pták (1958).
[§] This results is again due to Frobenius (1912).

**Theorem 2.6.** *Let $A$ be an $n \times n$ weakly cyclic matrix of index $k > 1$. Then $A^{jk}$ is completely reducible for every $j \geq 1$, i.e., there exists an $n \times n$ permutation matrix $P$ such that*

(2.39) $\qquad PA^{jk}P^T = \begin{bmatrix} C_1^j & & & & \\ & C_2^j & & & \\ & & \ddots & & \\ & & & & C_k^j \end{bmatrix}, \qquad j \geq 1.$

*where each diagonal submatrix $C_i$ is square, and*

$$\rho(C_1) = \rho(C_2) = \cdots = \rho(C_k) = \rho^k(A).$$

*Moreover, if the matrix $A$ is non-negative, irreducible, and cyclic of index $k$, then each submatrix $C_i$ is primitive.*

*Proof.* If $A$ is weakly cyclic of index $k > 1$, the direct computation of the powers of $PAP^T$ of (2.33') shows that the representation of $PA^{jk}P^T$ of (2.39) is valid for every $j \geq 1$. Again from (2.33'), the matrices $C_1$ of (2.39) turn out to be products of $k$ factors of the form $A_{i,i-1}A_{i-1,i-2} \cdots A_{i+1,i}$, which can be shown to have the same spectral radii. The remainder of this proof is considered in Exercise 4 below.

## EXERCISES

1. Let $A \geq O$ be an irreducible $n \times n$ matrix. If $A$ has exactly $d \geq 1$ diagonal entries positive, then $A$ is *primitive* and it is known (Holladay and Varga (1958)) that

$$\gamma(A) \leq 2n - d - 1.$$

Prove this result for the special case when $d = 1$.

2. Let $A \geq O$ be a primitive $n \times n$ matrix, and suppose that $A$ is *symmetrically non-negative*, i.e., $a_{i,j} > 0$ if and only if $a_{j,i} > 0$. Show that

$$\gamma(A) \leq 2n - 2.$$

(*Hint*: Apply Lemma 2.8 to $A^2$.)

**3.** If $A \geq O$ is an irreducible $n \times n$ cyclic matrix of index $k > 1$, prove that $k = 2$ if $A$ is symmetric. More generally, if $A \neq O$ is any weakly cyclic complex matrix, of index $k > 1$ (in the sense of Definition 2.3), prove that $k = 2$ if $A$ has the property that $a_{i,j} \neq 0$ if and only if $a_{j,i} \neq 0$.

**4.** If the matrix $A$ of Theorem 2.6 is a non-negative irreducible cyclic matrix of index $k > 1$, prove that the matrices $C_i$, $1 \leq i \leq k$, of (2.39) are primitive. (*Hint*: If some $C_i$ is cyclic, how many eigenvalues of $A$ have modulus $\rho(A)$?)

**5.** Using the notation of Exercise 2, Sec. 2.1, it is known that both the sequences $\{\bar{\lambda}_r\}_{r=0}^{\infty}$ and $\{\underline{\lambda}_r\}_{r=0}^{\infty}$ converge to $\rho(A)$ for an arbitrary initial vector $\mathbf{x}^{(0)}$ if and only if the irreducible matrix $A \geq O$ is primitive. Prove this result under the simplifying assumption that the eigenvectors of the $n \times n$ matrix $A$ span the vector space $V_n(C)$.

**6.** If $A \geq O$ is a primitive $n \times n$ matrix, prove that

$$\lim_{m \to \infty} [\text{tr } (A^m)]^{1/m} = \rho(A), \text{ where tr } (B) \equiv \sum_{i=1}^{n} b_{i,i} \text{ if } B = (b_{i,j}).$$

What is the corresponding result for cyclic matrices?

**7.** Let $A \geq O$ be an irreducible $n \times n$ matrix, and assume that its characteristic polynomial $\phi(t) = \det (tI - A)$ is one of the form

$$\phi(t) = t^n + \alpha_1 t^{n_1} + \alpha_2 t^{n_2} + \cdots + \alpha_m t^{n_m},$$

where each $\alpha_i$ is nonzero, $1 \leq i \leq m$, and $n > n_1 > \cdots > n_m \geq 0$. If $\nu$ is the greatest common divisor of the differences $n - n_1$, $n_1 - n_2$, $\cdots$, $n_{m-1} - n_m$, prove that $A$ is cyclic of index $\nu$ if $\nu > 1$, or that $A$ is primitive if $\nu = 1$ (Frobenius (1912)).

**8.** Show that there exist different partitionings of the matrix

$$B = \begin{bmatrix} 0 & 0 & 0 & i \\ 0 & 0 & 0 & 0 \\ 0 & 1 & 0 & 0 \\ 0 & 0 & 0 & 0 \end{bmatrix}$$

such that $B$ is simultaneously weakly cyclic of indices 2, 3, and 4. What is the characteristic polynomial for $B$?

**9.** Let $A \geq O$ be an irreducible $n \times n$ matrix. Prove that there exists an $n \times n$ matrix $B$ such that

$$\lim_{m \to \infty} \left( \frac{A}{\rho(A)} \right)^m = B.$$

if and only if $A$ is primitive. Moreover, if $A$ is primitive, prove that each row of $B$ is just a scalar multiple of any other row of $B$.

**10.** Let the $n \times n$ matrix $B$ be simultaneously weakly cyclic of indices $p$ and $q$, where $p > 1$ and $q > 1$ are relatively prime. Show that all eigenvalues of $B$ are zero.

**11.** Let $A \geq O$ be an $n \times n$ irreducible matrix that is weakly cyclic of index $k > 1$. Prove that $A$ is cyclic of index $mk > 1$, where $m$ is some positive integer. Construct an example where $m > 1$.

**12.** Let the $n \times n$ matrix $A$ be weakly cyclic of index $k > 1$, and let $\mathbf{z}$ be an eigenvalue of $A$ corresponding to the nonzero eigenvalue $\lambda$. Using $(2.33')$, directly construct $k - 1$ eigenvectors $\mathbf{w}_l$ such that

$$A\mathbf{w}_l = \lambda \exp\left(\frac{2\pi i l}{k}\right) \mathbf{w}_l, \qquad l = 1, 2, \cdots, k - 1.$$

**13.** An $n \times n$ circulant matrix $A$ is a matrix of the form

$$A = \begin{bmatrix} \alpha_0 & \alpha_1 & \alpha_2 & \cdots & \alpha_{n-2} & \alpha_{n-1} \\ \alpha_{n-1} & \alpha_0 & \alpha_1 & \cdots & \alpha_{n-3} & \alpha_{n-2} \\ \vdots & & & & & \vdots \\ \alpha_2 & \alpha_3 & \alpha_4 & \cdots & \alpha_0 & \alpha_1 \\ \alpha_1 & \alpha_2 & \alpha_3 & \cdots & \alpha_{n-1} & \alpha_0 \end{bmatrix}.$$

Show that the eigenvalues $\lambda_j$ of $A$ can be expressed as

$$\lambda_j = \alpha_0 + \alpha_1 \phi_j + \alpha_2 \phi_j^2 + \cdots + \alpha_{n-1} \phi_j^{n-1}, \qquad 0 \leq j \leq n - 1,$$

where $\phi_j = \exp(2\pi i j / n)$.

## 2.3. REDUCIBLE MATRICES

In the previous sections of this chapter we have investigated the Perron-Frobenius theory of non-negative irreducible square matrices. We now consider extensions of these basic results which make no assumption of irreducibility. Actually, such extensions follow easily by continuity arguments, since any non-negative square matrix can be made into a non-negative irreducible (even primitive) square matrix simply by replacing each zero entry of the matrix by an arbitrarily small $\epsilon > 0$. Nonetheless, we shall give the structure for non-negative reducible matrices, as this will be of later use to us.

Let $A$ be a reducible $n \times n$ matrix. By Definition 1.5, there exists an $n \times n$ permutation matrix $P_1$ such that

$$(2.40) \qquad P_1 A P_1^T = \begin{bmatrix} A_{1,1} & A_{1,2} \\ & \\ O & A_{2,2} \end{bmatrix},$$

where $A_{1,1}$ is an $r \times r$ submatrix and $A_{2,2}$ is an $(n - r) \times (n - r)$ submatrix, where $1 \le r < n$. We again ask if $A_{1,1}$ and $A_{2,2}$ are irreducible, and if not, we reduce them in the manner we initially reduced the matrix $A$ in (2.40). Thus, there exists an $n \times n$ permutation matrix $P$ such that

$$(2.41) \qquad P A P^T = \begin{bmatrix} R_{1,1} & R_{1,2} & \cdots & R_{1,m} \\ O & R_{2,2} & \cdots & R_{2,m} \\ \vdots & & \ddots & \vdots \\ O & O & \cdots & R_{m,m} \end{bmatrix},$$

where each square submatrix $R_{j,j}$, $1 \le j \le m$, is either irreducible or a $1 \times 1$ null matrix (see Definition 1.5). We shall say that (2.41) is the *normal form*† of a reducible matrix $A$. Clearly, the eigenvalues of $A$ are the eigenvalues of the square submatrices $R_{j,j}$, $1 \le j \le m$. With (2.41), we prove the following generalization of Theorem 2.1.

**Theorem 2.7.** *Let $A \ge O$ be an $n \times n$ matrix. Then,*

1. *$A$ has a non-negative real eigenvalue equal to its spectral radius. Moreover, this eigenvalue is positive unless $A$ is reducible and the normal form (2.41) of $A$ is strictly upper triangular.*‡
2. *To $\rho(A)$, there corresponds an eigenvector $\mathbf{x} \ge \mathbf{0}$.*
3. *$\rho(A)$ does not decrease when any entry of $A$ is increased.*

*Proof.* If $A$ is irreducible, the conclusions follow immediately from Theorem 2.1. If $A$ is reducible, assume that $A$ is in the normal form of (2.41). If any submatrix $R_{j,j}$ of (2.41) is irreducible, then $R_{j,j}$ has a positive eigenvalue equal to its spectral radius. Similarly, if $R_{j,j}$ is a $1 \times 1$ null matrix, its single eigenvalue is zero. Clearly, $A$ then has a non-negative real eigenvalue equal to its spectral radius. If $\rho(A) = 0$, then each $R_{j,j}$ of (2.41) is a $1 \times 1$ null matrix, which proves that the matrix of (2.41) is strictly upper triangular. The remaining statements of this theorem follow by applying a continuity argument to the result of Theorem 2.1.

† This terminology differs slightly from that of Gantmakher (1959, p. 90). We remark that this normal form of a reducible matrix is not necessarily unique. Gantmakher (1959) shows that this normal form is unique, up to permutations by blocks of (2.41).

‡ An $n \times n$ matrix $T = (t_{i,j})$ is *strictly upper triangular* only if $t_{i,j} = 0$ for all $i \ge j$.

Using the notation of Definition 2.1, we include the following generalization of Lemma 2.3.

**Theorem 2.8.** *Let $A$ and $B$ be two $n \times n$ matrices with $O \leq |B| \leq A$. Then,*

$$\rho(B) \leq \rho(A).$$

*Proof.* If $A$ is irreducible, then we know from Lemma 2.3 and its corollary that $\rho(B) \leq \rho(A)$. On the other hand, if $A$ is reducible, we note that the property $O \leq |B| \leq A$ is invariant under similarity transformations by permutation matrices. Putting $A$ into its normal form (2.41), we now simply apply the argument above to the submatrices $|R_{j,j}(B)|$ and $R_{j,j}(A)$ of the matrices $|B|$ and $A$, respectively.

## EXERCISES

**1.** Construct the normal form (2.41) of the matrix

$$A = \begin{bmatrix} 1 & 8 & 0 & 0 & 1 \\ 0 & 0 & 1 & 0 & 0 \\ 0 & 0 & 0 & 2 & 0 \\ 0 & 4 & 0 & 0 & 0 \\ 1 & 0 & 0 & 0 & 0 \end{bmatrix}.$$

What is the value of $\rho(A)$?

**2.** If $A = (a_{i,j}) \geq O$ is an $n \times n$ matrix, and $\mathbf{x}$ is any vector with positive components $x_1, x_2, \cdots, x_n$, prove the analog of Theorem 2.2:

$$\min_{1 \leq i \leq n} \left( \frac{\left| \sum_{j=1}^{n} a_{i,j} x_j \right|}{x_i} \right) \leq \rho(A) \leq \max_{1 \leq i \leq n} \left( \frac{\left| \sum_{j=1}^{n} a_{i,j} x_j \right|}{x_i} \right).$$

**\*3.** Let $A \geq O$ be an $n \times n$ matrix such that if $\mathbf{x} > \mathbf{0}$, then $A\mathbf{x} > \mathbf{0}$. Defining

$$\mathbf{x}^{(r)} = A\mathbf{x}^{(r-1)} = \cdots = A^r \mathbf{x}^{(0)},$$

where $\mathbf{x}^{(0)}$ is an arbitrary vector with positive components, let $\underline{\lambda}_r$ and and $\bar{\lambda}_r$ be defined as in Exercise 2 of Sec. 2.1. Show that

$$\underline{\lambda}_0 \leq \underline{\lambda}_1 \leq \cdots \leq \rho(A) \leq \cdots \leq \bar{\lambda}_1 \leq \bar{\lambda}_0.$$

When do both of these sequences converge to $\rho(A)$ for an arbitrary initial positive vector $\mathbf{x}^{(0)}$?

**4.** Let $a_1, a_2, \cdots, a_n$ be $n$ arbitrary complex numbers, and let $z_j$ be any zero of the polynomial

$$p_n(z) \equiv z^n - a_1 z^{n-1} - a_2 z^{n-2} - \cdots - a_n.$$

If $r^*$ is the largest non-negative real zero of the polynomial

$$\hat{p}_n(z) \equiv z^n - |a_1| z^{n-1} - |a_2| z^{n-2} - \cdots - |a_n|,$$

prove (Perron (1907)) that

$$\max_{1 \leq j \leq n} |z_j| \leq r^*.$$

(*Hint*: Use the fact that the *companion* or *Frobenius matrix*

$$B = \begin{bmatrix} 0 & 0 & \cdots & 0 & a_n \\ 1 & 0 & \cdots & 0 & a_{n-1} \\ 0 & 1 & & 0 & a_{n-2} \\ \vdots & & & & \vdots \\ \vdots & & & & \vdots \\ \vdots & & & & \vdots \\ 0 & 0 & \cdots & 1 & a_1 \end{bmatrix}$$

is such that $\det (zI - B) = p_n(z)$, and apply Theorem 2.8.)

**5.** Using the notation of the previous exercise, prove that

$$\max_{1 \leq j \leq n} |z_j| \leq r^* \leq \max \{|a_n|, 1 + |a_{n-1}|, \cdots, 1 + |a_1|\}.$$

(*Hint*: Apply the result of Exercise 2 to the matrix $|B|$. See Wilf (1961)).

**6.** Show that the upper bound $\nu$ for the spectral radius of an arbitrary $n \times n$ complex matrix $A$ of Corollary 3 of Theorem 1.5 can be deduced directly from Theorem 2.8 and the result of the previous Exercise 2.

## 2.4. NON-NEGATIVE MATRICES AND DIRECTED GRAPHS

We recall that the concept of a directed graph of a matrix, as introduced in Sec. 1.4, gives us a geometrical method for deciding whether a particular matrix is reducible or irreducible. If a matrix is reducible, its normal form (Sec. 2.3) is also of interest to us. But it becomes clear that, as similarity transformations by permutation matrices merely permute the subscripts of the nodes of the associated directed graph, the permutation matrix that brings a reducible matrix to its normal form can also be *deduced* from the associated directed graph of the matrix.

It is interesting that directed graphs of matrices are also useful in determining whether non-negative irreducible matrices are primitive or cyclic of some index $k$. First, the directed graphs for the powers of a non-negative $n \times n$ matrix $A$ can be simply deduced from the directed graph $G(A)$ of the matrix $A$, since the directed graph for the matrix $A^r$, $r \geq 1$, is the directed graph obtained by considering all paths of $G(A)$ of length exactly $r$. In other words, for the path $\overrightarrow{P_i P_{l_1}}$, $\overrightarrow{P_{l_1} P_{l_2}}$, $\cdots$, $\overrightarrow{P_{l_{r-1}} P_{l_r=j}}$ determined from the graph $G(A)$, we directly connect the node $P_i$ to the node $P_j$ with a path directed toward $P_j$ of length unity for the directed graph $G(A^r)$. With this observation, we can interpret Theorems 2.5 and 2.6 as follows: If $A$ is primitive, then for the directed graph $G(A^r)$, $r$ sufficiently large, each node $P_i$ is connected to every node $P_j$ by paths of length unity. On the other hand, if $A$ is irreducible and cyclic of index $k > 1$, each graph $G(A^{rk})$, $r \geq 1$, is the union of $k$ disjoint strongly connected directed subgraphs. To illustrate this, the matrix $B$ of (1.7) is cyclic of index 2. From the graph $G(B)$ of (1.35), we find that

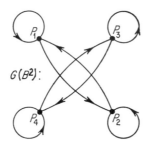

$$G(B^2):$$

showing geometrically that the directed graph of $B^2$ is the union of two disjoint strongly connected directed subgraphs. But, with the directed graphs for the powers of $A$, we can now give a simple graphical method for determining whether a given non-negative irreducible matrix $A$ is primitive or cyclic of some index $k > 1$.†

**Theorem 2.9.** *Let* $A = (a_{i,j}) \geq O$ *be an irreducible* $n \times n$ *matrix, with* $G(A)$ *as its directed graph. For each node* $P_i$ *of* $G(A)$, *consider all closed paths connecting* $P_i$ *to itself. If* $S_i$ *is the set of all the lengths* $m_i$ *of these closed paths, let* $k_i$ *be the greatest common divisor of the lengths, i.e.,*

$$(2.42) \qquad k_i = g.c.d._{m_i \epsilon S_i} \{m_i\}, \qquad 1 \leq i \leq n.$$

*Then,* $k_1 = k_2 = \cdots = k_n = k$, *where* $k = 1$ *when* $A$ *is primitive, and* $k > 1$ *when* $A$ *is cyclic of index* $k$.

† Due to Romanovsky (1936).

*Proof.* If $A^r \equiv (a_{i,j}^{(r)})$, let $m_1$ and $m_2$ be two positive integers in the set $S_i$. By definition, $a_{i,i}^{(m_1)}$ and $a_{i,i}^{(m_2)}$ are positive real numbers. Since

$$a_{i,i}^{(m_1+m_2)} = \sum_{l=1}^{n} a_{i,l}^{(m_1)} a_{l,i}^{(m_2)} \geq a_{i,i}^{(m_1)} a_{i,i}^{(m_2)} > 0,$$

it follows that $(m_1 + m_2) \in S_i$, proving that each set $S_i$ is a set of positive integers *closed under addition*. As is well known, the set $S_i$ contains all but a finite number of multiples of its greatest common divisor $k_i$.[†] If some $k_i = 1$, then $a_{i,i}^{(p)} > 0$ for all $p$ sufficiently large; from (2.33) we see that $A$ cannot be cyclic, and thus $A$ is primitive. But if $A$ is primitive, $a_{j,j}^{(p)} > 0$ for every $1 \leq j \leq n$ and all $p$ sufficiently large, so that

$$k_1 = k_2 = \cdots = k_n = 1.$$

Now, assume that some $k_i > 1$. As this implies that $a_{i,i}^{(p)}$ is not positive for all $p$ sufficiently large, then $A$ is cyclic of some index $\nu > 1$. From the proof of Theorem 2.6, we have that the only powers of the matrix $A$ which can have positive diagonal entries are of the form $A^{j\nu}$, and using the fact that the square diagonal submatrices $C_i$, $1 \leq i \leq k$, of (2.39) are *primitive*, then $A^{j\nu}$ has all its diagonal entries positive for all $j$ sufficiently large. Thus,

$$\nu = k_1 = k_2 = \cdots = k_n,$$

which completes the proof.

To illustrate these results, consider the non-negative matrices $B_i$ defined by

$$(2.43) \qquad B_1 = \begin{bmatrix} 0 & 0 & 1 & 0 \\ 0 & 0 & 0 & 1 \\ 0 & 1 & 0 & 0 \\ 1 & 0 & 0 & 0 \end{bmatrix}, \quad B_2 = \begin{bmatrix} 0 & 0 & 1 & 0 \\ 0 & 0 & 0 & 1 \\ 1 & 0 & 0 & 0 \\ 0 & 1 & 0 & 0 \end{bmatrix},$$

$$(2.43') \qquad B_3 = \begin{bmatrix} 0 & 1 & 0 & 0 \\ 0 & 0 & 1 & 0 \\ 0 & 0 & 0 & 1 \\ 1 & 1 & 0 & 0 \end{bmatrix}, \quad B_4 = \begin{bmatrix} 1 & 1 & 0 & 0 \\ 0 & 0 & 1 & 0 \\ 0 & 0 & 0 & 1 \\ 1 & 0 & 0 & 0 \end{bmatrix},$$

[†] For example, see Kemeny and Snell (1960, p. 6).

whose directed graphs are respectively

(2.44)    $G(B_1)$:

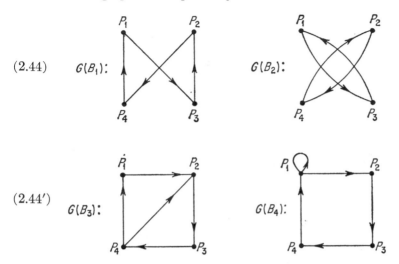

(2.44′)    $G(B_3)$:

It is easy to deduce *by inspection* of the above directed graphs that these matrices are irreducible with the exception of $B_2$.

We now apply the result of Theorem 2.9 to the particular matrices $B_1$, $B_3$, and $B_4$ of (2.43) and (2.43′). For the matrix $B_1$, inspection of the graph $G(B_1)$ of (2.44) shows that

$$k_1 = g.c.d. \{4, 8, 12, \cdots\} = 4,$$

so that $B_1$ is cyclic of index 4. For the matrix $B_3$, we find closed paths connecting $P_4$ to itself of lengths 3 and 4, whose greatest common divisor is unity. Thus, $k_4 = 1$ and we conclude that $B_3$ is primitive. For the matrix $B_4$, it is obvious that there is a loop, a closed path of length one, connecting $P_1$ to itself, so that $k_1 = 1$, and thus $B_4$ is also primitive.

With Theorem 2.9, it is convenient now to introduce the almost obvious notions of cyclic and primitive finite directed graphs. For this we make the following

DEFINITION 2.4. If $G$ is a strongly connected finite directed graph, then $G$ is a *cyclic graph of index* $k > 1$, or a *primitive graph* if the greatest common divisor of all the lengths of its closed paths is, respectively, $k > 1$ or $k = 1$.

These simple results from our excursion into graph theory will be very useful to us in Chapter 4 and 6. In Chapter 6 for example, we consider approximating differential equations by difference equations. These difference equations, a system of linear equations, have an associated directed graph, and the examination of this directed graph will be helpful in deciding when particular theoretical results for various iterative methods are rigorously applicable.

## EXERCISES

1. Construct the directed graph for the matrix $A$ of Exercise 1, Sec. 2.3. In the normal form for this reducible matrix, determine whether the non-null diagonal submatrices $R_{j,j}$ are cyclic or primitive.

2. Consider the $n \times n$ matrix

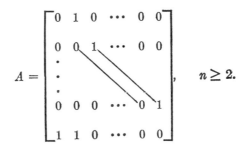

$$A = \begin{bmatrix} 0 & 1 & 0 & \cdots & 0 & 0 \\ 0 & 0 & 1 & \cdots & 0 & 0 \\ \vdots & & & & & \\ 0 & 0 & 0 & \cdots & 0 & 1 \\ 1 & 1 & 0 & \cdots & 0 & 0 \end{bmatrix}, \quad n \geq 2.$$

By considering the directed graph $G(A)$ of $A$, prove that $A$ is irreducible and primitive. Moreover, prove that its index of primitivity $\gamma(A)$ is such that

$$\gamma(A) = n^2 - 2n + 2.$$

Thus, the inequality of (2.38) is best possible (Wielandt (1950)).

3. With the explicit determination of the matrix $A^{\gamma(A)}$ for, say, $n = 5$, where $A$ is defined in the previous exercise, obtain bounds for $\rho(A)$ by applying Theorem 2.2 to $A^{\gamma(A)}$.

4. Let

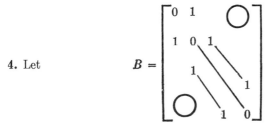

$$B = \begin{bmatrix} 0 & 1 & & \\ 1 & 0 & 1 & \\ & 1 & & 1 \\ & & 1 & 0 \end{bmatrix}$$

be an $n \times n$ matrix, $n \geq 2$. By means of graph theory, show that $B$ is irreducible and cyclic of index 2. More generally, if

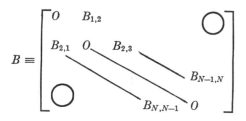

where the null diagonal submatrices are square, show that $B$ is weakly cyclic of index 2.

**5.** Let $A = (a_{i,j})$ be a non-negative $n \times n$ matrix. Prove (Ullman (1952)) that $\rho(A) > 0$ if and only if the directed graph $G(A)$ has at least one closed path. (*Hint*: Use Theorem 2.7.)

**6.** Let $B$ be an irreducible $n \times n$ complex matrix. Prove that $G(B)$ is a cyclic graph of index $p$ ($> 1$) if and only if $B$ is weakly cyclic of index $p$.

## BIBLIOGRAPHY AND DISCUSSION

**2.1.** The method of proof of the Perron-Frobenius Theorem 2.1 as given in Sec. 2.1 is basically due to Wielandt (1950), and is similar in part to the corresponding proofs given in Gantmakher (1959) and Debreu and Herstein (1953). It should be pointed out that there are other proofs of this important theorem. For example, Alexandroff and Hopf (1935, p. 480) essentially proved the weaker Theorem 2.8 by means of the Brouwer Fixpoint Theorem. For extensions of this result using topological facts closely related to the Brouwer Fixpoint Theorem, see Fan (1958). Another proof of the Perron-Frobenius theorem has been given by Householder (1958), who uses properties of general matrix norms to establish this result. Also, Putnam (1958) and Karlin (1959a) show how the Perron-Frobenius theorem and its generalizations are a consequence of the Vivanti-Pringsheim theorem for power series. See also Bellman (1960, p. 299).

For other recent proofs of the Perron-Frobenius theorem for matrices, see Ullman (1952), A. Brauer (1957b), and H. Samelson (1957).

The idea that a non-negative matrix carries the positive hyperoctant of vectors into itself has given rise to several important generalizations of the Perron-Frobenius Theorem to operators defined on more general spaces. For recent results, see Krein and Rutman (1948), G. Birkhoff (1957), H. Schaefer (1960), Mewborn (1960), Bonsall (1960), and references given therein.

The inclusion Theorem 2.2, giving upper- and lower-bound estimates for the spectral radius $\rho(A)$ of a non-negative irreducible matrix, is an improvement of Wielandt (1950) on the original result of Collatz (1942a). Moreover, it should be stated that Collatz (1942a) also shows in the same paper that if $A = (a_{i,j})$ is a real symmetric $n \times n$ matrix, and $\mathbf{u}$ is any vector none of whose components are zero, then the interval

$$\min_{1 \leq i \leq n} \left( \frac{\left| \sum_{j=1}^{n} a_{i,j} u_j \right|}{u_i} \right) \leq x \leq \max_{1 \leq i \leq n} \left( \frac{\left| \sum_{j=1}^{n} a_{i,j} u_j \right|}{u_i} \right)$$

contains at least one eigenvalue of $A$. For further results on inclusion theorems for $\rho(A)$ where $A \geq O$ is an $n \times n$ irreducible matrix with improved upper- and lower-bound estimates, see A. Brauer (1957a), W. Lederman (1950), A. Ostrowski (1952) and (1961), and A. Ostrowski and Hans Schneider (1960). For an extension of Collatz's inclusion theorem for $\rho(A)$ for more general operators, see Mewborn (1960).

Another min-max characterization of the spectral radius of a non-negative irreducible matrix has been given by Birkhoff and Varga (1958). See Exercise 7.

Non-negative square matrices $A = (a_{i,j})$ satisfying

$$\sum_{j=1}^{n} a_{i,j} = \sigma, \qquad 1 \leq i \leq n,$$

as in Lemma 2.5, are called *generalized stochastic matrices* for $\sigma > 0$. In the special case when $\sigma = 1$, these matrices are called *stochastic matrices*. Such matrices have applications in several branches of mathematics. See Romanovsky (1936), Bellman (1960), and Kemeny and Snell (1960).

**2.2.** Historically, Frobenius (1912) in his basic paper used the terms *primitive* and *imprimitive* to classify non-negative irreducible matrices. Later, Romanovsky (1936) introduced the term *cyclic of index k* for imprimitive matrices. In proving Theorem 2.5, we have used the method of proof of Herstein (1954). For generalizations of Theorem 2.3 to more general operators, see Rotha (1961).

For the index of primitivity $\gamma(A)$ of a primitive matrix, the first results for bounds for $\gamma(A)$ were obtained by Frobenius (1912), and Wielandt (1950) stated the result of (2.38). For more recent results, see Holladay and Varga (1958), D. Rosenblatt (1957), Pták (1958), and Marík and Pták (1960).

It is interesting that properties of both primitive and cyclic matrices have nontrivial applications to the iterative solution of matrix equations. In fact, we shall show (Chapter 5) that the result of Theorem 2.6 on the complete reducibility of certain powers of a weakly cyclic matrix gives rise to a newer, but nevertheless basic, iterative method for solving elliptic difference equations.

**2.4.** The use of directed graphs in matrix theory is not new, but recently several authors (*cf.* Harary (1959) and Parter (1961a)) have applied their geometrical aspects to the classical problems of Gaussian elimination and the determination of eigenvalues of large matrices. Specific use of directed graphs will be made in Chapter 6 in analyzing the matrix properties of matrix equations derived from elliptic (and parabolic) partial differential equations.

# CHAPTER 3

# BASIC ITERATIVE METHODS AND
# COMPARISON THEOREMS

## 3.1. THE POINT JACOBI, GAUSS-SEIDEL, AND SUCCESSIVE OVERRELAXATION ITERATIVE METHODS

Let $A = (a_{i,j})$ be an $n \times n$ complex matrix, and let us seek the solution of the system of linear equations

$$(3.1) \qquad \sum_{j=1}^{n} a_{i,j} x_j = k_i, \qquad 1 \leq i \leq n,$$

which we write in matrix notation as

$$(3.1') \qquad A\mathbf{x} = \mathbf{k},$$

where $\mathbf{k}$ is a given column vector. The solution vector $\mathbf{x}$ exists and is unique if and only if $A$ is *nonsingular*, and this solution vector is given explicitly by

$$(3.2) \qquad \mathbf{x} = A^{-1}\mathbf{k}.$$

We assume throughout that the matrix $A$ is nonsingular, and moreover that its diagonal entries $a_{i,i}$ are all nonzero complex numbers.

We can express the matrix $A$ as the matrix sum

$$(3.3) \qquad A = D - E - F,$$

where $D = \operatorname{diag} \{a_{1,1}, a_{2,2}, \cdots, a_{n,n}\}$, and $E$ and $F$ are respectively strictly lower and upper triangular $n \times n$ matrices, whose entries are the negatives of the entries of $A$ respectively below and above the main diagonal of $A$. We can write $(3.1')$ as

$$(3.4) \qquad D\mathbf{x} = (E + F)\mathbf{x} + \mathbf{k}.$$

**56**

Since the diagonal entries $a_{i,i}$ of $A$ are nonzero, we can carry out the following iterative method derived from (3.4):

$$(3.5) \qquad a_{i,i} x_i^{(m+1)} = - \sum_{\substack{j=1 \\ j \neq i}}^{n} a_{i,j} x_j^{(m)} + k_i, \qquad 1 \leq i \leq n, \ m \geq 0,$$

where the $x_i^{(0)}$'s are initial estimates of the components of the unique solution of $\mathbf{x}$ of (3.1'). Clearly, we can write (3.5) as

$$(3.6) \qquad x_i^{(m+1)} = - \sum_{\substack{j=1 \\ j \neq i}}^{n} \left( \frac{a_{i,j}}{a_{i,i}} \right) x_j^{(m)} + \frac{k_i}{a_{i,i}}, \qquad 1 \leq i \leq n, \ m \geq 0.$$

In matrix notation, (3.5) becomes

$$(3.5') \qquad D\mathbf{x}^{(m+1)} = (E + F)\mathbf{x}^{(m)} + \mathbf{k}, \qquad m \geq 0,$$

and since $D$ is a nonsingular matrix, the matrix analogue of (3.6) is

$$(3.6') \qquad \mathbf{x}^{(m+1)} = D^{-1}(E + F)\mathbf{x}^{(m)} + D^{-1}\mathbf{k}, \qquad m \geq 0.$$

We shall call this iterative method the *point Jacobi* or *point total-step iterative method*† and we call the matrix

$$(3.7) \qquad B = D^{-1}(E + F)$$

the *point Jacobi matrix*‡ associated with the matrix $A$.

Examination of the Jacobi iterative method of (3.6') shows that in general one must save all the components of the vector $\mathbf{x}^{(m)}$ while computing the components of the vector $\mathbf{x}^{(m+1)}$. It would intuitively seem very attractive to use the latest estimates $x_i^{(m+1)}$ of the components $x_i$ of the unique solution vector $\mathbf{x}$ in all subsequent computations, and this results in the iterative method

$$(3.8) \qquad a_{i,i} x_i^{(m+1)} = - \sum_{j=1}^{i-1} a_{i,j} x_j^{(m+1)} - \sum_{j=i+1}^{n} a_{i,j} x_j^{(m)} + k_i,$$

$$1 \leq i \leq n, \ m \geq 0.$$

This iterative method has the computational advantage that it does not require the simultaneous storage of the two approximations $x_i^{(m+1)}$ and

† This iterative method has many different names. It was originally considered by Jacobi (1845), but Geiringer (1949) calls it the *method of simultaneous displacements*, and Keller (1958) calls it the *Richardson iterative method*. It should be noted that *total-step method* is a translation of the German word *Gesamtschrittverfahren*.

‡ The term *Jacobi matrix* unfortunately has more than one meaning in the literature. Here, we do not mean in general a tridiagonal matrix.

$x_i^{(m)}$ in the course of computation as does the point Jacobi iterative method. In matrix notation, (3.8) becomes

$$(3.8')\qquad\qquad (D - E)\mathbf{x}^{(m+1)} = F\mathbf{x}^{(m)} + \mathbf{k}, \qquad m \geq 0,$$

and as $D - E$ is a nonsingular lower triangular matrix, we can write (3.8') equivalently as

$$(3.9)\qquad\qquad \mathbf{x}^{(m+1)} = (D - E)^{-1}F\mathbf{x}^{(m)} + (D - E)^{-1}\mathbf{k}, \qquad m \geq 0.$$

We shall call this iterative method the *point Gauss-Seidel* or *point single-step iterative method*,† and we call the matrix

$$(3.10)\qquad\qquad C = (D - E)^{-1}F$$

the *point Gauss-Seidel matrix* associated with the matrix $A$.

We now introduce a third basic iterative method for solving the system of linear equations of (3.1), a method closely related to the point Gauss-Seidel iterative method. Starting directly with the Gauss-Seidel method of (3.8), we define first the components of the auxiliary vector iterates $\bar{\mathbf{x}}^{(m)}$ from

$$(3.11)\qquad a_{i,i}\bar{x}_i^{(m+1)} = - \sum_{j=1}^{i-1} a_{i,j}x_j^{(m+1)} - \sum_{j=i+1}^{n} a_{i,j}x_j^{(m)} + k_i,$$

$$1 \leq i \leq n, \quad m \geq 0.$$

The actual components $x_i^{(m+1)}$ of this iterative method are then defined from

$$(3.12)\qquad x_i^{(m+1)} = x_i^{(m)} + \omega\{\bar{x}_i^{(m+1)} - x_i^{(m)}\} = (1 - \omega)x_i^{(m)} + \omega\bar{x}_i^{(m+1)}.$$

The quantity $\omega$ is called the *relaxation factor*, and from (3.12), $x_i^{(m+1)}$ is a weighted mean of $x_i^{(m)}$ and $\bar{x}_i^{(m+1)}$, the weights depending only on $\omega$. Only for $0 \leq \omega \leq 1$ are these weights non-negative, and we shall say that the use of $\omega > 1$ ($<1$) corresponds to *overrelaxation* (*underrelaxation*). We now combine the two equations (3.11) and (3.12) into the single equation

$$(3.13)\qquad a_{i,i}x_i^{(m+1)} = a_{i,i}x_i^{(m)}$$

$$+ \omega\left\{ - \sum_{j=1}^{i-1} a_{i,j}x_j^{(m+1)} - \sum_{j=i+1}^{n} a_{i,j}x_j^{(m)} + k_i - a_{i,i}x_i^{(m)}\right\}$$

in which the auxiliary vector iterates $\bar{\mathbf{x}}^{(m)}$ do not appear. Note that this method, like the point Gauss-Seidel method, also needs but one vector

† This iterative method is also called the *method of successive displacements* by Geiringer (1949), and in certain cases it is called the *Liebmann method* (see Frankel (1950)). Again it should be noted that *single-step method* is a translation of the German word *Einzelschrittverfahren*.

approximation in the course of computation. In matrix notation, this can be written as

$$(3.13') \qquad (D - \omega E)\mathbf{x}^{(m+1)} = \{(1 - \omega) D + \omega F\}\mathbf{x}^{(m)} + \omega\mathbf{k}, \qquad m \geq 0,$$

and as $D - \omega E$ is nonsingular for any choice of $\omega$, then with $L \equiv D^{-1}E$, $U \equiv D^{-1}F$, this takes the form

$$(3.14) \qquad \mathbf{x}^{(m+1)} = (I - \omega L)^{-1}\{(1 - \omega)I + \omega U\}\mathbf{x}^{(m)}$$

$$+ \omega(I - \omega L)^{-1} D^{-1}\mathbf{k}.$$

It would be more precise to call this iterative method the *point successive over-* or *underrelaxation iterative method* to show its dependence on the relaxation factor $\omega$, but for brevity we simply call this method the *point successive overrelaxation iterative method*,† and we call the matrix

$$(3.15) \qquad \mathcal{L}_\omega \equiv (I - \omega L)^{-1}\{(1 - \omega)I + \omega U\}$$

the *point successive relaxation matrix*. Selecting $\omega = 1$ shows that this iterative method reduces exactly to the point Gauss-Seidel iterative method, and the matrix $C$ defined in (3.10) is thus equal to $\mathcal{L}_1$. As yet, we do not know what effect the relaxation parameter $\omega$ has on the spectral radius or spectral norm of the matrix $\mathcal{L}_\omega$.

To each of the three iterative methods described, we can associate error vectors $\boldsymbol{\varepsilon}^{(m)}$ defined by

$$(3.16) \qquad \boldsymbol{\varepsilon}^{(m)} = \mathbf{x}^{(m)} - \mathbf{x}, \qquad m \geq 0,$$

where $\mathbf{x}$ is the unique vector solution of (3.1'), and from the very definition of each iterative method, such as (3.14), we can express the error vectors $\boldsymbol{\varepsilon}^{(m)}$ in each case as

$$(3.17) \qquad \boldsymbol{\varepsilon}^{(m)} = M\boldsymbol{\varepsilon}^{(m-1)} = \cdots = M^m\boldsymbol{\varepsilon}^{(0)}, \qquad m \geq 0,$$

where $M$ is the corresponding iteration matrix for the specific method. As we know (Theorem 1.4), the error vectors $\boldsymbol{\varepsilon}^{(m)}$ of these iterative methods tend to the zero vector for all $\boldsymbol{\varepsilon}^{(0)}$ if and only if the spectral radius $\rho(M)$ of the matrix $M$ is less than unity.

To illustrate two of these various iterative methods, we now consider a simple numerical example based on the 4 × 4 real, symmetric and posi-

---

† This iterative method is also called the *accelerated Liebmann* method by Frankel (1950) and many subsequent authors. Kahan (1958) calls it the *extrapolated Gauss-Seidel* method. It is often called the method of *systematic overrelaxation*.

tive definite matrix $A$ derived in Sec. 1.2. Specifically, we seek to solve the matrix equation $A\mathbf{x} = \mathbf{k}$, where

$$A = \begin{bmatrix} 1 & 0 & -\frac{1}{4} & -\frac{1}{4} \\ 0 & 1 & -\frac{1}{4} & -\frac{1}{4} \\ -\frac{1}{4} & -\frac{1}{4} & 1 & 0 \\ -\frac{1}{4} & -\frac{1}{4} & 0 & 1 \end{bmatrix}, \qquad \mathbf{k} = \frac{1}{2} \begin{bmatrix} 1 \\ 1 \\ 1 \\ 1 \end{bmatrix}.$$

The unique vector solution of this matrix problem is obviously

$$x_1 = x_2 = x_3 = x_4 = 1.$$

If our initial vector approximation is $\mathbf{x}^{(0)} = \mathbf{0}$, then the initial vector $\boldsymbol{\varepsilon}^{(0)}$ has all its components equal to $-1$. For this case, the point Jacobi matrix iterative method, the special case $\omega = 1$ of the point successive overrelaxation iterative method, then it can also be verified that

$$(3.18) \quad B = \frac{1}{4} \begin{bmatrix} 0 & 0 & 1 & 1 \\ 0 & 0 & 1 & 1 \\ 1 & 1 & 0 & 0 \\ 1 & 1 & 0 & 0 \end{bmatrix}, \qquad \mathcal{L}_\omega = \begin{bmatrix} \sigma & 0 & \tau & \tau \\ 0 & \sigma & \tau & \tau \\ \sigma\tau & \sigma\tau & 2\tau^2 + \sigma & 2\tau^2 \\ \sigma\tau & \sigma\tau & 2\tau^2 & 2\tau^2 + \sigma \end{bmatrix},$$

where $\tau = \omega/4$, $\sigma = 1 - \omega$. If $\boldsymbol{\alpha}^{(m)}$ denotes the errors for the point Jacobi iterative method, and $\boldsymbol{\beta}^{(m)}$ denotes the errors for the point Gauss-Seidel iterative method, the special case $\omega = 1$ of the point successive overrelaxation iterative method, then it can also be verified that

$$(3.19) \quad \boldsymbol{\alpha}^{(m)} = B^m \boldsymbol{\varepsilon}^{(0)} = \frac{-1}{2^m} \begin{bmatrix} 1 \\ 1 \\ 1 \\ 1 \end{bmatrix}, \qquad \boldsymbol{\beta}^{(m)} = \mathcal{L}_1^m \boldsymbol{\varepsilon}^{(0)} = \frac{-1}{4^m} \begin{bmatrix} 2 \\ 2 \\ 1 \\ 1 \end{bmatrix},$$

$$m \geq 1.$$

Thus, as

$$\| \boldsymbol{\alpha}^{(m)} \| = \frac{1}{2^{m-1}} > \| \boldsymbol{\beta}^{(m)} \| = \frac{\sqrt{10}}{4^m} \qquad \text{for } \textit{all } m \geq 1,$$

we conclude that the vector norms $\| \boldsymbol{\beta}^{(m)} \|$ tend more rapidly to zero than do the vector norms $\| \, \boldsymbol{\alpha}^{(m)} \, \|$ for this particular choice of $\boldsymbol{\varepsilon}^{(0)}$. This in turn leads us to suspect that the point Gauss-Seidel iterative method is in some cases "faster" than the point Jacobi iterative method. The next section is concerned with a precise mathematical formulation of what is meant by one iterative method's being faster than another.

## EXERCISES

**1.** For the special case $\omega = \frac{1}{2}$ of (3.18), determine the vector norms

$$\| \, \mathcal{L}_{1/2}^m \boldsymbol{\varepsilon}^{(0)} \, \|, \qquad 4 \geq m \geq 1,$$

where $\boldsymbol{\varepsilon}^{(0)}$ again has all its components equal to $-1$. How do these vector norms compare with the vector norms $\| \, \boldsymbol{\alpha}^{(m)} \, \|$ and $\| \, \boldsymbol{\beta}^{(m)} \, \|$ given from (3.19)?

**2.** For $\omega = \frac{1}{2}$, the matrix $\mathcal{L}_\omega$ of (3.18) is a non-negative irreducible matrix. Find upper and lower bounds for $\rho(\mathcal{L}_{1/2})$.

**3.** Let $A$ be the $5 \times 5$ matrix defined by

$$A = \begin{bmatrix} 2 & -1 & 0 & 0 & 0 \\ -1 & 2 & -1 & 0 & 0 \\ 0 & -1 & 2 & -1 & 0 \\ 0 & 0 & -1 & 2 & -1 \\ 0 & 0 & 0 & -1 & 1 \end{bmatrix}.$$

Derive its associated point Jacobi matrix, point Gauss-Seidel matrix, and point successive overrelaxation matrix.

a. Prove that the point Jacobi matrix is non-negative, irreducible, cyclic of index 2, and convergent.
b. Prove that the point Gauss-Seidel matrix is non-negative, reducible, and convergent. Upon constructing its normal form (2.41), show that one of the diagonal submatrices of the normal form is primitive.
c. Prove that the point successive overrelaxation matrix is non-negative, convergent, and primitive for $0 < \omega < 1$. Show that $\gamma(\mathcal{L}_\omega)$, the index of primitivity, is independent of $\omega$ in this range.

## 3.2. AVERAGE RATES OF CONVERGENCE

We now suppose that we have a general iterative method

$$(3.20) \qquad \mathbf{x}^{(m+1)} = M\mathbf{x}^{(m)} + \mathbf{g}, \qquad m \geq 0,$$

where $M$ is an $n \times n$ complex matrix. If $(I - M)$ is nonsingular, then there exists a unique solution of $(I - M)\mathbf{x} = \mathbf{g}$, and if as before the error vectors $\boldsymbol{\varepsilon}^{(m)}$ for the vector iterates of (3.20) are defined as $\boldsymbol{\varepsilon}^{(m)} = \mathbf{x}^{(m)} - \mathbf{x}$, then

$$(3.21) \qquad \boldsymbol{\varepsilon}^{(m)} = M\boldsymbol{\varepsilon}^{(m-1)} = \cdots = M^m\boldsymbol{\varepsilon}^{(0)}, \qquad m \geq 0.$$

Using matrix and vector norms, Theorem 1.2 shows that

$$(3.22) \qquad || \, \boldsymbol{\varepsilon}^{(m)} \, || \leq || \, M^m \, || \cdot || \, \boldsymbol{\varepsilon}^{(0)} \, ||, \qquad m \geq 0,$$

with the equality possible for each $m$ for some vectors $\boldsymbol{\varepsilon}^{(0)}$. Thus, if $\boldsymbol{\varepsilon}^{(0)}$ is not the null vector, then $|| \, M^m \, ||$ gives us a sharp upper-bound estimate for the ratio $|| \, \boldsymbol{\varepsilon}^{(m)} \, ||/|| \, \boldsymbol{\varepsilon}^{(0)} \, ||$ of the Euclidean norms of $|| \, \boldsymbol{\varepsilon}^{(m)} \, ||$ and $|| \, \boldsymbol{\varepsilon}^{(0)} \, ||$ for $m$ *iterations*. We are emphasizing here the fact that we are considering such ratios for all $m \geq 0$. Since the initial vectors $\boldsymbol{\varepsilon}^{(0)}$ are unknown in practical problems, then $|| \, M^m \, ||$ serves as a basis of comparison of different iterative methods.

DEFINITION 3.1. Let $A$ and $B$ be two $n \times n$ complex matrices. If, for some positive integer $m$, $|| \, A^m \, || < 1$, then

$$(3.23) \qquad R(A^m) \equiv - \ln \left[ (|| \, A^m \, ||)^{1/m} \right] = \frac{- \ln || \, A^m \, ||}{m}$$

is the *average rate of convergence for m iterations* of the matrix $A$. If $R(A^m) < R(B^m)$, then $B$ is *iteratively faster*† *for m iterations* than $A$.

In terms of actual computations, the significance of the average rate of convergence $R(A^m)$ is the following. Clearly, the quantity

$$(3.24) \qquad \sigma \equiv \left( \frac{|| \, \boldsymbol{\varepsilon}^{(m)} \, ||}{|| \, \boldsymbol{\varepsilon}^{(0)} \, ||} \right)^{1/m}$$

is the *average* reduction factor per iteration for the successive error norms. But if $|| \, A^m \, || < 1$, then by definition

$$(3.25) \qquad \sigma \leq (|| \, A^m \, ||)^{1/m} = e^{-R(A^m)}$$

where $e$ is the base of the natural logarithms. In other words, $R(A^m)$ is the exponential decay rate for a sharp upper bound of the average error reduction $\sigma$ per iteration in this $m$ step iterative process. Defining $N_m \equiv$

---

† We remark that, because of norm inequalities, it may result for a particular vector $\boldsymbol{\varepsilon}$ that $|| \, B^m\boldsymbol{\varepsilon} \, || < || \, A^m\boldsymbol{\varepsilon} \, ||$, while $|| \, A^m \, || < || \, B^m \, ||$. Nevertheless $A$ is iteratively faster, by definition, than $B$ for $m$ iterations.

$(R(A^m))^{-1}$, we see from the previous inequality that

$$\sigma^{N_m} \leq 1/e,$$

so that $N_m$ is a *measure* of the number of iterations required to reduce the norm of the initial error vector by a factor $e$.

Thus far, we have not brought into the discussion of the comparison of iterative methods the spectral radius of the iteration matrix $M$ of (3.20). If $A_1$ and $A_2$ are both Hermitian (or normal), then

$$\| A_i^m \| = (\rho(A_i))^m,$$

and thus, if $\rho(A_1) < \rho(A_2) < 1$, then

$$\| A_1^m \| < \| A_2^m \| < 1 \qquad \text{for } all \quad m \geq 1.$$

Unfortunately, although we know that $\| A^m \| \to 0$ as $m \to \infty$ for any convergent matrix, the spectral norms $\| A^m \|$ can behave quite erratically. In terms of Definition 3.1, it is entirely possible for a matrix $A$ to be iteratively faster than matrix $B$ for $m_1$ iterations, but iteratively slower for $m_2 \neq m_1$ iterations. To illustrate this, consider the two convergent matrices

$$(3.26) \qquad A = \begin{bmatrix} \alpha & 4 \\ 0 & \alpha \end{bmatrix}, \qquad B = \begin{bmatrix} \alpha & 0 \\ 0 & \beta \end{bmatrix}, \qquad 0 < \alpha < \beta < 1.$$

It is readily verified that

$$(3.27) \qquad A^m = \begin{bmatrix} \alpha^m & 4m\alpha^{m-1} \\ 0 & \alpha^m \end{bmatrix}, \qquad B^m = \begin{bmatrix} \alpha^m & 0 \\ 0 & \beta^m \end{bmatrix},$$

and from this, we obtain

$$(3.28) \qquad \| A^m \| = \left\{ \alpha^{2m} + 8m^2\alpha^{2m-2} \left[ 1 + \left( 1 + \frac{\alpha^2}{4m^2} \right)^{1/2} \right] \right\}^{1/2},$$

and

$$(3.28') \qquad \| B^m \| = \beta^m.$$

Thus, for $\alpha$ sufficiently close to unity, we see from (3.28) that the norms $\| A^m \|$ are initially *increasing* for $m \geq 1$, and $\| A^m \| > \| B^m \|$ for small values of $m$. On the other hand, for $m \to \infty$ it is clear that $\| A^m \| < \| B^m \|$.

For large powers of the matrix $A$, more precise information is available about $\| A^m \|$. We begin with

**Lemma 3.1.** *Let $J$ be an upper bi-diagonal complex $p \times p$ matrix of the form*†

(3.29)
$$
J = \begin{bmatrix}
\lambda & 1 & & \\
 & \lambda & 1 & \\
 & & \ddots & 1 \\
 & & & \lambda
\end{bmatrix},
$$

*where $\lambda \neq 0$. Then,*‡

(3.30)   $\| J^m \| \sim \dbinom{m}{p-1} [\rho(J)]^{m-(p-1)}, \qquad m \to \infty.$

*Proof.* As in the similar proof of Theorem 1.4,

$$
\binom{m}{p-1}
$$

is again the binomial coefficient, and, as was indicated in Sec. 1.3, it can be verified inductively that

(3.31)
$$
J^m = \begin{bmatrix}
\lambda^m & \dbinom{m}{1}\lambda^{m-1} & \cdots & \dbinom{m}{p-1}\lambda^{m-(p-1)} \\
0 & \lambda^m & \cdots & \dbinom{m}{p-2}\lambda^{m-(p-2)} \\
\vdots & & & \vdots \\
0 & 0 & \cdots & \lambda^m
\end{bmatrix},
$$

for $m \geq p - 1$. As $|\lambda| = \rho(J) > 0$ by hypothesis, we can divide through by

$$
\binom{m}{p-1}\lambda^{m-(p-1)}
$$

---

† In the notation of Wedderburn (1934), the matrix $J$ has elementary divisors of degree $p$.

‡ By $g(m) \sim h(m)$ as $m \to \infty$, we mean that $g(m)/h(m) \to 1$ as $m \to \infty$.

in (3.31), which defines

(3.32)
$$K_m = \frac{J^m}{\binom{m}{p-1}\lambda^{m-(p-1)}},$$

where $K_m = (k_{i,j}^{(m)})$ is a $p \times p$ matrix such that for all $1 \leq i, j \leq p$, $k_{i,j}^{(m)} = 0(1/m)$, as† $m \to \infty$, with the sole exception of the entry $k_{1,p}^{(m)}$ which is unity. Forming $K_m^* K_m = (l_{i,j}^{(m)})$, it is clear that all the terms of this product are $0(1/m)$, with the exception of the term $l_{p,p}^{(m)} = 1 + 0(1/m)$, and from the fact that the eigenvalues of a matrix are continuous functions of the entries of the matrix, it follows that $\| K_m \| \to 1$ as $m \to \infty$, which establishes (3.30).

**Theorem 3.1.** *Let $A$ be an arbitrary $n \times n$ complex matrix such that $\rho(A) > 0$. Then,*

(3.33)
$$\| A^m \| \sim \nu \cdot \binom{m}{p-1}[\rho(A)]^{m-(p-1)}, \qquad m \to \infty,$$

*where $p$ is the largest order of all diagonal submatrices $J_r$ of the Jordan normal form of $A$ with $\rho(J_r) = \rho(A)$, and $\nu$ is a positive constant.*

*Proof.* Let $S$ be a nonsingular $n \times n$ matrix such that $A = SJS^{-1}$, where $J$ is the Jordan normal form of the matrix $A$. We can suppose that the Jordan matrix $J$ is such that its diagonal submatrices $J_l$, $1 \leq l \leq r$, which have the form given in (3.29), are arranged in order of nondecreasing spectral radius, and $J_r$, the final diagonal submatrix, is such that $|\lambda| = \rho(J_r) = \rho(A)$, and the order of $J_r$ is not exceeded by any other submatrix $J_l$ where $\rho(J_l) = \rho(A)$. If the order of $J_r$ is $p$, then the matrix

$$\frac{J^m}{\binom{m}{p-1}[\rho(A)]^{m-(p-1)}}$$

tends to the $n \times n$ matrix $M$ all of whose entries are zero, except for the certain entries of modulus unity such as in the $(n - p + 1)$st row and $n$th column. Precisely, the total number of such entries of modulus unity of $M$ is equal to the number of diagonal submatrices $J_l$ for which $\rho(J_l) = \rho(A)$

---

† By $g(m) = 0(1/m)$ as $m \to \infty$, we mean that $| mg(m) | \leq \sigma$ for all $m$ sufficiently large, where $\sigma$ is a positive constant.

and the order of $J_l$ is equal to the order of $J_r$. Thus,

$$(3.34) \qquad \left\| \frac{A^m}{\binom{m}{p-1}[\rho(A)]^{m-(p-1)}} \right\| \to \| SMS^{-1} \| \equiv \nu, \qquad m \to \infty,$$

which completes the proof.

The constant $\nu = \| SMS^{-1} \|$ can be estimated as follows: Let the constant $\Lambda$ be defined by $\Lambda = \| S \| \cdot \| S^{-1} \|$.† As $A^m = SJ^mS^{-1}$ and $J^m = S^{-1}A^mS$, then using norm inequalities it follows from the definition of $\Lambda$ that

$$\frac{\| J^m \|}{\Lambda} \leq \| A^m \| \leq \Lambda \| J^m \|,$$

and thus,

$$(3.35) \qquad \frac{1}{\Lambda} \leq \nu \leq \Lambda.$$

To illustrate this last result, consider the matrix $A$ of (3.26). In this case, $p = 2$, $\alpha = \rho(A)$, and it can be verified that the choice

$$S = \begin{bmatrix} 4 & 0 \\ 0 & 1 \end{bmatrix}$$

is such that $A = SJS^{-1}$ where $J$ is a proper Jordan normal form of $A$ as required in Theorem 3.1. But as

$$SMS^{-1} = \begin{bmatrix} 0 & 4 \\ 0 & 0 \end{bmatrix},$$

then the spectral norm of this matrix is 4, and thus $\nu = 4$. As $\| S \| = 4$, $\| S^{-1} \| = 1$, then $\Lambda = 4$ for this example. We have also incidentally indicated that equality is possible in (3.35).

Returning to the topic of rates of convergence, we see from (3.30) that if a matrix $A$ is *convergent*, then the spectral norms of all sufficiently high

---

† Since $\| \mathbf{x} \| = \| S \cdot S^{-1} \mathbf{x} \| \leq \| S \| \cdot \| S^{-1} \| \cdot \| \mathbf{x} \|$, then the constant

$$\Lambda = \| S \| \cdot \| S^{-1} \| \geq 1.$$

This constant is called the *condition number* of the matrix $S$.

powers of the matrix are *monotone decreasing*. In fact, from (3.33) we can actually prove

**Theorem 3.2.** *Let $A$ be a convergent $n \times n$ complex matrix. For all $m$ sufficiently large, the average rate of convergence for $m$ iterations $R(A^m)$ satisfies*

$$(3.36) \qquad \lim_{m \to \infty} R(A^m) = -\ln \rho(A) \equiv R_\infty(A).$$

*Proof.* From the main result (3.33) of Theorem 3.1, taking logarithms we obtain,

$$R(A^m) = \frac{-\ln \| A^m \|}{m} \sim \frac{-\ln \nu}{m} - \frac{\ln \binom{m}{p-1}}{m} - \left( \frac{m - p + 1}{m} \right) \ln \rho(A).$$

As it is readily verified that

$$\frac{\ln \binom{m}{p-1}}{m} \to 0 \text{ as } m \to \infty,$$

then we obtain

$$\lim_{m \to \infty} R(A^m) = -\ln \rho(A) \equiv R_\infty(A),$$

the desired result.

We speak of the negative of the logarithm of the spectral radius of a convergent matrix $A$ as the *asymptotic rate of convergence* $R_\infty(A)$ for the matrix $A$. Since $\| A^m \| \geq (\rho(A))^m$ for all $m \geq 1$, we have immediately the

**Corollary.** *If $A$ is an arbitrary convergent $n \times n$ complex matrix, then*

$$(3.37) \qquad\qquad R_\infty(A) \geq R(A^m)$$

*for any positive integer $m$ for which $\| A^m \| < 1$.*

It should be said that the asymptotic rate of convergence† $R_\infty(A)$ for a convergent matrix $A$ is by far the simplest practical measure of rapidity of convergence of a matrix which is in common use. However, its indiscriminate use can give quite misleading information. As an example, let us consider again the convergent matrix $A$ of (3.26). If $\alpha = 0.99$, then $R_\infty(A) = 0.01005$, so that its reciprocal $N_\infty$ is 99.50. This would indicate that approximately only one hundred iterations would be necessary to

---

† What we have defined as the asymptotic rate of convergence $R_\infty(A)$ is what **Young** (1954a) calls the *rate of convergence*.

reduce the Euclidean norm of any initial error vector by a factor of $e$. For this particular example, however, $\|A^m\|$ is initially increasing, and $\|A^m\| \geq 1$ for *all* $m < 805$. In order to have $\|A^m\| \leq 1/e$, it is necessary that $m \geq 918$.

## EXERCISES

1. Prove directly from the construction of the matrix $M$ of Theorem 3.1 that $\|M\| = 1 = \|M\|_1 = \|M\|_\infty$. (See Exercises 1 and 2 of Sec. 1.3.)

2. Prove, similarly, that (3.33) of Theorem 3.1 is valid for the matrix norms $\|A^m\|_1$ and $\|A^m\|_\infty$.

3. To indicate how slowly $R(A^m)$ can converge to $R_\infty(A)$, show that for the matrix $A$ of (3.26) with $\alpha = 0.99$, the least positive integer $m$ for which

$$1 \geq \frac{R(A^m)}{R_\infty(A)} \geq 0.99$$

is approximately $10^5$.

## 3.3. THE STEIN-ROSENBERG THEOREM

With the definition of *asymptotic rate of convergence*, we now compare the asymptotic rates of convergence, or equivalently the spectral radii, of the point Jacobi matrix and the point Gauss-Seidel matrix in the case where the point Jacobi matrix has non-negative real entries. This fundamental comparison based on the Perron-Frobenius theory is due to Stein and Rosenberg (1948).

To begin with, let $B$ be any $n \times n$ matrix with zero diagonal entries. We express the matrix $B$ as

$$(3.38) \qquad\qquad B = L + U,$$

where $L$ and $U$ are, respectively, $n \times n$ strictly lower and upper triangular matrices. Abstractly, we can call such matrices $B$ *Jacobi matrices*. The particular example in Sec. 3.1 of a point Jacobi matrix certainly satisfied these hypotheses. As the matrix $(I - \omega L)$ is nonsingular for all real $\omega$, we similarly define

$$(3.39) \qquad\qquad \mathcal{L}_\omega = (I - \omega L)^{-1}\{\omega U + (1 - \omega)I\}$$

as the associated *successive overrelaxation matrix*. By means of the Perron-Frobenius theory, we now *geometrically* characterize the spectral radius of the Gauss-Seidel matrix $\mathcal{L}_1$ in terms of the intersection of two curves.

DEFINITION 3.2. Let $L$ and $U$ be respectively strictly lower and upper triangular $n \times n$ matrices. Then, $m(\sigma) \equiv \rho(\sigma L + U)$, and $n(\sigma) \equiv \rho(L + \sigma U)$, $\sigma \geq 0$.

From the definitions of the functions $m(\sigma)$ and $n(\sigma)$, it is obvious that $m(0) = n(0) = 0$, $m(1) = n(1) = \rho(L + U) = \rho(B)$, and

$$(3.40) \qquad n(\sigma) = \sigma m(1/\sigma), \qquad \sigma > 0.$$

**Lemma 3.2.** *Let $B = L + U$ be a non-negative $n \times n$ matrix. If $\rho(B) > 0$, then the functions $m(\sigma)$ and $n(\sigma)$ are both strictly increasing for $\sigma \geq 0$. If $\rho(B) = 0$, then $m(\sigma) = n(\sigma) = 0$ for all $\sigma \geq 0$.*

*Proof.* For convenience we shall assume here that the non-negative matrix $B$ is irreducible so that $\rho(B) > 0$. (For the reducible case, see Exercise 1 of this section). Thus, neither $L$ nor $U$ is the null matrix. For $\sigma > 0$, the non-negative matrix $M(\sigma) = \sigma L + U$ has the same directed graph as does $B$, so that $M(\sigma)$ is also irreducible, and we see from the Perron-Frobenius Theorem 2.1 that increasing $\sigma$ *increases* the spectral radius of $M(\sigma)$. The same argument shows that $n(\sigma)$ is also a strictly increasing function, which completes the proof for the irreducible case.

For $\rho(B) > 0$, we represent this result graphically in Figure 3.

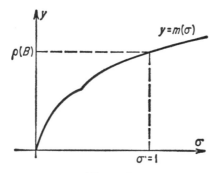

**Figure 3**

We now consider the Gauss-Seidel matrix $\mathcal{L}_1 = (I - L)^{-1}U$, where the matrix $B = L + U$ is assumed to be non-negative and irreducible. As $L$ is strictly lower triangular,

$$(I - L)^{-1} = I + L + L^2 + \cdots + L^{n-1},$$

and from this it follows that $\mathcal{L}_1$ is also a non-negative matrix. Thus, there exists a nonzero vector $\mathbf{x} \geq \mathbf{0}$ such that

$$(3.41) \qquad (I - L)^{-1}U\mathbf{x} = \lambda\mathbf{x}, \qquad \text{where} \qquad \lambda = \rho(\mathcal{L}_1) \geq 0.$$

From the non-negative irreducible character of $B$, it can be verified that

$\lambda$ is indeed positive, and $\mathbf{x}$ is a positive vector†. Thus, we can write (3.41) equivalently as

$$(3.42) \qquad (\lambda L + U)\mathbf{x} = \lambda \mathbf{x} \quad \text{and} \quad \left(L + \frac{1}{\lambda} U\right)\mathbf{x} = \mathbf{x}.$$

Since

$$(\lambda L + U) \quad \text{and} \quad \left(L + \frac{1}{\lambda} U\right)$$

are non-negative irreducible matrices and $\mathbf{x} > \mathbf{0}$, we deduce from Theorem 2.1 and Definition 3.2 that

$$(3.42') \qquad m(\lambda) = \lambda \quad \text{and} \quad n(1/\lambda) = 1.$$

If $\rho(B) = 1$, then $n(1) = 1$, and it follows from the monotone increasing behavior of $n(\sigma)$ in Lemma 3.2 that $\lambda = 1$, and conversely.

Now, assume that $0 < \rho(B) < 1$. Since $n(1) = \rho(B) < 1$ and $n(1/\lambda) = 1$, the monotone behavior of $n(\sigma)$ again shows us that $1/\lambda > 1$, or $0 < \lambda < 1$. But as $m(\sigma)$ is also strictly increasing and $m(1) = \rho(B)$, it follows that $0 < \lambda < \rho(B) < 1$. Similarly, if $\rho(B) > 1$, then $\lambda = \rho(\mathcal{L}_1) > \rho(B) > 1$. Thus, we have proved the following theorem of Stein and Rosenberg for the case that $B \geq O$ is irreducible.

**Theorem 3.3.** *Let the Jacobi matrix $B \equiv L + U$ be a non-negative $n \times n$ matrix with zero diagonal entries, and let $\mathcal{L}_1$ be the Gauss-Seidel matrix, the special case $\omega = 1$ of (3.39). Then, one and only one of the following mutually exclusive relations is valid:*

1. $\rho(B) = \rho(\mathcal{L}_1) = 0$.

2. $0 < \rho(\mathcal{L}_1) < \rho(B) < 1$.

3. $1 = \rho(B) = \rho(\mathcal{L}_1)$.

4. $1 < \rho(B) < \rho(\mathcal{L}_1)$.

*Thus, the Jacobi matrix $B$ and the Gauss-Seidel matrix $\mathcal{L}_1$ are either both convergent, or both divergent.*

As an immediate corollary, we have the

**Corollary.** *If the non-negative Jacobi matrix $B$ is such that $0 < \rho(B) < 1$, then*

$$(3.43) \qquad R_\infty(\mathcal{L}_1) > R_\infty(B).$$

† See Exercise 6.

Thus, the Stein-Rosenberg theorem gives us our first *comparison theorem* for two different iterative methods. Interpreted in a more practical way, not only is the point Gauss-Seidel iterative method computationally more *convenient* to use (because of storage requirements) than the point Jacobi iterative method, but it is also asymptotically *faster* when the Jacobi matrix $B$ is non-negative with $0 < \rho(B) < 1$.

It is interesting to note that Theorem 3.3 has a geometrical interpretation. If we plot the curves $y_1(\sigma) = m(\sigma)$ and $y_2(\sigma) = \sigma$ as a function of $\sigma > 0$, then from (3.42'), these curves intersect in a point whose abscissa is $\rho(\mathcal{L}_1)$, which is shown in Figure 4.

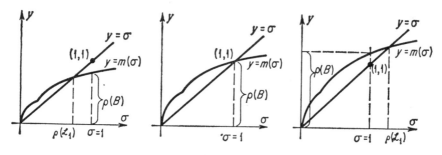

**Figure 4**

Returning to the example of Sec. 3.1, the $4 \times 4$ convergent point Jacobi matrix $B$ of (3.18) certainly satisfies the hypotheses of the Corollary, and thus the asymptotic rate of convergence of the point Gauss-Seidel iterative method is greater than that of the point Jacobi iterative method, as suggested in (3.19). It would now be of interest to compare the average rate of convergence of the Jacobi iterative method for $m$ iterations with that of the Gauss-Seidel iterative method for *each* $m \geq 0$, in the case that the Jacobi matrix $B$ is non-negative and $0 < \rho(B) < 1$. If we again examine the simple example of (3.18), we see that the matrix $B$ of (3.18) is real and symmetric, so that

$$\| B^m \| = (\rho(B))^m = \frac{1}{2^m}.$$

On the other hand, by direct computation from (3.18),

$$\mathcal{L}_1^m = \frac{1}{2 \cdot 4^m} \left[ \begin{array}{cc|cc} 0 & 0 & 2 & 2 \\ 0 & 0 & 2 & 2 \\ \hline 0 & 0 & 1 & 1 \\ 0 & 0 & 1 & 1 \end{array} \right], \qquad m \geq 1,$$

so that

$$(\mathcal{L}_1^m)^* \cdot (\mathcal{L}_1^m) = \frac{1}{4^{2m+1}} \left[ \begin{array}{cc|cc} 0 & 0 & 0 & 0 \\ 0 & 0 & 0 & 0 \\ \hline 0 & 0 & 10 & 10 \\ 0 & 0 & 10 & 10 \end{array} \right], \quad m \geq 1.$$

Thus,

$$|| \mathcal{L}_1^m || = \frac{\sqrt{5}}{4^m}, \quad m \geq 1,$$

and we see that $|| \mathcal{L}_1^m || > || B^m ||$ for $m = 1$. In fact, if we had chosen our initial error vector as

$$\varepsilon^{(0)} = \sqrt{2} \left[ \begin{array}{c} 0 \\ 0 \\ 1 \\ 1 \end{array} \right],$$

which has the same Euclidean norm as the choice $\varepsilon^{(0)}$ of Sec. 3.1, then

$$|| \varepsilon^{(0)} || = 2, \quad || B\varepsilon^{(0)} || = 1,$$

and

$$|| \mathcal{L}_1 \varepsilon^{(0)} || = \frac{\sqrt{5}}{2} > || B\varepsilon^{(0)} ||.$$

We reach the surprising conclusion in this case that for just *one* iteration, the point Jacobi method is iteratively *faster* than the point Gauss-Seidel iterative method. This simple counterexample shows that we *cannot* extend the inequality (3.43) of the asymptotic rates of convergence given by the Stein-Rosenberg theorem to average rates of convergence for all $m \geq 1$ even in the non-negative case. However, in Chapter 5, using semi-iterative techniques, we shall see how, such comparisons can be made for all $m \geq 1$.

Thus far our comparisons of the Jacobi and Gauss-Seidel iterative methods have been restricted to the case where the Jacobi matrix is non-negative, and here our fundamental results from the Perron-Frobenius theory were the basis for a comparison of these iterative methods. Unfortunately, for the case when the Jacobi matrix $B$ is not a non-negative matrix, it is not possible to prove that the inequalities relating the spectral radii of the two iterative methods in Theorem 3.3 carry over to the general case. Specifically, one can find examples (see Exercise 5 of this section) where one iterative method is convergent while the other is divergent, so that in the general case, neither of the two methods can be universally favored.

In the case of diagonally dominant complex matrices, however, we can apply Theorem 3.3 to prove that both the associated point Jacobi and Gauss-Seidel matrices are convergent.

**Theorem 3.4.** *Let* $A = (a_{i,j})$ *be a strictly or irreducibly diagonally dominant* $n \times n$ *complex matrix. Then, both the associated point Jacobi and the point Gauss-Seidel matrices are convergent, and the iterative methods of* (3.5) *and* (3.8) *for the matrix problem* $A\mathbf{x} = \mathbf{k}$ *are convergent for any initial vector approximation* $\mathbf{x}^{(0)}$.

*Proof.* From Definition 1.7 and Theorem 1.8, we know that the matrix $A$ is nonsingular and that the diagonal entries of $A$ are necessarily nonzero. From (3.7), the associated point Jacobi matrix $B = (b_{i,j})$ derived from the matrix $A$ is such that

$$(3.44) \qquad b_{i,i} = \begin{cases} 0, & i = j \\ \dfrac{-a_{i,j}}{a_{i,i}}, & i \neq j \end{cases}.$$

Again, from Definition 1.7 it follows that

$$\sum_{j=1}^{n} |b_{i,j}| < 1 \qquad \text{for all} \quad 1 \leq i \leq n$$

if $A$ is strictly diagonally dominant, or

$$\sum_{j=1}^{n} |b_{i,j}| \leq 1 \qquad \text{for all} \quad 1 \leq i \leq n$$

with strict inequality for at least one $i$ if $A$ is irreducibly diagonally dominant. Recalling from Chapter 2 the notation that $|B|$ is the non-negative matrix whose entries are $|b_{i,j}|$, these row sums tells us that $\rho(|B|) < 1$. Now, applying Theorem 2.8, we conclude that $\rho(B) \leq \rho(|B|) < 1$, so that the point Jacobi iterative method is necessarily convergent. Similarly, if $B = L + U$ and $\mathcal{L}_1 = (I - L)^{-1}U$, then

$$\rho(\mathcal{L}_1) \leq \rho\{(I - |L|)^{-1}|U|\}.$$

But as $\rho(|B|) < 1$, we have from Theorem 3.3 that

$$\rho\{(I - |L|)^{-1}|U|\} \leq \rho(|B|) < 1;$$

thus we conclude that the point Gauss-Seidel iterative method is also convergent, which completes the proof.

## EXERCISES

**1.** Prove Lemma 3.2, assuming that the matrix $B$ is reducible. (*Hint*: Apply the proof of Lemma 3.2 for the irreducible case to the normal reduced form (2.41) of $M(\sigma) = \sigma L + U$.)

**2.** Let

$$B = \begin{bmatrix} 0 & \alpha & 0 & 0 & 0 \\ \alpha & 0 & 0 & 0 & 0 \\ 0 & 0 & 0 & 0 & \beta \\ 0 & 0 & \beta & 0 & 0 \\ 0 & 0 & 0 & \beta & 0 \end{bmatrix}, \qquad \text{where } 0 < \alpha < \beta < 1,$$

and determine the function $m(\sigma)$. Show that $m(\sigma)$ in this case, as in the illustration of Figure 3, has a discontinuous derivative at some point $\sigma^* > 0$.

**3.** If the $n \times n$ matrix $B = L + U \geq O$ is such that $\rho(B) > 0$, prove that $m(\sigma) = \rho(\sigma L + U) \sim K\sigma^\nu$ as $\sigma \to 0$, where $K > 0$, and

$$\frac{1}{n} \leq \nu \leq \frac{n-1}{n}.$$

(*Hint.* If

$$\det \{\lambda I - \sigma L - U\} \equiv \sum_{j=0}^{n} a_j(\sigma)\lambda^{n-j},$$

show that each $a_j(\sigma)$ is a polynomial of degree $(j-1)$ in $\sigma$.)

**4.** Complete the proof of Theorem 3.3 for the case when $B$ is reducible.

**5. a.** Let
$$B \equiv \begin{bmatrix} 0 & -2 & 2 \\ -1 & 0 & -1 \\ -2 & -2 & 0 \end{bmatrix}.$$

Show that $\rho(B) < 1$ while $\rho(\mathcal{L}_1) > 1$.

**b.** Let
$$B \equiv \tfrac{1}{2}\begin{bmatrix} 0 & 1 & -1 \\ -2 & 0 & -2 \\ 1 & 1 & 0 \end{bmatrix}.$$

Show that $\rho(B) > 1$ while $\rho(\mathcal{L}_1) < 1$. (Collatz (1942b)).

**c.** Construct similar examples using $4 \times 4$ matrices.

**6.** Let $B = L + U$ be a non-negative irreducible $n \times n$ Jacobi matrix, $n \geq 2$. Prove that the spectral radius of the Gauss-Seidel matrix $\mathcal{L}_1 = (I - L)^{-1}U$ is positive and that its associated eigenvector has positive components.

**7.** Let $B = L + U$ be a non-negative irreducible $n \times n$ Jacobi matrix, $n \geq 2$. Prove that the Gauss-Seidel matrix $\mathcal{L}_1 = (I - L)^{-1}U$ is reducible. Moreover, show that there is but one diagonal submatrix of the reduced normal form of $\mathcal{L}_1$ with spectral radius $\rho(\mathcal{L}_1)$. Finally, show that $\mathcal{L}_1 \mathbf{x} > 0$ whenever $\mathbf{x} > 0$.

**8.** If the $n \times n$ Jacobi matrix $B = L + U \geq O$ is such that $\rho(B) > 0$, prove that

$$m(x) < m(xy) < ym(x)$$

for all $x > 0$, $y > 1$.

**\*9.** If the $n \times n$ Jacobi matrix $B = L + U \geq O$ is such that $\rho(B) > 0$, prove, only using Theorem 3.3, that $n(\sigma)$ is unbounded as $\sigma \to +\infty$. Similarly, prove that $m(\sigma)$ is unbounded as $\sigma \to +\infty$ (Kahan (1958)).

**10.** If the $n \times n$ Jacobi matrix $B = L + U$ is non-negative and irreducible, prove that the curves $y_1 = m(\sigma)$ and $y_2 = \sigma$ intersect in a *unique* point with positive abscissa.

## 3.4. THE OSTROWSKI-REICH THEOREM

In the previous section, the Jacobi and Gauss-Seidel iterative methods were compared in terms of their spectral radii in only the non-negative case, and we saw how the matrix property of being strictly or irreducibly diagonally dominant was a sufficient condition for the convergence of the point Jacobi and point Gauss-Seidel iterative methods. Whereas $\rho(M) < 1$ is both a necessary and sufficient condition that $M$ be a convergent matrix, in this section we examine results of Ostrowski (1954) and Reich (1949) which give different necessary and sufficient conditions for the convergence of the successive overrelaxation and Gauss-Seidel iterative methods in a form which will be very useful in future chapters. We begin with the following result of Kahan (1958).

**Theorem 3.5.** *Let $B$ be any $n \times n$ complex matrix with zero diagonal entries which we write in the form $B = L + U$, where $L$ and $U$ are respectively strictly lower and upper triangular matrices. If*

$$\mathcal{L}_\omega \equiv (I - \omega L)^{-1}\{\omega U + (1 - \omega)I\},$$

*then for all real $\omega$,*

(3.45) $$\rho(\mathcal{L}_\omega) \geq |\omega - 1|,$$

*with equality only if all eigenvalues of $\mathcal{L}_\omega$ are of modulus $|\omega - 1|$.*

*Proof.* Let $\phi(\lambda) = \det (\lambda I - \mathcal{L}_\omega)$ be the characteristic polynomial of $\mathcal{L}_\omega$. Since $L$ is strictly lower triangular, then $(I - \omega L)$ is nonsingular, and as $\det (I - \omega L) = 1$, then

$$0 = \phi(\lambda) = \det (I - \omega L) \cdot \det (\lambda I - \mathcal{L}_\omega)$$

$$= \det \{(\lambda + \omega - 1)I - \omega \lambda L - \omega U\}.$$

The constant term $\sigma$ of $\phi(\lambda)$, the product of the eigenvalues of $\mathcal{L}_\omega$, is obtained by setting $\lambda = 0$ in this expression. Thus,

$$(3.46) \qquad \sigma = \prod_{i=1}^{n} (-\lambda_i(\omega)) = \det \{(\omega - 1)I - \omega U\} = (\omega - 1)^n.$$

Therefore,

$$(3.47) \qquad \rho(\mathcal{L}_\omega) = \max_i |\lambda_i(\omega)| \geq |\omega - 1|,$$

with equality possibly only if all eigenvalues of $\rho(\mathcal{L}_\omega)$ are of modulus $|\omega - 1|$, which completes the proof.

If in the general case our goal is to determine the real value of $\omega$ which minimizes $\rho(\mathcal{L}_\omega)$, then the above result of Kahan shows that we need only consider the open interval $0 < \omega < 2$ for values of $\omega$ such that $\mathcal{L}_\omega$ is convergent.

In order to solve iteratively the matrix equation $A\mathbf{x} = \mathbf{k}$, we now assume that the $n \times n$ matrix $A$ is Hermitian and can be expressed in the form

$$(3.48) \qquad A = D - E - E^*,$$

where $D$ and $E$ are $n \times n$ matrices and $D$ is Hermitian and *positive definite*. We also assume that the matrix $D - \omega E$ is nonsingular for all $0 \leq \omega \leq 2$. From the matrix equation $A\mathbf{x} = \mathbf{k}$ and (3.48), we can write

$$(3.49) \qquad \omega(D - E)\mathbf{x} = \omega E^*\mathbf{x} + \omega \mathbf{k},$$

and adding $(1 - \omega) D\mathbf{x}$ to both sides gives

$$(3.50) \qquad (D - \omega E)\mathbf{x} = \{\omega E^* + (1 - \omega)D\}\mathbf{x} + \omega \mathbf{k},$$

which serves to introduce the iterative method

$$(3.51) \qquad (D - \omega E)\mathbf{x}^{(m+1)} = \{\omega E^* + (1 - \omega)D\}\mathbf{x}^{(m)} + \omega \mathbf{k}, \qquad m \geq 0,$$

which in principle can be carried out since $(D - \omega E)$ is by assumption nonsingular for $0 \leq \omega \leq 2$. But upon letting $D^{-1}E \equiv L$ and $D^{-1}E^* \equiv U$, this iterative method takes the more familiar form

$$(3.51') \qquad \mathbf{x}^{(m+1)} = \mathcal{L}_\omega \mathbf{x}^{(m)} + \omega(D - \omega E)^{-1}\mathbf{k}, \qquad m \geq 0,$$

where $\mathcal{L}_\omega = (I - \omega L)^{-1}\{\omega U + (1 - \omega)I\}$, so that it is the successive overrelaxation iterative method. In contrast with Sec. 3.3, however, note that the matrices $L$ and $U$ defined here need not be respectively strictly lower and upper triangular matrices. The important theorem of Ostrowski (1954) regarding the convergence of the iterative method (3.51) is

**Theorem 3.6.** *Let $A = D - E - E^*$ be an $n \times n$ Hermitian matrix, where $D$ is Hermitian and positive definite, and $D - \omega E$ is nonsingular for $0 \leq \omega \leq 2$. Then, $\rho(\mathcal{L}_\omega) < 1$ if and only if $A$ is positive definite and $0 < \omega < 2$.*

*Proof.* If $\varepsilon_0$ is an arbitrary nonzero initial error vector, then the error vectors for the successive overrelaxation iterative method are defined by

$$(3.52) \qquad \varepsilon_{m+1} = \mathcal{L}_\omega \varepsilon_m, \qquad m \geq 0,$$

or equivalently,

$$(3.53) \qquad (D - \omega E)\varepsilon_{m+1} = (\omega E^* + (1 - \omega)D)\varepsilon_m, \qquad m \geq 0.$$

Also, let $\delta_m \equiv \varepsilon_m - \varepsilon_{m+1}$ for all $m \geq 0$. With the relation $A = D - E - E^*$, (3.53) can be written as

$$(3.54) \qquad (D - \omega E)\delta_m = \omega A \varepsilon_m, \qquad m \geq 0,$$

and

$$(3.55) \qquad \omega A \varepsilon_{m+1} = (1 - \omega)D\delta_m + \omega E^*\delta_m, \qquad m \geq 0.$$

If we premultiply these equations respectively by $\varepsilon_m^*$ and $\varepsilon_{m+1}^*$, and use the fact that the matrices $A$ and $D$ are Hermitian, these relations can be combined, after some manipulation,† into the single equation

$$(3.56) \qquad (2 - \omega)\delta_m^* D\delta_m = \omega\{\varepsilon_m^* A \varepsilon_m - \varepsilon_{m+1}^* A \varepsilon_{m+1}\}, \qquad m \geq 0.$$

In this form, it is now easy to establish the desired conclusion. First, assuming $A$ to be positive definite and $0 < \omega < 2$, choose $\varepsilon_0$ to be an

† See Exercise 4.

eigenvector of $\mathcal{L}_\omega$. Thus, $\varepsilon_1 = \mathcal{L}_\omega \varepsilon_0 = \lambda \varepsilon_0$, and $\delta_0 = (1 - \lambda) \varepsilon_0$. With this particular choice of $\varepsilon_0$, (3.56) reduces to

$$(3.56') \qquad \left(\frac{2 - \omega}{\omega}\right) | 1 - \lambda |^2 \varepsilon_0^* D \varepsilon_0 = (1 - | \lambda |^2) \varepsilon_0^* A \varepsilon_0.$$

Now, $\lambda$ cannot be unity, for then $\delta_0$ would be the null vector, and thus from (3.54), $A \varepsilon_0 = 0$. As $\varepsilon_0$ is by definition nonzero, this contradicts the assumption that $A$ is positive definite. With $0 < \omega < 2$, both sides of (3.56') are positive, which proves that $| \lambda | < 1$. Thus, $\mathcal{L}_\omega$ is convergent.

Now, assume that $\mathcal{L}_\omega$ is convergent. It necessarily follows that $0 < \omega < 2$ from Theorem 3.5, and for any nonzero vector $\varepsilon_0$, the vectors $\varepsilon_m$ tend to zero as $m$ increases. Thus, $\varepsilon_m^* A \varepsilon_m$ also approaches zero. As $D$ is positive definite, we deduce from (3.56) that

$$(3.57) \qquad \varepsilon_m^* A \varepsilon_m = \varepsilon_{m+1}^* A \varepsilon_{m+1} + \left(\frac{2 - \omega}{\omega}\right) \delta_m D \delta_m \geq \varepsilon_{m+1}^* A \varepsilon_{m+1},$$

$$m \geq 0.$$

If $A$ is not positive definite, we could find a nonzero vector $\varepsilon_0$ such that $\varepsilon_0^* A \varepsilon_0 \leq 0$. But no eigenvalue of $\mathcal{L}_\omega$ can be unity. Hence, $\delta_0 = (I - \mathcal{L}_\omega) \varepsilon_0$ is not the null vector, which gives us the strict inequality $\varepsilon_1^* A \varepsilon_1 < \varepsilon_0^* A \varepsilon_0 \leq 0$. The nonincreasing property of (3.57) now contradicts the fact that $\varepsilon_m^* A \varepsilon_m$ tends to zero, which completes the proof.

The special case of $\omega = 1$, first proved by Reich (1949), is given in

**Corollary 1.** *Let $A = D - E - E^*$ be an $n \times n$ Hermitian matrix where $D$ is Hermitian and positive definite and $(D - E)$ is nonsingular. Then, the Gauss-Seidel iterative method, $\omega = 1$ of (3.51) is convergent if and only if $A$ is positive definite.*

As an application of Theorem 3.6, let the $n \times n$ matrix $A$ be partitioned into the form

$$(3.58) \qquad A - \begin{bmatrix} A_{1,1} & A_{1,2} & \cdots & A_{1,N} \\ A_{2,1} & A_{2,2} & \cdots & A_{2,N} \\ \cdot & & & \cdot \\ \cdot & & & \cdot \\ \cdot & & & \cdot \\ A_{N,1} & A_{N,2} & \cdots & A_{N,N} \end{bmatrix},$$

where the diagonal blocks $A_{i,i}$, $1 \leq i \leq N$, are square and nonvoid.† From

† By nonvoid we mean that each submatrix $A_{i,i}$ is an $r_i \times r_i$ matrix where $r_i \geq 1$.

this partitioning of the matrix $A$, we define the matrices

$$(3.59) \qquad D = \begin{bmatrix} A_{1,1} & & & \mathbf{O} \\ & A_{2,2} & & \\ & & \ddots & \\ \mathbf{O} & & & A_{N,N} \end{bmatrix},$$

$$E = - \begin{bmatrix} 0 & 0 & \cdots & 0 \\ A_{2,1} & 0 & \cdots & 0 \\ \vdots & & \ddots & \vdots \\ A_{N,1} & A_{N,2} & \cdots & 0 \end{bmatrix},$$

$$F = - \begin{bmatrix} 0 & A_{1,2} & \cdots & A_{1,N} \\ 0 & 0 & \cdots & A_{2,N} \\ \vdots & & \ddots & \vdots \\ 0 & 0 & \cdots & 0 \end{bmatrix},$$

where the matrices $E$ and $F$ are respectively lower and upper triangular matrices, and $A = D - E - F$. Assuming $A$ to be Hermitian, it follows that $D$ is also Hermitian, and $E^* = F$. If we further assume that $D$ is positive definite, then, from the form of the matrices $D$ and $E$, it follows that $D - \omega E$ is nonsingular for all values of $\omega$. If the column vectors $\mathbf{x}$ and $\mathbf{k}$ of the matrix problem $A\mathbf{x} = \mathbf{k}$ are partitioned relative to the partitioning of (3.58), then the matrix problem can be written as

$$(3.60) \qquad \begin{bmatrix} A_{1,1} & A_{1,2} & \cdots & A_{1,N} \\ A_{2,1} & A_{2,2} & \cdots & A_{2,N} \\ \vdots & & & \vdots \\ A_{N,1} & A_{N,2} & \cdots & A_{N,N} \end{bmatrix} \begin{bmatrix} X_1 \\ X_2 \\ \vdots \\ X_N \end{bmatrix} = \begin{bmatrix} K_1 \\ K_2 \\ \vdots \\ K_N \end{bmatrix},$$

or equivalently

$$(3.60') \qquad \sum_{j=1}^{N} A_{i,j} X_j = K_i, \qquad 1 \le i \le N.$$

Since $D$ is nonsingular, each diagonal submatrix $A_{i,i}$ is nonsingular, and we can define the associated *block*† (or *partitioned*) *successive overrelaxation iterative method* to be

$$(3.61) \qquad A_{i,i}X_i^{(m+1)} = A_{i,i}X_i^{(m)} + \omega\left\{ -\sum_{j=1}^{i-1} A_{i,j}X_j^{(m+1)} - \sum_{j=i+1}^{N} A_{i,j}X_j^{(m)} \right.$$

$$\left. +K_i - A_{i,i}X_i^{(m)} \right\}, \qquad m \geq 0,$$

where we have assumed that matrix equations such as $A_{i,i}X_i = G_i$ can be solved directly‡ for $X_i$, given $G_i$. As is readily verified, this iterative procedure can be written as

$$(3.62) \quad (D - \omega E)\mathbf{x}^{(m+1)} = \{\omega F + (1 - \omega)D\}\mathbf{x}^{(m)} + \omega\mathbf{k}, \qquad m \geq 0,$$

and applying Theorem 3.6, we have immediately

**Corollary 2.** *Let $A$ be an $n \times n$ Hermitian matrix partitioned as in (3.58), and let the matrices $D$, $E$, and $F$ be defined as in (3.59). If $D$ is positive definite, then the block successive overrelaxation method of (3.62) is convergent for all $\mathbf{x}^{(0)}$ if and only if $0 < \omega < 2$ and $A$ is positive definite.*

We remark that if the matrix $A$ is Hermitian and positive definite, *any* partitioning of the matrix $A$ as in (3.58) directly guarantees the positive definite Hermitian character of the matrix $D$ of (3.59). Thus, the block Gauss-Seidel iterative method of (3.62) is necessarily convergent.

### EXERCISES

**1.** Let the real symmetric matrix $A$ be given explicitly by

$$A = \begin{bmatrix} 5 & 2 & 2 \\ 2 & 5 & 3 \\ 2 & 3 & 5 \end{bmatrix}.$$

Verifying first that $A$ is positive definite, determine the spectral radius $\rho_1$ of its associated point Gauss-Seidel matrix (see Sec. 3.1). If $A$ is partitioned

---

† If each diagonal submatrix $A_{i,i}$ of (3.58) is a $1 \times 1$ matrix, this "block" interative method reduces to what we have previously called a *point iterative method*.

‡ Such practical considerations, as to which matrix equations can be efficiently solved directly, will be discussed in Chapter 6.

in the manner of (3.58) so that $A = D_2 - E_2 - E_2^*$ where

$$D_2 = \begin{bmatrix} 5 & 2 & 0 \\ 2 & 5 & 0 \\ \hline 0 & 0 & 5 \end{bmatrix}, \qquad E_2 = \begin{bmatrix} 0 & 0 & 0 \\ 0 & 0 & 0 \\ \hline -2 & -3 & 0 \end{bmatrix},$$

show that the spectral radius $\rho_2$ of the block Gauss-Seidel matrix $(D_2 - E_2)^{-1}E_2^*$ is such that $1 > \rho_2 > \rho_1$. Thus, the block methods introduced in this section do *not* necessarily give improvements over the corresponding point methods.

2. Let $A$ be the $4 \times 4$ matrix of (1.6). If $A$ is partitioned in the manner of (3.58) so that $A = D - E - E^*$ where

$$D = \begin{bmatrix} 1 & 0 & -\frac{1}{4} & 0 \\ 0 & 1 & -\frac{1}{4} & 0 \\ -\frac{1}{4} & -\frac{1}{4} & 1 & 0 \\ \hline 0 & 0 & 0 & 1 \end{bmatrix}, \qquad E = \begin{bmatrix} 0 & 0 & 0 & 0 \\ 0 & 0 & 0 & 0 \\ 0 & 0 & 0 & 0 \\ \hline \frac{1}{4} & \frac{1}{4} & 0 & 0 \end{bmatrix},$$

determine the spectral radius $\rho_2$ of the block Gauss-Seidel matrix $(D - E)^{-1}E^*$. If $\rho_1$ is the spectral radius of the corresponding point Gauss-Seidel matrix, show that $1 > \rho_1 > \rho_2$. From this example, we conclude that block iterative methods *can* be asymptotically faster than their corresponding point iterative methods.

3. Let $A$ be a strictly or irreducibly diagonally dominant $n \times n$ Hermitian matrix with positive real diagonal entries. If this matrix is arbitrarily partitioned in the manner of (3.58), and if the matrices $D$, $E$, and $F$ are defined as in (3.59), prove that the block successive overrelaxation iterative method of (3.62) is convergent for any $0 < \omega < 2$.

4. Verify (3.56).

## 3.5. STIELTJES MATRICES AND M-MATRICES

At the conclusion of Sec. 3.4 we stated that any partitioning of a positive definite Hermitian matrix of the form (3.58) gives rise to a convergent block Gauss-Seidel iterative method, and we are ultimately interested in comparing the asymptotic rates of convergence of iterative methods derived from different partitionings of the matrix $A$. The first exercise of Sec. 3.4 in essence shows that taking "larger" blocks does not always result in larger asymptotic rates of convergence. In Sec. 3.6, however, we shall show that, under suitable conditions, iterative methods based on partitionings with larger blocks of the matrix $A$ *do* have larger asymptotic rates of convergence. As these "suitable conditions" for the matrix $A$ insure that the matrix $A^{-1}$ is a non-negative matrix†, the purpose of this section is to introduce the related concepts of a *Stieltjes matrix* and an *M-matrix*, whose inverses are indeed non-negative matrices. We shall see also in Chapter 6 that such matrices arise naturally in the numerical solution of elliptic partial differential equations.

Historically, Stieltjes (1887) proved that if $A$ is a real symmetric and positive definite $n \times n$ matrix with all its off-diagonal entries negative, then $A^{-1} > O$. Later, Frobenius (1912) proved the stronger result that if $B > O$ is an $n \times n$ matrix, and $\alpha$ is a real number with $\alpha > \rho(B)$, then the matrix $\alpha I - B$ is nonsingular, and $(\alpha I - B)^{-1} > O$. We now consider more recent generalizations of Ostrowski (1937) and Fan (1958) dealing with matrix properties which insure that the inverse of a matrix has only non-negative entries. We begin with

**Theorem 3.7.** *If $M$ is an arbitrary complex $n \times n$ matrix with $\rho(M) < 1$, then $I - M$ is nonsingular, and*

$$(3.63) \qquad (I - M)^{-1} = I + M + M^2 + \cdots,$$

*the series on the right converging. Conversely, if the series on the right converges, then $\rho(M) < 1$.*

*Proof.* First, assume that $\rho(M) < 1$. If $\mu$ is an eigenvalue of $M$, then $1 - \mu$ is an associated eigenvalue of $I - M$, and, as $\rho(M) < 1$, $I - M$ is nonsingular. From the identity

$$I - (I - M)(I + M + M^2 + \cdots + M^r) = M^{r+1},$$

we have, upon premultiplying by $(I - M)^{-1}$, that

$$(I - M)^{-1} - (I + M + M^2 + \cdots + M^r) = (I - M)^{-1}M^{r+1}.$$

Thus

$$(3.64) \qquad \| (I - M)^{-1} - (I + M + M^2 + \cdots + M^r) \|$$
$$\leq \| (I - M)^{-1} \| \cdot \| M^{r+1} \|,$$

for all $r \geq 0$. As $M$ is convergent, it follows from Theorem 1.4 that $\| M^{r+1} \|$ tends to zero with increasing $r$. Thus, the series of (3.63) is convergent and

† The simple example of the $4 \times 4$ matrix $A$ of (1.6) shows (1.9) that $A^{-1} > O$. Opposed to this, it is easy to verify that the matrix $A$ of Exercise 1 of Sec. 3.4 does not share this property.

is equal to $(I - M)^{-1}$. Conversely, if the series converges, let $\mu$ be an eigenvalue of $M$, corresponding to the eigenvector $\mathbf{x}$. Then

$$(I + M + M^2 + \cdots)\mathbf{x} = (1 + \mu + \mu^2 + \cdots)\mathbf{x}.$$

Thus, the convergence of the matrix series implies the convergence of the series $1 + \mu + \mu^2 + \cdots$ for any eigenvalue $\mu$ of $M$. However, as is well known, for this series of complex numbers to converge, it is necessary that $|\mu| < 1$ for all eigenvalues of $M$, and thus $\rho(M) < 1$, completing the proof.

The result of Theorem 3.7 is the basis for

**Theorem 3.8.** *If $A \geq O$ is an $n \times n$ matrix, then the following are equivalent*:

1. $\alpha > \rho(A)$.

2. $\alpha I - A$ *is nonsingular, and* $(\alpha I - A)^{-1} \geq O$.

*Proof.* Assuming part 1, it is then obvious that the matrix $M = \alpha I - A$ is nonsingular. Writing $M = \alpha(I - A/\alpha)$, then

$$M^{-1} = \frac{1}{\alpha}\left(I - \frac{A}{\alpha}\right)^{-1},$$

and as $A/\alpha$ has spectral radius less than unity by hypothesis, applying Theorem 3.7 we have that

$$(3.65) \qquad M^{-1} = \frac{1}{\alpha}\left\{ I + \frac{A}{\alpha} + \frac{A^2}{\alpha^2} + \cdots \right\}.$$

Since $A \geq O$, so are its powers. Thus $M^{-1} \geq O$, proving that part 1 implies part 2. Assuming part 2, let $\mathbf{x} \geq \mathbf{0}$ be an eigenvector of $A$ with $A\mathbf{x} = \rho(A)\mathbf{x}$. Thus,

$$(\alpha I - A)\mathbf{x} = (\alpha - \rho(A))\mathbf{x},$$

and as the matrix $(\alpha I - A)$ is nonsingular, $\alpha - \rho(A) \neq 0$. Now,

$$(\alpha I - A)^{-1}\mathbf{x} = \frac{\mathbf{x}}{(\alpha - \rho(A))},$$

and since $\mathbf{x} \geq \mathbf{0}$, $(\alpha I - A)^{-1} \geq O$, it follows that $\alpha > \rho(A)$. Thus, part 2 implies part 1, completing the proof.

As in the past, the use of the matrix property of irreducibility often enables us to strengthen certain results. With irreducibility, we can strengthen Theorem 3.8 as follows. Its proof, following along familiar lines now, is left as an exercise.

**Theorem 3.9.** *If $A \geq O$ is an $n \times n$ matrix, then the following are equivalent*:

1. $\alpha > \rho(A)$ *and $A$ is irreducible.*
2. $\alpha I - A$ *is nonsingular, and* $(\alpha I - A)^{-1} > O$.

The following theorem uses Theorem 3.8 as its basis.

**Theorem 3.10.** *If $A = (a_{i,j})$ is a real $n \times n$ matrix with $a_{i,j} \leq 0$ for all $i \neq j$, then the following are equivalent:*

1. *$A$ is nonsingular, and $A^{-1} \geq O$.*
2. *The diagonal entries of $A$ are positive real numbers, and letting $D$ be the diagonal matrix whose diagonal entries $d_{i,i}$ are defined as*

$$(3.66) \qquad\qquad d_{i,i} = 1/a_{i,i}, \qquad 1 \leq i \leq n,$$

*then the matrix $B = I - DA$ is non-negative and convergent.*

*Proof.* Assuming part 1, let $A^{-1} = (r_{i,j})$. As $A$ has nonpositive off-diagonal entries, then

$$(3.67) \qquad r_{i,i}a_{i,i} - \sum_{\substack{j=1 \\ j \neq i}}^{n} r_{i,j} \, | \, a_{j,i} \, | = 1, \qquad 1 \leq i \leq n.$$

Thus, $A^{-1} \geq O$ implies that all the diagonal entries of $A$ are positive real numbers. With the definition of the diagonal entries of the diagonal matrix $D$ in (3.66), it follows that the matrix $B = I - DA$ is non-negative. Moreover, as $I - B = DA$, then $I - B$ is nonsingular and

$$(3.68) \qquad\qquad A^{-1}D^{-1} = (I - B)^{-1},$$

and since $A^{-1}$ and $D^{-1}$ are both non-negative matrices, it follows that $(I - B)^{-1} \geq O$. Thus, from Theorem 3.8, $\rho(B) < 1$, and part 1 implies part 2. Assuming part 2, we have from Theorem 3.8 that $(I - B)$ is nonsingular, and $(I - B)^{-1} \geq O$, which, from (3.68), implies part 1, completing the proof.

Finally, the analog of Theorem 3.10, obtained through considerations of irreducibility, is given in Theorem 3.11. The proof is similar to those given above and is again left as an exercise.

**Theorem 3.11.** *If $A = (a_{i,j})$ is a real $n \times n$ matrix with $a_{i,j} \leq 0$ for all $i \neq j$, then the following are equivalent:*

1. *$A$ is nonsingular, and $A^{-1} > O$.*
2. *The diagonal entries of $A$ are positive real numbers. If $D$ is the diagonal matrix defined by (3.66), then the matrix $B = I - DA$ is non-negative, irreducible, and convergent.*

With the above results, we can prove the following corollaries.

**Corollary 1.** *If $A = (a_{i,j})$ is a real, irreducibly diagonally dominant $n \times n$ matrix with $a_{i,j} \leq 0$ for all $i \neq j$, and $a_{i,i} > 0$ for all $1 \leq i \leq n$, then $A^{-1} > O$.*

*Proof.* From Theorem 3.4, we have that the point Jacobi matrix $\tilde{B}$ (see (3.44)) associated with the matrix $A$ is convergent. From the hypotheses of this corollary, this point Jacobi matrix $\tilde{B}$ is non-negative and irreducible. But as the diagonal entries of $A$ are positive, it follows that the matrix $B$ of Theorem 3.11 is *precisely* the point Jacobi matrix $\tilde{B}$, and the result follows now from Theorem 3.11.

**Corollary 2.** *If $A = (a_{i,j})$ is a real, symmetric and nonsingular $n \times n$ irreducible matrix, where $a_{i,j} \leq 0$ for all $i \neq j$, then $A^{-1} > O$ if and only if $A$ is positive definite.*

*Proof.* Let $A^{-1} = (r_{i,j})$. As $A$ has nonpositive off-diagonal entries, then $A^{-1} \geq O$ implies again from (3.67) that the diagonal entries of $A$ are positive real numbers. But similarly, $A$'s being positive definite also implies this. With the positivity of the diagonal entries of the matrix $A$, the $n \times n$ matrix $B$ defined in Theorem 3.11 is evidently non-negative and irreducible and is just the point Jacobi matrix associated with $A$. Employing Theorem 3.11, $A^{-1} > O$ if and only if $\rho(B) < 1$, and from the Stein-Rosenberg Theorem 3.3, we conclude then that $A^{-1} > O$ if and only if the point Gauss-Seidel matrix is convergent. Remembering that the diagonal entries of $A$ are positive, then from Reich's Corollary 1 of Theorem 3.6, $A^{-1} > O$ if and only if $A$ is positive definite, which completes the proof.

We now make some definitions which we shall use frequently in subsequent sections.

DEFINITION 3.3. A real $n \times n$ matrix $A = (a_{i,j})$ with $a_{i,j} \leq 0$ for all $i \neq j$ is an *M-matrix* if $A$ is nonsingular, and $A^{-1} \geq O$.

DEFINITION 3.4. A real $n \times n$ matrix $A = (a_{i,j})$ with $a_{i,j} \leq 0$ for all $i \neq j$ is a *Stieltjes matrix* if $A$ is symmetric and positive definite.

With this last definition and Corollary 2 above, we immediately have

**Corollary 3.** *If $A$ is a Stieltjes matrix, then it is also an M-matrix. If $A$ is in addition irreducible, then $A^{-1} > O$.*

As a final result in this section, we include

**Theorem 3.12.** *Let $A$ be an $n \times n$ M-matrix, and let $C$ be any matrix obtained from $A$ by setting certain off-diagonal entries of the matrix $A$ to zero. Then, $C$ is also an M-matrix.*

*Proof.* Since the diagonal entries of the matrices of $A$ and $C$ are identical, the diagonal matrix $D$ defined in (3.66) is the same for the matrices $A$ and $C$. If $B_A \equiv I - DA$, and $B_C \equiv I - DC$, we have by hypothesis that $0 \le B_C \le B_A$, and as $\rho(B_A) < 1$ from Theorem 3.10, we conclude from the Theorem 2.8 that $B_C$ is also convergent. Thus, from Theorem 3.10, $C$ is an $M$-matrix.

As a consequence of this theorem, note that any principal minor of an $M$-matrix is also an $M$-matrix.

As a simple example of the above results, consider the real $4 \times 4$ matrix $A$ first described in (1.6). As we have seen, this matrix is symmetric, irreducible, and positive definite. Applying Corollary 2 above, we conclude that $A^{-1} > 0$, which of course agrees with the expression for $A^{-1}$ in (1.9). In terms of Definition 3.4, this matrix is evidently an irreducible Stieltjes matrix. As a second example, considering the matrix

$$A = \begin{bmatrix} 2 & -1/2 & 0 & 0 & 0 \\ -3/2 & 2 & -1/2 & 0 & 0 \\ 0 & -3/2 & 2 & -1/2 & 0 \\ 0 & 0 & -3/2 & 2 & -1/2 \\ 0 & 0 & 0 & -3/2 & 2 \end{bmatrix}.$$

We deduce by inspection that $A$ is an irreducibly diagonally dominant matrix, and applying Corollary 1, we necessarily have that $A^{-1} > 0$. Thus, $A$ is evidently an irreducible $M$-matrix.

## EXERCISES

1. Prove the following generalization of Theorem 3.7. If all the eigenvalues of an $n \times n$ complex matrix lie in the interior of the circle of convergence of the power series

$$\phi(z) = \sum_{k=0}^{\infty} c_k z^k,$$

then the matrix power series

$$\phi(A) \equiv \sum_{k=0}^{\infty} c_k A^k$$

converges. (*Hint*: assume that $A$ is in its Jordan normal form). As a consequence, conclude that

$$\exp (A) \equiv I + A + \frac{A^2}{2} + \cdots$$

is defined for any $n \times n$ complex matrix.

2. In Theorem 3.8, show that the following property (Fan (1958)) is equivalent to properties 1 and 2 of Theorem 3.8:

   3. *There exists a vector* $\mathbf{u} > \mathbf{0}$ *such that* $(\alpha I - A)\mathbf{u} > \mathbf{0}$.

3. Prove Theorem 3.9.

4. In Theorem 3.10, show that the following property is equivalent to properties 1 and 2 of Theorem 3.10.

   3. *If* $\lambda$ *is any eigenvalue of* $A$, *then* $\mathrm{Re}\,\lambda > 0$.

5. Prove Theorem 3.11.

6. An $n \times n$ matrix $A$ is *monotone* (Collatz (1952)) if for any vector $\mathbf{r}$, $A\mathbf{r} \geq \mathbf{0}$ implies $\mathbf{r} \geq \mathbf{0}$. If the off-diagonal entries of $A$ are nonpositive, show that $A$ is monotone if and only if $A$ is an $M$-matrix.

7. Let $A$ be an $n \times n$ *monotone* matrix whose off-diagonal entries are nonpositive. If $\mathbf{x}_1$ and $\mathbf{x}_2$ are two vectors with

$$A\mathbf{x}_1 \geq \mathbf{k}, \qquad A\mathbf{x}_2 \leq \mathbf{k},$$

prove, using the exercise above, that

$$\mathbf{x}_1 \geq \mathbf{x} \geq \mathbf{x}_2,$$

where $\mathbf{x}$ is the unique solution of $A\mathbf{x} = \mathbf{k}$. (Collatz (1952)).

8. Prove that a symmetric $M$-matrix is a Stieltjes matrix.

9. Let $A$ and $B$ be two $n \times n$ $M$-matrices, where $A \geq B$. Prove that $B^{-1} \geq A^{-1}$.

## 3.6. REGULAR SPLITTINGS OF MATRICES

With the definitions of a Stieltjes matrix and an $M$-matrix, we return to the problem of comparing the asymptotic rates of convergence of iterative methods defined from different partitionings of a given matrix. Abstractly, if we seek to solve the matrix equation $A\mathbf{x} = \mathbf{k}$ where $A$ is a given nonsingular $n \times n$ matrix, we consider expressing the matrix $A$ in the form

$$(3.69) \qquad\qquad A = M - N,$$

where $M$ and $N$ are also $n \times n$ matrices. If $M$ is nonsingular, we say that this expression represents a *splitting* of the matrix $A$, and associated with this splitting is an iterative method

(3.70) $$M\mathbf{x}^{(m+1)} = N\mathbf{x}^{(m)} + \mathbf{k}, \qquad m \geq 0,$$

which we write equivalently as

(3.70') $$\mathbf{x}^{(m+1)} = M^{-1}N\mathbf{x}^{(m)} + M^{-1}\mathbf{k}, \qquad m \geq 0.$$

We now point out that *all* iterative methods considered thus far can be described from this point of view. For example, if the diagonal entries of the matrix $A = (a_{i,j})$ are all nonzero, and we express the matrix $A$ as in (3.3) as the matrix sum

$$A = D - E - F$$

where $D = \mathrm{diag}\,\{a_{1,1}, a_{2,2}, \cdots, a_{n,n}\}$ and $E$ and $F$ are respectively strictly lower and upper triangular $n \times n$ matrices, then the following choices

$$M = D; \qquad\qquad N = E + F,$$

(3.71) $$M = D - E; \qquad N = F,$$

$$M = \frac{1}{\omega}(D - \omega E); \qquad N = \frac{1}{\omega}(\omega F + (1 - \omega)D), \qquad \omega \neq 0,$$

give in (3.70) respectively the point Jacobi, point Gauss-Seidel, and point successive overrelaxation iterative methods. Similar remarks are valid for the block iterative methods of Sec. 3.4. While we may have been inclined to call the iterative method of (3.70) a generalized Jacobi iterative method, we see that such splittings (3.69) of the matrix $A$ in reality cover all iterative methods previously described.

Special properties of the matrices $M$ and $N$ defining a splitting of the matrix $A$ have been of considerable use in Sec. 3.3. There we in essence assumed that $D$ is a positive diagonal matrix, and that $E + F$ is a nonnegative matrix. Clearly, for $0 < \omega \leq 1$ this guarantees that the different splittings of (3.71) share the common properties that $M^{-1} \geq O$ and $N \geq O$, which brings us to

DEFINITION 3.5.  For $n \times n$ real matrices $A$, $M$, and $N$, $A = M - N$ is a *regular splitting of the matrix* $A$ if $M$ is nonsingular with $M^{-1} \geq O$, and $N \geq O$.

In terms of the iteration matrix $M^{-1}N$ of (3.70'), we can from (3.69) express this matrix product in the form

(3.72) $$M^{-1}N = (A + N)^{-1}N,$$

and as $A$ is nonsingular by assumption, we can write this as

$$(3.73) \qquad M^{-1}N = (I + G)^{-1}G, \qquad G \equiv A^{-1}N.$$

Because of this relation, if $\mathbf{x}$ is any eigenvector of $G$ corresponding to the eigenvalue $\tau$, then as

$$G\mathbf{x} = \tau\mathbf{x}, \qquad (I + G)^{-1}G\mathbf{x} = \frac{\tau}{1 + \tau}\mathbf{x},$$

it follows that $\mathbf{x}$ is also an eigenvector of $M^{-1}N$ corresponding to the eigenvalue $\mu$ given by

$$(3.74) \qquad \mu = \frac{\tau}{1 + \tau}.$$

Conversely, if $\mu$ is any eigenvalue of $(I + G)^{-1}G$ with $(I + G)^{-1}G\mathbf{z} = \mu\mathbf{z}$, then

$$G\mathbf{z} = \mu(I + G)\mathbf{z}.$$

From this expression, it follows that $\mu$ *cannot* be unity, so that we can write

$$G\mathbf{z} = \left(\frac{\mu}{1 - \mu}\right)\mathbf{z} \equiv \tau\mathbf{z},$$

which is again the eigenvalue relationship of (3.74). (See also Exercise 1.)

**Theorem 3.13.** *If $A = M - N$ is a regular splitting of the matrix $A$ and $A^{-1} \geq O$, then*

$$(3.75) \qquad \rho(M^{-1}N) = \frac{\rho(A^{-1}N)}{1 + \rho(A^{-1}N)} < 1.$$

*Thus, the matrix $M^{-1}N$ is convergent, and the iterative method of (3.70) converges for any initial vector $\mathbf{x}^{(0)}$.*

*Proof.* By Definition 3.5, we have that $M^{-1}N \geq O$. Thus, in determining $\rho(M^{-1}N)$, we can, by the Perron-Frobenius theory of non-negative matrices (Theorem 2.7), restrict our attention to those eigenvectors $\mathbf{x}$ of $M^{-1}N$ and $G$ which are non-negative. But from (3.73) and the hypotheses, $G$ is also a non-negative matrix, and therefore the associated eigenvalue $\tau$ of $G$ is necessarily non-negative. Clearly, $\tau/(1 + \tau)$ is a monotone function of $\tau$ for $\tau \geq 0$, and thus $\mu$ in (3.74) is maximized in modulus by choosing $\tau = \rho(G)$, which proves (3.75). With this expression, it is immediately obvious that $\rho(M^{-1}N) < 1$, and thus the matrix $M^{-1}N$ is convergent. As a consequence, the iterative method of (3.70) is convergent, completing the proof.

In terms of satisfying the hypothesis that $A^{-1} \geq O$ in Theorem 3.13, we now recall results from Sec. 3.5.

**Corollary.** *Let $A$ be a Stieltjes matrix, and let $A = M - N$ be a regular splitting of $A$ where $N$ is real and symmetric. Then,*

$$(3.76) \qquad \rho(M^{-1}N) \leq \frac{\rho(N) \cdot \rho(A^{-1})}{1 + \rho(N) \cdot \rho(A^{-1})} < 1.$$

*Proof.* By the hypotheses, we conclude that $A^{-1} \geq O$ and that the results of Theorem 3.13 are valid. As $A^{-1}$ and $N$ are both real symmetric matrices, then $\rho(A^{-1}) = \| A^{-1} \|$, and $\rho(N) = \| N \|$. Thus, as

$$\rho(A^{-1}N) \leq \| A^{-1}N \| \leq \| A^{-1} \| \cdot \| N \| = \rho(A^{-1}) \cdot \rho(N),$$

the first inequality of (3.76) follows directly from (3.75) and the monotonicity of $\tau/(1 + \tau)$ for $\tau \geq 0$.

As a means for generating regular splittings, the results of Sec. 3.5 are again useful.

**Theorem 3.14.** *Let $A = (a_{i,j})$ be an $n \times n$ M-matrix. If $M$ is any $n \times n$ matrix obtained by setting certain off-diagonal entries of $A$ to zero, then $A = M - N$ is a regular splitting of $A$ and $\rho(M^{-1}N) < 1$.*

*Proof.* This follows directly from Theorem 3.12 and Theorem 3.13.

We now assume that $A = M_1 - N_1 = M_2 - N_2$ are *two* regular splittings of $A$, and that we have the stronger hypothesis $A^{-1} > O$. We seek to compare the spectral radii of the two products $M_1^{-1}N_1$ and $M_2^{-1}N_2$.

**Theorem 3.15.** *Let $A = M_1 - N_1 = M_2 - N_2$ be two regular splittings of $A$, where $A^{-1} > O$. If $N_2 \geq N_1 \geq O$, equality excluded,† then*

$$(3.77) \qquad 1 > \rho(M_2^{-1}N_2) > \rho(M_1^{-1}N_1) > 0.$$

*Proof.* Using (3.75) of Theorem 3.13, $\rho(M^{-1}N)$ is monotone with respect to $\rho(A^{-1}N)$, and it suffices to prove that $\rho(A^{-1}N_2) > \rho(A^{-1}N_1) > 0$. Since $N_2 \geq N_1 \geq O$, equality excluded, and $A^{-1} > O$, we have that

$$A^{-1}N_2 \geq A^{-1}N_1 \geq O,$$

equality excluded. We now complete the proof of this theorem under the assumption that $A^{-1}N_1$ is irreducible, the reducible case being left as an exercise. But as some entry of the irreducible non-negative matrix $A^{-1}N_1$

---

† By this we mean that neither $N_1$ nor $N_2 - N_1$ is the null matrix.

is *increased* in passing to $A^{-1}N_2$, then from the Perron-Frobenius Theorem 2.1, we immediately conclude that

$$\rho(A^{-1}N_2) > \rho(A^{-1}N_1) > 0,$$

which completes the proof in the irreducible case.

In terms of asymptotic rates of convergence, Theorem 3.15 immediately gives us

**Corollary 1.** *If $A = M_1 - N_1 = M_2 - N_2$ are two regular splittings of $A$ where $A^{-1} > O$, and $N_2 \geq N_1 \geq O$, equality excluded, then*

$$(3.78) \qquad\qquad R_\infty(M_1^{-1}N_1) > R_\infty(M_2^{-1}N_2) > 0.$$

Thus, the theory of regular splittings leads to another basic comparison of asymptotic convergence rates between two iterative methods. For Stieltjes matrices, we have

**Corollary 2.** *Let $A$ be an irreducible $n \times n$ Stieltjes matrix and let $M_1$ and $M_2$ be $n \times n$ matrices, each obtained by setting certain off-diagonal entries of $A$ to zero. If $A \equiv M_1 - N_1$ and $A \equiv M_2 - N_2$, and $N_1 \geq N_2 \geq O$, equality excluded, then*

$$0 < \rho(M_2^{-1}N_2) < \rho(M_1^{-1}N_1) < 1.$$

*Proof.* This is an immediate consequence of Theorem 3.15 and Corollary 3 of Theorem 3.11.

The importance of this corollary lies in the following observation: If $A$ is an irreducible $n \times n$ Stieltjes matrix, we know that its associated point Gauss-Seidel iterative method and any block Gauss-Seidel iterative method, in the sense of Sec. 3.4, are convergent. As long as some one diagonal sub-matrix $A_{i,i}$ of the partitioned matrix $A$ of (3.58) contains a nonzero entry from *above* the main diagonal of $A$, the block Gauss-Seidel iterative method *must* have a larger asymptotic rate of convergence than that of the point Gauss-Seidel iterative method. Practically speaking, we must then ascertain that the work involved in directly solving smaller matrix equations does not overbalance the gain made in asymptotic rates of convergence. Such practical topics will be considered in Chapter 6.

We finally point out that the example given in Exercise 1 of Sec. 3.4 where a block iterative method was slower than a point method was such that the associated matrix $A$ was neither a Stieltjes matrix nor an $M$-matrix.

Some of the comparisons of asymptotic rates of convergence which we have previously given can be proved in special cases as direct applications of this regular splitting theory. For example, take the Stein-Rosenberg Theorem 3.3. If $A \equiv I - B$, where $B$ is a non-negative, irreducible, and convergent matrix with zero diagonal entries, then writing $B = L + U$, where $L$ and $U$ are respectively strictly lower and upper triangular matrices, we define

$$M_1 = I; \qquad N_1 = L + U,$$

$$M_2 = I - L; \qquad N_2 = U.$$

It is obvious that these are both regular splittings of $A$, and $A^{-1} > 0$. But as $N_1 \geq N_2 \geq 0$, equality excluded, we conclude directly from Theorem 3.15 that

$$0 < \rho\{(I - L)^{-1}U\} < \rho(L + U) < 1,$$

which is statement 2 of the Stein-Rosenberg Theorem 3.3. But a more interesting application is

**Theorem 3.16.** *Let $A = I - B$, where $B = L + U$ is a non-negative, irreducible, and convergent $n \times n$ matrix, and $L$ and $U$ are respectively strictly lower and upper triangular matrices. Then, the successive overrelaxation matrix $\mathcal{L}_\omega$ of* (3.39) *is convergent for all $0 < \omega \leq 1$. Moreover, if $0 < \omega_1 < \omega_2 \leq 1$ then*

$$(3.79) \qquad\qquad 0 < \rho(\mathcal{L}_{\omega_2}) < \rho(\mathcal{L}_{\omega_1}) < 1,$$

*and consequently*

$$(3.79') \qquad\qquad R_\infty(\mathcal{L}_{\omega_1}) < R_\infty(\mathcal{L}_{\omega_2}).$$

*Proof.* Defining $A = M_\omega - N_\omega$, where

$$M_\omega = \frac{1}{\omega}(I - \omega L) \quad \text{and} \quad N_\omega = \frac{1}{\omega}\{\omega U + (1 - \omega)I\},$$

it is evident from the hypotheses that $A = M_\omega - N_\omega$ is a regular splitting of $A$ for all $0 < \omega \leq 1$. If $0 < \omega_1 < \omega_2 \leq 1$, then $0 \leq N_{\omega_2} \leq N_{\omega_1}$, equality excluded, and as $A^{-1} > 0$, the result follows directly from Theorem 3.15.

Evidently, for non-negative irreducible and convergent Jacobi matrices, *underrelaxation* $(0 < \omega < 1)$ for the successive overrelaxation iterative method can only result in asymptotically slower iterative procedures, and as such, these are of little practical interest. Continuing in this case, if $\rho(\mathcal{L}_\omega)$ is *differentiable* as a function of $\omega$ for $\omega = 1$, and

$$(3.80) \qquad\qquad \frac{d}{d\omega}\rho(\mathcal{L}_\omega)\;\Big|_{\omega=1} < 0,$$

the above results would strongly suggest that the use of $\omega$'s greater than unity would decrease even further the spectral radius of $\rho(\mathcal{L}_\omega)$ beneath $\rho(\mathcal{L}_1)$. In other words, for $B \geq 0$ irreducible and convergent, *overrelaxation* $(\omega > 1)$ would appear to be the promising way of minimizing the spectral radius $\rho(\mathcal{L}_\omega)$ of the successive overrelaxation matrix. Indeed, the pioneering work of Young (1950) shows that this is true for a large class of matrices. We will consider this topic in detail in Chapters 4 and 5, where we shall see that the theory of cyclic and primitive matrices, as developed in Chapter 2, can be used as a basis for such a topic.

Before leaving this section, however, we consider a simple counter-example† to the notion that the minimum of $\rho(\mathcal{L}_\omega)$ occurs when $\omega > 1$, even in the case where the point Jacobi matrix is non-negative, irreducible, and convergent. Let

$$(3.81) \qquad B = \alpha \begin{bmatrix} 0 & 1 & 0 \\ 0 & 0 & 1 \\ 1 & 0 & 0 \end{bmatrix}, \qquad 0 < \alpha < 1.$$

The matrix $B$ is evidently non-negative, convergent, and irreducible, and the matrix $B$ is in fact in the form (2.33) of a matrix which is *cyclic of index* 3. By direct computation,

$$(3.82) \qquad \mathcal{L}_\omega = \begin{bmatrix} 1 - \omega & \alpha\omega & 0 \\ 0 & 1 - \omega & \alpha\omega \\ \alpha\omega(1 - \omega) & \alpha^2\omega^2 & 1 - \omega \end{bmatrix},$$

and it can be verified that $\rho(\mathcal{L}_\omega)$ *increases* as $\omega$ increases from unity. For this example, $\rho(\mathcal{L}_\omega)$ is not differentiable for $\omega = 1$, and

$$\min_\omega \rho(\mathcal{L}_\omega) = \rho(\mathcal{L}_1).$$

### EXERCISES

1. Let $G$ be any $n \times n$ complex matrix such that $I + G$ is nonsingular. Upon considering the Jordan normal form of $G$, prove that $\mu_k$ is an eigenvalue of multiplicity $n_k$ of $(I + G)^{-1}G$ if and only if the associated eigenvalue $\tau_k$ of $G$ of (3.74) is also of multiplicity $n_k$.

2. With the hypotheses of the Corollary of Theorem 3.13, show that

$$\rho(M^{-1}N) = \frac{\rho(N)\rho(A^{-1})}{1 + \rho(N)\rho(A^{-1})}$$

if the matrices $A$ and $N$ commute, i.e. $AN = NA$.

3. Prove Theorem 3.15 under the assumption that the matrix $A^{-1}N_1$ is reducible. (*Hint*: Use the normal form (2.41) of $A^{-1}N_1$).

† Due to Kahan (1958).

**4.** Let $B = L + U$ be a non-negative irreducible and convergent $n \times n$ matrix, and let $m(\sigma) \equiv \rho(\sigma L + U), \sigma \geq 0$, as in Definition 3.2. Prove that

a. If $\omega \neq 0$ and $\tau$ is a non-negative real number with

$$m(\tau) = \frac{\tau + \omega - 1}{\omega},$$

then $\tau$ is an eigenvalue of $\mathcal{L}_\omega$, and its corresponding eigenvector can be chosen to have positive components.

b. If $0 < \omega \leq 1$, prove that $\rho(\mathcal{L}_\omega)$ is the largest such value of $\tau$ for which

$$m(\tau) = \frac{\tau + \omega - 1}{\omega}.$$

c. if $0 < \omega < 1$, prove that $\mathcal{L}_\omega$ is primitive.

**5.** Let the $4 \times 4$ matrix $A$ be given by

$$A \equiv \begin{bmatrix} 4 & -1 & 0 & 0 \\ -1 & 4 & -1 & 0 \\ 0 & -1 & 4 & -1 \\ 0 & 0 & -1 & 4 \end{bmatrix},$$

and let $D_1 \equiv 4I$,

$$D_2 \equiv \begin{bmatrix} 4 & -1 & 0 & 0 \\ -1 & 4 & 0 & 0 \\ 0 & 0 & 4 & 0 \\ 0 & 0 & 0 & 4 \end{bmatrix}, \quad \text{and} \quad D_3 \equiv \begin{bmatrix} 4 & -1 & 0 & 0 \\ -1 & 4 & -1 & 0 \\ 0 & -1 & 4 & 0 \\ 0 & 0 & 0 & 4 \end{bmatrix}.$$

With $A = D_i - F_i$ defining the matrices $F_i, 1 \leq i \leq 3$, show that each is a regular splitting of $A$. Find $\rho(D_i^{-1}F_i)$ and $R_\infty(D_i^{-1}F_i)$ for $1 \leq i \leq 3$.

## BIBLIOGRAPHY AND DISCUSSION

**3.1.** For excellent bibliographical references to the older contributions of Gauss, Jacobi, and others on cyclic iterative methods, see Ostrowski (1956) and Bodewig (1959). The first paper to consider simultaneously the point Jacobi and point Gauss-Seidel iterative methods with sufficient conditions for their convergence was apparently the paper by von Mises and Pollaczek-Geiringer (1929). However, much earlier Nekrasov (1885 and 1892) and Mehmke (1892) examined the point Gauss-Seidel iterative method and considered sufficient conditions for its convergence.

The point successive overrelaxation iterative method was simultaneously considered by Frankel (1950) and Young (1950), who both pointed out the advantages gained in certain cases by using a fixed relaxation factor $\omega$ greater than unity. However, the idea that "over-relaxing" could be beneficial occurred much earlier, at least in the context of noncyclic iterative methods. Explicit mention of the use of "over-relaxation" in noncyclic iterative methods can be found in the well-known books of Southwell (1946), Shaw (1953), and Allen (1954).

**3.2.** The terminology concerning average rates of convergence for $m$ iterations is not uniform in the literature. Kahan (1958) defines the average rate of convergence for $m$ iterations as

$$\frac{-\ln \left[\| \, \boldsymbol{\varepsilon}^{(m)} \, \|/\| \, \boldsymbol{\varepsilon}^{(0)} \, \|\right]}{m},$$

where $\boldsymbol{\varepsilon}^{(0)}$ is the initial error vector and $\boldsymbol{\varepsilon}^{(m)}$ is the error vector after $m$ iterations. This, of course, is bounded above by the quantity $R(A^m)$ of Definition 3.1. Young (1954a) defines $-\ln \rho(A)$ as the rate of convergence of the matrix $A$, and it is clear that $R(A^m) = -\ln \rho(A)$ for all $m$ if $A$ is Hermitian or normal.

The results of this section on average rates of convergence are essentially restatements of results by Werner Gautschi (1953a, 1953b, and 1954), and Ostrowski (1956), who found bounds for the norm of the matrix $A^p$, as a function of $p$, for general matrix norms. For generalizations, see Marcus (1955) and Kato (1960). The particular result of Theorem 3.2, which can be restated as $\| \, A^m \, \|^{1/m} \to \rho(A)$, is true more generally for bounded linear operators on a Banach space. For example, see Dunford and Schwartz, p. 567 (1958).

The quantity $\Lambda \equiv \| \, S \, \| \cdot \| \, S^{-1} \, \|$, the *condition number* for the nonsingular matrix $S$, which is important in the study of rounding errors of direct inversions of matrices and matrix equations, can be traced back at least to Turing (1948), although it is implicit in the papers of von Neumann and Goldstine (1947), and Goldstein and von Neumann (1951). In the case that $S$ is Hermitian or normal, this reduces to the *P-condition number* of $S$, which has been considered by Todd (1950, 1954, and 1958). For recent generalizations, see Bauer (1960).

**3.3.** The proof of the Stein-Rosenberg Theorem 3.3 is due in part to Kahan (1958) and differs from the original proof of Stein and Rosenberg (1948). Both proofs are nevertheless based on the Perron-Frobenius theory of non-negative matrices.

Many sufficient conditions for the convergence of the point Jacobi and point Gauss-Seidel iterative methods have been repeatedly derived. If the matrix $A$ is strictly diagonally dominant, von Mises and Geiringer (1929) proved that the point Jacobi method is convergent. See also Wittmeyer (1936). Collatz (1942b) proved, with the same assumption, that both the point Jacobi and point Gauss-Seidel iterative methods are convergent. Later, Geiringer (1949) proved that this result was also true for irreducibly diagonally dominant matrices, and these combined conclusions are given in Theorem 3.4. For other results dealing with sufficient conditions for the convergence of one or both of these iterative methods, see Collatz (1950), Sassenfeld (1951), and Weissinger (1952).

**3.4.** It was known for some time that if the matrix $A$ is real, symmetric and positive definite, then either the point or block Gauss-Seidel iterative method is convergent (von Mises and Geiringer (1929)). The converse was first established by Reich (1949) and later by Geiringer (1949) and Schmeidler (1949). Theorem 3.6, which includes these results as the special case $\omega = 1$, is a less general form of the result by Ostrowski (1954). See also Householder (1958) and Weissinger (1951 and 1953). The proof we have given for Theorem 3.6 is based on Ostrowski's (1954) use of quadratic forms, but different proofs are given by Householder (1958), who uses basic properties of matrix norms, and Reich (1949). For extensions to nearly symmetric matrices, see Stein (1951) and Ostrowski (1954).

In Exercise 1 of Sec. 3.4, an example is given where a point method is iteratively faster than a block iterative method. Such an example was first pointed out by Arms and Gates (1956), although the use of block methods has long been practically advocated (von Mises and Geiringer (1929) and Southwell (1946)). Such block methods apparently go back to Gerling (1843).

**3.5.** The number of people who have contributed to the literature of matrices whose inverses have non-negative real entries is quite large. Beginning with the early results of Stieltjes (1887) and Frobenius (1912), Ostrowski (1937 and 1956) gave the definition and related results for an $M$-matrix. From Theorem 3.10 and Exercise 4, we see that there are several equivalent definitions of an $M$-matrix. For other equivalent definitions of an $M$-matrix see the excellent paper by Fan (1958), and references given therein. For other related results, see de Rham (1952), Kotelyanskii (1952), Egerváry (1954), and Birkhoff and Varga (1958). In Exercise 6, the related work of Collatz (1952) on monotone matrices is given. See also Collatz (1955 and 1960, pp. 42–47).

**3.6.** The definition of a regular splitting of a matrix is due to Varga (1960a), and we have largely followed the exposition given there. However, these results are closely related to results of Householder (1958). Fiedler and Pták (1956) establish results similar to those of Theorem 3.15 but without strict inequality between spectral radii. See also Juncosa and Mulliken (1960), and Kahan (1958).

# CHAPTER 4

# SUCCESSIVE OVERRELAXATION
# ITERATIVE METHODS

## 4.1. p-Cyclic Matrices

The point successive overrelaxation iterative method of Chapter 3 was simultaneously introduced by Frankel (1950) and Young (1950). Whereas Frankel considered the special case of the numerical solution of the Dirichlet problem for a rectangle and showed for this case that the point successive overrelaxation iterative method, with suitably chosen relaxation factor, gave substantially larger (by an order of magnitude) asymptotic rates of convergence than those for the point Jacobi and point Gauss-Seidel iterative methods, Young (1950) and (1954a) showed that these conclusions held more generally for matrices satisfying his definition of *property A*, and that these results could be rigorously applied to the iterative solution of matrix equations arising from discrete approximations to a large class of elliptic partial differential equations for general regions. Then, Arms, Gates, and Zondek (1956) with their definition of *property $A^\pi$* generalized Young's results. In so doing, they enlarged the class of matrix equations to which the basic results of Young and Frankel on the successive overrelaxation iterative method could be rigorously applied.

These important matrix properties mentioned above are fundamentally related to the work of Frobenius on cyclic matrices. By giving a further generalization of these matrix properties, we can bring out in detail their relationship to weakly cyclic matrices, as given in Sec. 2.2. Also, the use of graph theory as introduced in Sec. 1.4 will aid us in determining when these matrix properties are present.

Suppose that we seek to solve the matrix equation

$$(4.1) \qquad\qquad A\mathbf{x} = \mathbf{k},$$

where $A = (a_{i,j})$ is a given $n \times n$ complex matrix, $n \geq 2$, which is partitioned in the following form:

(4.2)
$$A = \begin{bmatrix} A_{1,1} & A_{1,2} & \cdots & A_{1,N} \\ A_{2,1} & A_{2,2} & \cdots & A_{2,N} \\ \cdot & & & \cdot \\ \cdot & & & \cdot \\ \cdot & & & \cdot \\ A_{N,1} & A_{N,2} & \cdots & A_{N,N} \end{bmatrix}$$

as in (3.58), where the diagonal submatrices $A_{i,i}$, $1 \leq i \leq N$, are square. We now assume that all the diagonal submatrices $A_{i,i}$ are nonsingular, so that the block diagonal matrix $D$

(4.3)
$$D = \begin{bmatrix} A_{1,1} & & & O \\ & A_{2,2} & & \\ & & \ddots & \\ O & & & A_{N,N} \end{bmatrix}$$

derived from the matrix $A$ of (4.2) is also nonsingular. The $n \times n$ matrix $B$ defined by

(4.4)
$$B \equiv -D^{-1}A + I$$

is the *block Jacobi matrix* of $A$, corresponding to the partitioning of the matrix $A$ in (4.2).

Of special interest are the particular matrices

(4.5)
$$A_1 = \begin{bmatrix} A_{1,1} & 0 & 0 & \cdots & 0 & A_{1,p} \\ A_{2,1} & A_{2,2} & 0 & \cdots & 0 & 0 \\ \cdot & & \ddots & & & \cdot \\ \cdot & & & \ddots & & \cdot \\ \cdot & & & & \ddots & \cdot \\ 0 & 0 & 0 & \cdots & A_{p,p-1} & A_{p,p} \end{bmatrix}, \quad p \geq 2,$$

and

(4.6)
$$A_2 = \begin{bmatrix} A_{1,1} & A_{1,2} & & & O \\ A_{2,1} & A_{2,2} & A_{2,3} & & \\ & \ddots & \ddots & \ddots & \\ & & & A_{N-1,N-1} & A_{N-1,N} \\ O & & & A_{N,N-1} & A_{N,N} \end{bmatrix},$$

and we say that $A_2$ is a *block tri-diagonal matrix*. The particular matrices $A_1$ and $A_2$ of (4.5) and (4.6) give rise to block Jacobi matrices of the form

$$(4.7) \qquad B_1 = \begin{bmatrix} O & O & O & \cdots & O & B_{1,p} \\ B_{2,1} & O & O & \cdots & O & O \\ O & B_{3,2} & O & \cdots & O & O \\ \vdots & & & & & \\ O & O & O & \cdots & B_{p,p-1} & O \end{bmatrix},$$

and

$$(4.8) \qquad B_2 = \begin{bmatrix} O & B_{1,2} & & & \\ B_{2,1} & O & B_{2,3} & & \\ & & \ddots & B_{N-1,N} \\ & & B_{N,N-1} & O \end{bmatrix},$$

where the matrices $B_1$ and $B_2$ are partitioned relative to the partitioning of the matrices $A_1$ and $A_2$, respectively. But the matrices $B_1$ and $B_2$ have interesting properties. First of all, the matrix $B_1$ is in the normal form (2.33′) of a matrix which is *weakly cyclic of index p*. (See Definition 2.3 of Sec. 2.2). Similarly, we can permute the *blocks* of the partitioned matrix $B_2$ to show that $B_2$ is *weakly cyclic of index* 2. (See Exercise 4 of Sec. 2.4.) This leads us to

DEFINITION 4.1. If the block Jacobi matrix $B$ of (4.4) for the matrix $A$ of (4.2) is weakly cyclic of index $p(\geq 2)$, then $A$ is *p-cyclic*, relative to the partitioning of (4.2).

We point out that if the diagonal submatrices $A_{i,i}$ of (4.2) are $1 \times 1$ matrices, then the property that the matrix $A$ is 2-cyclic in the sense of Definition 4.1 is equivalent to Young's *property A*. Similarly, the statement that the matrix $A$ of (4.2) is 2-cyclic in the sense of Definition 4.1 with no restrictions on the order of the diagonal submatrices $A_{i,i}$ is implied by *property $A^\pi$* of Arms, Gates, and Zondek.

It is interesting to point out that by choosing $N = 2$ in (4.2), the matrix $B$ of (4.4) takes the simple form

$$B = \begin{bmatrix} O & B_{1,2} \\ B_{2,1} & O \end{bmatrix},$$

which is obviously weakly cyclic of index 2. In other words, *any* partitioning of the matrix $A$ of (4.2) with $N = 2$, for which the diagonal submatrices are nonsingular, is such that the matrix $A$ is then 2-cyclic.

Our first use of graph theory in this chapter† is directed toward the question of deciding *when* the given matrix $A$ of (4.2) is $p$-cyclic, or equivalently, when its block Jacobi matrix $B$ of (4.4) is weakly cyclic of index $p$. To this end, we consider the *block directed graph* $G_\pi(B)$ of the $n \times n$ matrix $B$. Instead of taking $n$ modes for the directed graph of $B$, as in Sec. 1.4, we now take $N$ nodes $P_{\pi(1)}, P_{\pi(2)}, \cdots, P_{\pi(N)}$, where $N$ is the number of diagonal submatrices of the partitioned matrix $A$ of (4.2). By analogy with our definitions of directed graphs of matrices in Sec. 1.4, if the submatrix $B_{i,j}$ of the partitioned matrix $B$ has at least one nonzero entry, we connect the node $P_{\pi(i)}$ to the node $P_{\pi(j)}$ by means of a path directed from $P_{\pi(i)}$ to $P_{\pi(j)}$. To illustrate this,‡ the block directed graph of the matrix $B_1$ of (4.7) is, for $p = 5$,

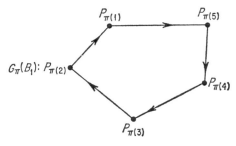

**Figure 5**

Similarly, the block directed graph of the matrix $B_2$ of (4.8) is, for $N = 5$,

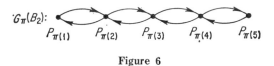

**Figure 6**

It is now clear that the greatest common divisor of the lengths of closed paths of the first block directed graph is $p = 5$, and thus, using the terminology of Def. 2.4, the first block directed graph is a *cyclic graph* of index $p = 5$. Similarly, the second block directed graph is a cyclic graph of index 2. The introduction of block directed graphs permits us to describe $p$-cyclic matrices in geometrical terms.

---

† See also Sec. 4.4.
‡ We have assumed here that all the distinguished submatrices $B_{j,k}$ of (4.7) and (4.8) have at least one nonzero entry.

**Theorem 4.1.** *Let the block directed graph of the partitioned Jacobi matrix $B$ of (4.4) be strongly connected. Then, the matrix $A$ of (4.2) is $p$-cyclic if the block directed graph of the matrix $B$ is a cyclic graph of index $p$.*

*Proof.* Similar to Exercise 6 of Sec. 2.4, if the block directed graph $G_\pi(B)$ of $B$ is strongly connected, then $G_\pi(B)$ is a cyclic graph of index $p(\geq 2)$ only if the partitioned matrix $B$ is weakly cyclic of index $p$.

This simple idea on the relationship between block directed graphs and $p$-cyclic matrices will be quite useful in Chapter 6 when matrix properties of difference equations derived from partial differential equations are considered.

Since the block Jacobi matrix $B$ of (4.4) must necessarily have zero diagonal entries, we can again express the matrix $B$ as $B = L + U$, where $L$ and $U$ are respectively strictly lower and upper triangular matrices.

DEFINITION 4.2. If the matrix $A$ of (4.2) is $p$-cyclic, then the matrix $A$ is *consistently ordered†* if all the eigenvalues of the matrix

$$B(\alpha) \equiv \alpha L + \alpha^{-(p-1)} U,$$

derived from the Jacobi matrix $B = L + U$, are independent of $\alpha$, for $\alpha \neq 0$. We then also say that the matrix $B$ is consistently ordered. Otherwise the matrices $A$ and $B$ are inconsistently ordered.

Let us consider the matrices $A_1$ and $A_2$ of (4.5) and (4.6). With

$$(4.9) \quad B_1(\alpha) = \begin{bmatrix} 0 & 0 & 0 & \cdots & 0 & \alpha^{-(p-1)}B_{1,p} \\ \alpha B_{2,1} & 0 & 0 & \cdots & 0 & 0 \\ 0 & \alpha B_{3,2} & 0 & \cdots & 0 & 0 \\ \vdots & & & & & \vdots \\ 0 & 0 & 0 & \cdots & \alpha B_{p,p-1} & 0 \end{bmatrix}$$

it can be verified by direct multiplication that $B_1^p(\alpha) = B_1^p$ for all $\alpha \neq 0$. Thus, as the eigenvalues of $B_1(\alpha)$ are independent of $\alpha$, we conclude that the matrix $A_1$ of (4.5) is a *consistently ordered $p$-cyclic matrix*. Similarly, if

$$B_2(\alpha)\mathbf{x} = \lambda\mathbf{x}$$

† For an eigenvalue-independent definition of the concept of a consistent ordering, see Exercise 9.

where $\mathbf{x} \neq \mathbf{0}$, we can partition the vector $\mathbf{x}$ relative to the partitioning of (4.8), so that this is equivalent to the set of equations

$$(4.10) \qquad \alpha B_{j,j-1} X_{j-1} + \frac{1}{\alpha} B_{j,j+1} X_{j+1} = \lambda X_j, \qquad 1 \leq j \leq N,$$

where $B_{1,0}$ and $B_{N,N+1}$ are null matrices. Defining $Z_j \equiv (1/\alpha^{j-1}) X_j$, $1 \leq j \leq N$, (4.10) takes the form

$$(4.11) \qquad B_{j,j-1} Z_{j-1} + B_{j,j+1} Z_{j+1} = \lambda Z_j, \qquad 1 \leq j \leq N,$$

so that any eigenvalue $\lambda$ of $B_2(\alpha)$ is, for any $\alpha \neq 0$, an eigenvalue of $B_2$, proving that $A_2$ is a consistently ordered 2-cyclic matrix. In other words, assuming nonsingular diagonal submatrices, *any block tri-diagonal matrix of the form* (4.6) *is a consistently ordered 2-cyclic matrix.*

While not all $p$-cyclic matrices $A$ are consistently ordered (see Exercise 4), it is not difficult to find an $n \times n$ permutation matrix $P$ which permutes the entries of $A$ by blocks such that $PAP^T$ is $p$-cyclic and consistently ordered. In fact, from Definition 2.3, there exists a permutation matrix $P$ which permutes the entries of the block Jacobi matrix $B$ by blocks such that $PBP^T$ is weakly cyclic of index $p$ and of the form (4.7), and is thus consistently ordered. We point out, however, that such permutation matrices $P$ giving rise to consistent orderings are *not* unique.

The basic reason for introducing the concepts of a $p$-cyclic matrix, and consistent orderings for such matrices, is that the knowledge of these matrix properties greatly facilitates the theoretical analysis of the successive overrelaxation iterative method, which will be treated in Sec. 4.2 and Sec. 4.3. Fortunately, as we shall see in Chapter 6, these matrix properties are not difficult to obtain practically.

The distinction between consistent and inconsistent orderings is theoretically important when one considers, as in Sec. 4.4, the effect that orderings have on asymptotic rates of convergence. The nature of consistent orderings will be more fully explored for the case $p = 2$ in Sec. 4.4, where again we will find graph theory to be both convenient and instructive.

The final result of this section is a type of determinantal invariance for consistently ordered, weakly cyclic matrices of index $p$.

**Theorem 4.2.** *Let $B = L + U$ be a consistently ordered, weakly cyclic matrix of index $p$. Then, for any complex constants $\alpha$, $\beta$, and $\gamma$,*

$$(4.12) \qquad \det \{\gamma I - \alpha L - \beta U\} = \det \{\gamma I - (\alpha^{p-1}\beta)^{1/p}(L + U)\}.$$

*Proof.* This is almost an immediate consequence of Definition 4.2. As we know, for any complex constants $\alpha$, $\beta$, and $\gamma$,

$$\det\{\gamma I - (\alpha L + \beta U)\} = \prod_{i=1}^{n} (\gamma - \sigma_i);$$

(4.13)

$$\det\{\gamma I - (\alpha^{p-1}\beta)^{1/p}(L + U)\} = \prod_{i=1}^{n} (\gamma - \tau_i),$$

where the $\sigma_i$'s and $\tau_i$'s are, respectively, eigenvalues of the matrices $\alpha L + \beta U$ and $(\alpha^{p-1}\beta)^{1/p}(L + U)$. We proceed now to show that the sets

$$\{\sigma_i\}_{i=1}^{n} \quad \text{and} \quad \{\tau_i\}_{i=1}^{n}$$

are identical. If either $\alpha$ or $\beta$ is zero, then $\alpha L + \beta U$ is strictly triangular, which implies that each $\sigma_i$ is zero. For this case, it is obvious that each $\tau_i$ is also zero, and thus the two determinants of (4.13) are equal. If $\alpha$ and $\beta$ are both different from zero, let

$$(\alpha/\beta)^{1/p} \equiv \nu.$$

Thus, we can write

$$\alpha L + \beta U = (\alpha^{p-1}\beta)^{1/p}\{\nu L + \nu^{-(p-1)}U\}.$$

Using the hypothesis that $B = L + U$ is a consistently ordered weakly cyclic matrix of index $p$, we can set $\nu = 1$, and we conclude that the eigenvalue sets $\{\sigma_i\}_{i=1}^{n}$ and $\{\tau_i\}_{i=1}^{n}$ are again identical, which proves that the determinants of (4.13) are equal, completing the proof.

### EXERCISES

1. Let $A$ be the partitioned matrix of (4.2), where the diagonal submatrices are square and nonsingular. Prove that $A$ is $p$-cyclic if there exist $p$ disjoint nonempty sets $S_m$, $m = 0, 1, 2, \cdots, p - 1$ of $W$, the set of the first $N$ positive integers, such that

$$\bigcup_{m=0}^{p-1} S_m = W,$$

and if $A_{i,j} \neq 0$, then either $i = j$, or if $i \in S_m$, then $j \in S_{m-1}$, subscripts taken modulo $p$.

2. Let $A$ satisfy the hypotheses of Exercise 1. Prove that $A$ is $p$-cyclic if there exists a vector $\gamma^T = (\gamma_1, \gamma_2, \cdots, \gamma_N)$ with integral components such that if $A_{i,j} \neq 0$, then $\gamma_j - \gamma_i$ is equal to $p - 1$ or $-1$, and for each integer $r$, $0 \leq r \leq p - 1$, there exists some $\gamma_j \equiv r \pmod{p}$. The vector $\gamma^T$ is said to be an *ordering vector* for $A$. Find the ordering vectors for the matrices $A_1$ and $A_2$ of (4.5) and (4.6).

**3.** For the $4 \times 4$ matrix $A$ of $(1.6)$, partitioned so that its diagonal submatrices are all $1 \times 1$ matrices, prove that

a. $A$ is 2-cyclic;

b. $A$ is consistently ordered.

**4.** Prove that the matrix

$$A = \begin{bmatrix} 1 & -\frac{1}{4} & 0 & -\frac{1}{4} \\ -\frac{1}{4} & 1 & -\frac{1}{4} & 0 \\ 0 & -\frac{1}{4} & 1 & -\frac{1}{4} \\ -\frac{1}{4} & 0 & -\frac{1}{4} & 1 \end{bmatrix},$$

partitioned so that its diagonal submatrices are all $1 \times 1$ matrices, is 2-cyclic, but inconsistently ordered.

**5.** Show that the matrix $B$ of $(3.81)$, partitioned so that all its diagonal submatrices are $1 \times 1$ matrices, is cyclic of index 3, but inconsistently ordered.

**6.** Let the matrix $A$, partitioned in the form $(4.2)$, be expressed as $D - E - F$, where $D$ is the block diagonal matrix of $(4.3)$, and the matrices $E$ and $F$ are respectively strictly lower and upper triangular matrices. If $A$ is a consistently ordered $p$-cyclic matrix, and $D$ is nonsingular, prove that

$$\det \{D - \alpha E - \beta F\} = \det \{D - (\alpha^{p-1}\beta)^{1/p}(E + F)\},$$

where $\alpha$ and $\beta$ are arbitrary complex constants.

**7.** Show that the matrix

$$A = \begin{bmatrix} 20 & -4 & -4 & -1 & 0 & 0 \\ -4 & 20 & -1 & -4 & 0 & 0 \\ -4 & -1 & 20 & -4 & -4 & -1 \\ -1 & -4 & -4 & 20 & -1 & -4 \\ 0 & 0 & -4 & -1 & 20 & -4 \\ 0 & 0 & -1 & -4 & -4 & 20 \end{bmatrix},$$

partitioned so that its diagonal submatrices are all $1 \times 1$ matrices, is not $p$-cyclic for any $p \geq 2$. On the other hand, if $A$ is partitioned so that its diagonal submatrices are all $2 \times 2$ matrices, show that $A$ is a consistently ordered 2-cyclic matrix.

**\*8.** Consider the matrix $A_2$ of (4.6), which was shown to be a consistently ordered 2-cyclic matrix when its diagonal submatrices are nonsingular. Show (Friedman (1957)) that *any* permutation of rows and columns by blocks of $A$ leaves invariant this property of being consistently ordered and 2-cyclic. [*Hint*: Follow the approach of (4.10)].

**9.** Let $A$ satisfy the hypotheses of Exercise 1. If $\gamma^T$ is an ordering vector of Exercise 2, so that $A$ is $p$-cyclic, then $A$ is consistently ordered if and only if the following conditions hold. If $A_{i,j} \neq 0$, and $i \neq j$, then:

a. If $j > i$, then $\gamma_j - \gamma_i = p - 1$.

b. If $i > j$, then $\gamma_j - \gamma_i = -1$.

**10.** While the block directed graph of the Jacobi matrix $B$ for the partitioned matrix

$$
A = 
\begin{bmatrix}
8 & -1 & 0 & -1 & -1 & -1 \\
-1 & 8 & 0 & -1 & -1 & -1 \\
-1 & 1 & 8 & -1 & 8 & -8 \\
8 & -8 & -1 & 8 & -1 & 1 \\
-1 & -1 & -1 & 0 & 8 & -1 \\
-1 & -1 & -1 & 0 & -1 & 8
\end{bmatrix}
$$

is *not* a cyclic graph of index 2, show that $B$ is nevertheless weakly cyclic of index 2. What does this imply about the converse of Theorem 4.1?

## 4.2. The Successive Overrelaxation Iterative Method for p-Cyclic Matrices

With the partitioned matrix $A$ of (4.2), and its associated nonsingular block diagonal matrix $D$ of (4.3), as in Sec. 3.4 we now let the respectively strictly lower and upper triangular matrices $E$ and $F$ be defined from the decomposition $A = D - E - F$, where these matrices are explicitly given in (3.59). In terms of splittings of the matrix $A$, writing $A = M_1 - N_1$ where $M_1 = D$, $N_1 = E + F$ gives rise to the *block Jacobi iterative method*

$$(4.14) \qquad D\mathbf{x}^{(m+1)} = (E + F)\mathbf{x}^{(m)} + \mathbf{k}, \qquad m \geq 0.$$

Similarly, the splitting $A = M_2 - N_2$ where

$$M_2 = \frac{1}{\omega}(D - \omega E), \qquad N_2 = \frac{1}{\omega}\{(\omega F + (1 - \omega)D\}$$

for $\omega \neq 0$ gives rise to the *block successive overrelaxation iterative method*

$$(4.15) \qquad (D - \omega E)\mathbf{x}^{(m+1)} = (\omega F + (1 - \omega)D)\mathbf{x}^{(m)} + \omega\mathbf{k}, \qquad m \geq 0.$$

With $L \equiv D^{-1}E$ and $U \equiv D^{-1}F$, these iterative methods can be written simply as

$$(4.14') \qquad \mathbf{x}^{(m+1)} = (L + U)\mathbf{x}^{(m)} + D^{-1}\mathbf{k}, \qquad m \geq 0$$

and

$$(4.15') \qquad (I - \omega L)\mathbf{x}^{(m+1)} = \{\omega U + (1 - \omega)I\}\mathbf{x}^{(m)} + \omega D^{-1}\mathbf{k}, \qquad m \geq 0.$$

Assuming that the matrix $A$ is a consistently ordered $p$-cyclic matrix, the main result of this section is a *functional relationship* (4.18) between the eigenvalues $\mu$ of the block Jacobi matrix

$$(4.16) \qquad\qquad B = L + U,$$

and the eigenvalues $\lambda$ of the block successive overrelaxation matrix

$$(4.17) \qquad \mathcal{L}_\omega = (I - \omega L)^{-1}\{\omega U + (1 - \omega)I\}.$$

That such a functional relationship actually exists is itself interesting, but the importance of this result† lies in the fact that it is the basis for the precise determination of the value of $\omega$ which minimizes $\rho(\mathcal{L}_\omega)$.

**Theorem 4.3.** *Let the matrix $A$ of (4.2) be a consistently ordered $p$-cyclic matrix with nonsingular diagonal submatrices $A_{i,i}$, $1 \leq i \leq N$. If $\omega \neq 0$, and $\lambda$ is a nonzero eigenvalue of the matrix $\mathcal{L}_\omega$ of (4.17) and if $\mu$ satisfies*

$$(4.18) \qquad\qquad (\lambda + \omega - 1)^p = \lambda^{p-1}\omega^p\mu^p,$$

*then $\mu$ is an eigenvalue of the block Jacobi matrix $B$ of (4.16). Conversely, if $\mu$ is an eigenvalue of $B$ and $\lambda$ satisfies (4.18), then $\lambda$ is an eigenvalue of $\mathcal{L}_\omega$.*

*Proof.* The eigenvalues of $\mathcal{L}_\omega$ are the roots of its characteristic polynomial

$$(4.19) \qquad\qquad \det\{\lambda I - \mathcal{L}_\omega\} = 0,$$

but as $(I - \omega L)$ is nonsingular, as in the proof of Theorem 3.5, we conclude that equation (4.19) is equivalent to

$$(4.20) \qquad \det\{(\lambda + \omega - 1)I - \lambda\omega L - \omega U\} = 0.$$

Now, let

$$(4.21) \qquad \phi(\lambda) \equiv \det\{(\lambda + \omega - 1)I - \lambda\omega L - \omega U\}.$$

† Originally due to Young (1950) for the case $p = 2$.

Upon applying Theorem 4.2 directly to (4.21), we simply obtain

$$(4.22) \qquad \phi(\lambda) = \det\{(\lambda + \omega - 1)I - \lambda^{(p-1)/p}\omega B\}.$$

Since $A$ is $p$-cyclic by hypothesis, then the matrix $B$, and hence $\lambda^{(p-1)/p}\omega B$, is weakly cyclic of index $p$. Thus we necessarily have, upon applying Romanovsky's Theorem 2.4,

$$(4.23) \qquad \phi(\lambda) = (\lambda + \omega - 1)^m \prod_{i=1}^{r} \{(\lambda + \omega - 1)^p - \lambda^{p-1}\omega^p\mu_i^p\}$$

where the $\mu_i$ are nonzero if $r \geq 1$. To prove the second part of this theorem, let $\mu$ be an eigenvalue of the block Jacobi matrix $B$, and let $\lambda$ satisfy (4.18). Then, evidently one of the factors of $\phi(\lambda)$ of (4.23) vanishes, proving that $\lambda$ is an eigenvalue of $\mathcal{L}_\omega$. To prove the first part of the theorem, let $\omega \neq 0$, and let $\lambda$ be an eigenvalue of $\mathcal{L}_\omega$, so that at least one factor of (4.23) vanishes. If $\mu \neq 0$, and $\mu$ satisfies (4.18), then $(\lambda + \omega - 1) \neq 0$. Thus,

$$(\lambda + \omega - 1)^p = \lambda^{p-1}\omega^p\mu_i^p$$

for some $i$, $1 \leq i \leq r$, where $\mu_i$ is nonzero. Combining this with (4.18), we have that

$$\lambda^{p-1}\omega^p(\mu^p - \mu_i^p) = 0,$$

and since $\lambda$ and $\omega$ are nonzero, then $\mu^p = \mu_i^p$. Taking $p$th roots, we have that

$$\mu = \mu_i e^{2\pi i r/p},$$

where $r$ if a non-negative integer satisfying $0 \leq r < p$. But from the weakly cyclic nature of the matrix $B$, it is evident (Theorem 2.4) that $\mu$ is also an eigenvalue of $B$. To conclude the proof, if $\omega \neq 0$, $\lambda$ is a nonzero eigenvalue of $\mathcal{L}_\omega$, and $\mu = 0$ satisfies (4.18), then we must show that $\mu = 0$ is an eigenvalue of $B$. But with these hypotheses, it is obvious from (4.22) that $\phi(\lambda) = \det B = 0$, proving that $\mu = 0$ is an eigenvalue of $B$, which completes the proof.

The choice of $\omega = 1$ is such that $\mathcal{L}_1$ reduces in particular to the block Gauss-Seidel matrix of Sec. 3.4. As an immediate consequence of Theorem 4.3, we have the following

**Corollary.** *Let the matrix $A$ of (4.2) be a consistently ordered $p$-cyclic matrix with nonsingular diagonal blocks $A_{i,i}$, $1 \leq i \leq N$. If $\mu$ is an eigenvalue of the block Jacobi matrix $B$ of (4.16), then $\mu^p$ is an eigenvalue of $\mathcal{L}_1$. If $\lambda$ is a nonzero eigenvalue of $\mathcal{L}_1$ and $\mu^p = \lambda$, then $\mu$ is an eigenvalue of $B$. Thus, the* *block Jacobi iterative method converges if and only if the block Gauss-Seidel iterative method converges, and if both converge, then*

$$(4.24) \qquad \rho(\mathcal{L}_1) = (\rho(B))^p < 1,$$

*and consequently*

$$(4.24') \qquad R_\infty(\mathcal{L}_1) = pR_\infty(B).$$

It is interesting to consider the result of the above corollary in relation to the similar result of the Stein-Rosenberg Theorem 3.3. With the basic assumption that the Jacobi matrix $B$ of (4.16) is a non-negative and convergent irreducible matrix, it follows from Theorem 3.3 that

$$0 < \rho(\mathcal{L}_1) < \rho(B) < 1,$$

so that $R_\infty(\mathcal{L}_1) > R_\infty(B)$. Note that the exact ratio value for the asymptotic rates of convergence is not determined. On the other hand, the result of (4.24') gives the *exact* value for this ratio with the basic, but obviously different, assumption that the matrix $A$ of (4.2) is a consistently ordered $p$-cyclic matrix.

It is instructive at this point to reconsider the numerical example of Sec. 3.1. There, both the point Jacobi and point Gauss-Seidel iterative methods were applied to a particular equation $A\mathbf{x} = \mathbf{k}$. As this matrix $A$ is a consistently ordered 2-cyclic matrix (see Exercise 3 of Sec. 4.1), we conclude *theoretically* that the asymptotic rate of convergence of the point Gauss-Seidel iterative method is *twice* that of the point Jacobi iterative method. But we can observe this *numerically* from (3.19), where the error vectors for these iterative methods are explicitly determined for a specific initial error vector.

## EXERCISES

**1.** Let the $p \times p$ matrix $A$, $p \geq 2$, be defined by

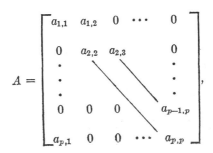

where the diagonal entries of $A$ are nonzero. Show that the analogous functional equation

$$(\lambda + \omega - 1)^p = \lambda \omega^p \mu^p$$

exists between the eigenvalues $\lambda$ of the associated point successive overrelaxation matrix and the eigenvalues $\mu$ of the point Jacobi matrix derived from $A$.

2. If $A$ is the matrix of the above exercise, show that for each $k = 1, 2, \cdots,$ $p - 1$, there exists a $p \times p$ permutation matrix $P_k$ such that the functional equation relating the eigenvalues $\lambda$ of the point successive overrelaxation matrix and the eigenvalues $\mu$ of the point Jacobi matrix $B$ derived from $P_k A P_k^T$ is of the form

$$(\lambda + \omega - 1)^p = \lambda^k \omega^p \mu^p.$$

If $B$ is convergent, which value of $k$ maximizes $R_\infty(\mathcal{L}_1)$?

## 4.3. Theoretical Determination of an Optimum Relaxation Factor

With the matrix $A$ of (4.2) a consistently ordered $p$-cyclic matrix with nonsingular diagonal submatrices $A_{i,i}$, $1 \leq i \leq N$, we now assume that the associated block Jacobi matrix $B$ of (4.16) is *convergent*. From the Corollary of Theorem 4.3, we then know that the block successive overrelaxation matrix $\mathcal{L}_\omega$ is convergent for $\omega = 1$, and by continuity it is also convergent for some interval in $\omega$ containing unity. This in turn implies that there is a comparable interval in $\omega$ for which the average rate of convergence $R(\mathcal{L}_\omega^m)$ is defined for all sufficiently large positive integers $m$. The problem that we should like to solve is the determination of a relaxation factor $\omega_m$ which maximizes the average rate of convergence $R(\mathcal{L}_\omega^m)$ as a function of $\omega$ for each positive integer $m$, provided that $R(\mathcal{L}_\omega^m)$ is defined, but the solution of this problem is complicated by the fact that $\omega_m$ is presumably a function of $m$. We now investigate the simpler problem of determining the optimum relaxation factor $\omega_b$ which *maximizes* the asymptotic rate of convergence $R_\infty(\mathcal{L}_\omega)$, or equivalently, we seek

$$(4.25) \qquad \min_\omega \rho(\mathcal{L}_\omega) = \rho(\mathcal{L}_{\omega_b}).$$

In particular, if the eigenvalues of $B^p$ are real and non-negative, it is known that the relaxation factor $\omega_b$ for which (4.25) is valid is *unique* and is precisely specified as the unique positive real root (less than $p/(p - 1)$) of the equation

$$(4.26) \qquad (\rho(B)\omega_b)^p = [p^p(p - 1)^{1-p}] \cdot (\omega_b - 1),$$

where $\rho(B)$ is the spectral radius of the associated block Jacobi matrix. For $p = 2$, $\omega_b$ can be expressed equivalently[†] as

$$(4.26') \qquad \omega_b = \frac{2}{1 + \sqrt{1 - \rho^2(B)}} = 1 + \left(\frac{\rho(B)}{1 + \sqrt{1 - \rho^2(B)}}\right)^2.$$

For simplicity, we shall *verify* that the relaxation factor $\omega_b$ of $(4.26')$ does indeed give the solution of $(4.25)$ for the special case $p = 2$. The method of proof we give in this case does, however, essentially carry over to the general case.[‡] If the eigenvalues of $B^2$ are non-negative real numbers, then we know from the weakly-cyclic-of-index-2 character (Definition 4.1) of the block Jacobi matrix $B$ that the nonzero eigenvalues $\mu_i$ of $B$ occur in $\pm$ pairs,[§] and thus,

$$-\rho(B) \le \mu_i \le \rho(B) < 1.$$

From Theorem 4.3, the eigenvalues $\lambda$ of the block successive overrelaxation matrix and the eigenvalues $\mu$ of the block Jacobi matrix are related through

$$(4.27) \qquad (\lambda + \omega - 1)^2 = \lambda \omega^2 \mu^2,$$

and at this point, we could begin finding the roots of this quadratic equation to determine the optimum parameter $\omega_b$. More in line with the development in Sec. 4.4, however, we take square roots throughout $(4.27)$, which, for $\omega \ne 0$, results in

$$(4.28) \qquad \frac{\lambda + \omega - 1}{\omega} = \pm\lambda^{1/2}\mu.$$

Defining the functions

$$(4.29) \qquad g_\omega(\lambda) \equiv \frac{\lambda + \omega - 1}{\omega}, \qquad \omega \ne 0,$$

and

$$(4.30) \qquad m(\lambda) \equiv \lambda^{1/2}\mu, \qquad 0 \le \mu \le \rho(B) < 1,$$

then $g_\omega(\lambda)$ is a straight line through the point $(1, 1)$, whose slope decreases monotonically with increasing $\omega$. It is now evident that $(4.28)$ can be

[†] The third member of $(4.26')$ is due to Young (1950).
[‡] See Exercise 4 of this section, which indicates the similarity with the general case.
[§] More precisely, if $\mu_i$ is a nonzero eigenvalue of $B$, so is $-\mu_i$. See Theorem 2.4.

geometrically interpreted as the intersection of the curves $g_\omega(\lambda)$ and $\pm m(\lambda)$, as illustrated in Figure 7 below.

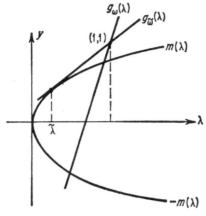

**Figure 7**

It is also evident that the largest abscissa of the two points of intersection *decreases* with increasing $\omega$ until $g_\omega(\lambda)$ becomes tangent to $m(\lambda)$, which occurs when

$$\tilde{\omega} = \frac{2}{1 + \sqrt{1 - \mu^2}},$$

and the abscissa $\tilde{\lambda}$ of the point of tangency is readily verified to be $\tilde{\lambda} = \tilde{\omega} - 1$. For $\omega > \tilde{\omega}$, the polynomial of (4.27) has two conjugate complex zeros of modulus $\omega - 1$, which now *increase* in modulus with increasing $\omega$. Thus, for this fixed eigenvalue $\mu$ of $B$, the value $\omega$ which *minimizes* the zero of largest modulus of (4.27) is $\tilde{\omega}$. Continuing, it is clear that the curve $\pm m(\lambda) = \pm\lambda^{1/2}\rho(B)$ is an *envelope* for all the curves $\pm\lambda^{1/2}\mu$, $0 \le \mu \le \rho(B)$, and we conclude, using the above argument, that

$$\min_\omega \rho(\mathcal{L}_\omega) = \rho(\mathcal{L}_{\omega_b}) = \omega_b - 1,$$

where $\omega_b$ is defined in (4.26′). This is the special case $p = 2$ of

**Theorem 4.4.** *Let the matrix $A$ of (4.2) be a consistently ordered p-cyclic matrix, with nonsingular diagonal submatrices $A_{i,i}$, $1 \le i \le N$. If all the eigenvalues of the pth power of the associated block Jacobi matrix $B$ are real and non-negative, and $0 \le \rho(B) < 1$, then with $\omega_b$ defined in (4.26),*

(4.31)

  1. $\rho(\mathcal{L}_{\omega_b}) = (\omega_b - 1)(p - 1)$;

  2. $\rho(\mathcal{L}_\omega) > \rho(\mathcal{L}_{\omega_b})$    *for all $\omega \ne \omega_b$.*

*Moreover, the block successive overrelaxation matrix $\mathcal{L}_\omega$ is convergent for all $\omega$ with $0 < \omega < p/(p-1)$.*

Using the definition of asymptotic rates of convergence of Sec. 3.2, we now compare the quantities $R_\infty(\mathcal{L}_{\omega_b})$ and $R_\infty(\mathcal{L}_1)$.

**Theorem 4.5.** *Let the matrix $A$ of (4.2) be a consistently ordered $p$-cyclic matrix, with nonsingular diagonal submatrices $A_{i,i}$, $1 \leq i \leq N$, and let the eigenvalues of the pth power of the associated block Jacobi matrix $B$ be real non-negative, with $0 \leq \rho(B) < 1$. As $\rho(B) \to 1-$,*

$$(4.32) \qquad R_\infty(\mathcal{L}_{\omega_b}) \sim \left(\frac{2p}{p-1}\right)^{1/2} [R_\infty(\mathcal{L}_1)]^{1/2}.$$

*Proof.* Since

$$R_\infty(\mathcal{L}_{\omega_b}) = -\ln \left[(\omega_b - 1)(p-1)\right]$$

from Theorem 4.4 and

$$R_\infty(\mathcal{L}_1) = -p \ln \rho(B)$$

from the Corollary of Sec. 4.2, the result follows upon making use of (4.26) by applying L'Hospital's rule twice to

$$(4.33) \qquad \lim_{\rho(B) \to 1} \frac{R_\infty(\mathcal{L}_{\omega_b})}{[R_\infty(\mathcal{L}_1)]^{1/2}}.$$

Combining the results of Theorem 4.5 and the Corollary of Theorem 4.3, we have the

**Corollary.** *With the assumptions of Theorem 4.5, if $R_\infty(B)$ denotes the asymptotic rate of convergence of the block Jacobi iterative method, then as $\rho(B) \to 1-$,*

$$(4.33') \qquad R_\infty(\mathcal{L}_{\omega_b}) \sim \left(\frac{2p^2}{p-1}\right)^{1/2} [R_\infty(B)]^{1/2}.$$

We remark that, by Theorem 4.5, the successive overrelaxation iterative method (with optimum relaxation factor $\omega_b$) rigorously applied to consistently ordered $p$-cyclic matrices gives a substantial improvement in the asymptotic rates of convergence of slowly convergent (spectral radius close to unity) block Jacobi and Gauss-Seidel iterative methods. To illustrate this, if $\rho(B) = 0.9999$, then

$$R_\infty(B) \doteq 0.0001 \quad \text{and} \quad R_\infty(\mathcal{L}_1) \doteq 0.0002,$$

whereas $R_\infty(\mathcal{L}_{\omega_b}) = 0.0283$, if $p = 2$. In this case, the ratio

$$\frac{R_\infty(\mathcal{L}_{\omega_b})}{R_\infty(\mathcal{L}_1)}$$

of asymptotic rates of convergence is greater than one hundred, which indicates that for comparable accuracy roughly one hundred times as many iterations of the Gauss-Seidel method are required as for the successive overrelaxation method. We further remark that although the coefficient $[2p^2/(p - 1)]^{1/2}$ of (4.33′) is strictly increasing as a function of $p$, the corresponding coefficient $[2p/(p - 1)]^{1/2}$ of (4.32) is strictly decreasing and is clearly maximized when the cyclic index is 2. This, as we shall see, is of considerable value, since the case $p = 2$ occurs naturally by virtue of symmetry in the numerical solution of many physical problems.

As an example of the difference in asymptotic rates of convergence that orderings alone can make, consider the $3 \times 3$ matrix $B$ of (3.81), with $\alpha = \rho(B) = 0.9999$. In Sec. 3.6, it was shown that

$$\min_\omega \rho(\mathcal{L}_\omega) = \rho(\mathcal{L}_1),$$

which in this case is $(0.9999)^{3/2} \doteq 0.99985$. Thus,

(4.34) $$\max_\omega R_\infty(\mathcal{L}_\omega) = R_\infty(\mathcal{L}_1) \doteq 0.00015.$$

But if $P$ is a particular $3 \times 3$ permutation matrix such that

(4.35) $$PBP^T = \tilde{B} = \alpha \begin{bmatrix} 0 & 0 & 1 \\ 1 & 0 & 0 \\ 0 & 1 & 0 \end{bmatrix},$$

then $\tilde{B}$ is a consistently ordered 3-cyclic matrix. Since all the eigenvalues of $\tilde{B}^3$ are $\alpha^3$ and hence positive, we can apply the Corollary above, and it follows that

(4.36) $$\max_\omega R_\infty(\tilde{\mathcal{L}}_\omega) = R_\infty(\tilde{\mathcal{L}}_{\omega_b}) \doteq 0.03,$$

a *significant* improvement over that of (4.34). Further results of the effect of orderings on asymptotic rates of convergence will be discussed in Sec. 4.4 for the special case $p = 2$.

The method of proof of Theorem 4.4 shows that the optimum relaxation factor $\omega_b$ is obtained by *coalescing* two positive eigenvalues of $\mathcal{L}_\omega$. This in turn results in the fact† that the Jordan normal form of $\mathcal{L}_{\omega_b}$ contains a single $2 \times 2$ submatrix $J$ of the form

$$J = \begin{bmatrix} \lambda & 1 \\ 0 & \lambda \end{bmatrix}$$

where $\lambda = \rho(\mathcal{L}_{\omega_b})$. The discussion in Sec. 3.2 on norms of powers of matrices is now relevant, and applying Theorems 3.1 and 4.4, we conclude that

$$(4.37) \qquad \| \mathcal{L}_{\omega_b}^m \| \sim \nu m [(\omega_b - 1)(p - 1)]^{m-1}, \qquad m \to \infty.$$

In Chapter 5 the goal is to minimize

$$\left\| \prod_{i=1}^m \mathcal{L}_{\omega_i} \right\|$$

for *each* $m \geq 0$, by an appropriate selection of a sequence of relaxation factors $\omega_i$.

Finally, in the numerical solution by iterative methods of practical problems, one generally is not given, nor can one precisely determine *a priori*, the spectral radius of the block Jacobi or the block Gauss-Seidel iteration matrices. Thus, in practice we would determine estimates of the optimum relaxation factor $\omega_b$ of (4.26) or (4.26′) from estimates of $\rho(B)$. In this regard, simple monotonicity arguments applied to (4.26) show that overestimates and underestimates of $\rho(B)$ give, respectively, overestimates and underestimates of $\omega_b$. But, from the nature of the determination of $\omega_b$ by the coalescing of two eigenvalues, it was originally shown by Young (1954a) that overestimating $\rho(B)$ by a small amount causes a smaller decrease in $R_\infty(\mathcal{L}_\omega)$ than does underestimating $\rho(B)$ by a comparable amount. This is best seen for the case $p = 2$ by plotting $\rho(\mathcal{L}_\omega)$ as a function of $\omega$ for $0 \leq \omega \leq 2$, as shown in Figure 8.

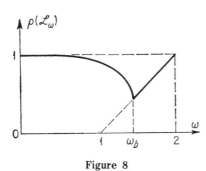

**Figure 8**

## EXERCISES

**1.** For $0 \leq \rho(B) < 1$, show that there exists a unique positive real root $\omega_b$ less than $p/(p - 1)$ of (4.26). Also, show that this unique positive real root can be expressed in the form (4.26′) when $p = 2$.

† See Young (1954a) for the case $p = 2$ and Varga (1959a) for the general case $p \geq 2$.

**2.** Consider the *divergent* Jacobi matrix

$$B = \begin{pmatrix} 0 & 1 \\ \alpha & 0 \end{pmatrix},$$

where $-\alpha = |r| > 1$. Using the fact that the matrix $B$ is a consistently ordered weakly cyclic matrix of index 2, show that the associated successive overrelaxation matrix $\mathcal{L}_\omega$ is *convergent* for some real $\omega$, $0 < \omega < 2$. Determine an optimum relaxation factor $\omega_b$ which maximizes the asymptotic rate of convergence $R_\infty(\mathcal{L}_\omega)$ for such convergent matrices $\mathcal{L}_\omega$.

**3.** Let the matrices $A$ and $B$ satisfy the hypotheses of Theorem 4.4 for $p = 2$. If the spectral radius of the Jacobi matrix $B$ satisfies $0 < \rho(B) < 1$, $\omega_b$ is defined as in (4.26'), and $\rho(\mathcal{L}_\omega)$ is the spectral radius of the successive overrelaxation method, prove that $d(\rho(\mathcal{L}_\omega))/d\omega$ becomes unbounded as $\omega$ increases to $\omega_b$.

**4.** It has been shown that the optimum relaxation factor $\omega_b$ for the case $p = 2$ of (4.26') is obtained by finding the value of $\omega$ for which the straight line $(\lambda + \omega - 1)/\omega$ is tangent to $\lambda^{1/2}\rho(B)$. Show that the general formula of (4.26) for $\omega_b$ is similarly obtained as the value of $\omega$ for which $(\lambda + \omega - 1)/\omega$ is tangent to $\lambda^{(p-1)/p}\rho(B)$.

**5.** For the $3 \times 3$ matrix (3.81) with $0 < \alpha < 1$, construct the normal form (2.41) of its corresponding Gauss-Seidel matrix $\mathcal{L}_1$. Similarly, construct the normal form (2.41) of the Gauss-Seidel matrix $\tilde{\mathcal{L}}_1$ obtained from the matrix $\tilde{B}$ of (4.35). Show that the spectral radii of $\mathcal{L}_1$ and $\tilde{\mathcal{L}}_1$ are obtained respectively from a cyclic (of index 2) submatrix and a primitive submatrix of these normal forms.

**\*6.** Let $B = L + U \geq O$ be an irreducible and convergent $n \times n$ matrix, where $L$ and $U$ are respectively strictly lower and upper triangular matrices. If the spectral radius of the reducible Gauss-Seidel matrix $(I - L)^{-1}U$ is obtained from a primitive submatrix of its normal form (2.41), prove that the value of $\omega$ which minimizes

$$\rho\{(I - \omega L)^{-1}(\omega U + (1 - \omega)I)\}$$

is necessarily *greater* than unity. (*Hint*: use Exercise 4 of Sec. 3.6 and continuity.)

## 4.4. Extensions of the p-Cyclic Theory of Matrices

In the previous sections of this chapter, properties of the successive overrelaxation method were investigated under the assumption that the matrix $A$ of (4.1) was a consistently ordered $p$-cyclic matrix, and it was found that a precise formula for the relaxation factor $\omega$ which maximizes $R_\infty(\mathcal{L}_\omega)$ could be derived, with certain additional assumptions. Of course, it is clear that the basic assumption that the matrix $A$ is a consistently ordered $p$-cyclic matrix allowed us to deduce the functional relationship (4.18) between the eigenvalues of the Jacobi matrix and the eigenvalues of the successive overrelaxation matrix, which was, so to speak, the stepping-stone to the precise determination of an optimum relaxation factor.

We now consider extensions of this theory,† relative to a particular set of matrices. With the matrix $A$ partitioned as in (4.2), where the diagonal submatrices are square and nonsingular, we assume that the block Jacobi matrix $B$ of (4.4) is an element of the set $S$ of

DEFINITION 4.3. The square matrix $B \in S$ if $B$ satisfies the following properties:

1. $B \geq O$ with zero diagonal entries.

2. $B$ is irreducible and convergent, i.e., $0 < \rho(B) < 1$.

3. $B$ is symmetric.

For any matrix $B \in S$, we can decompose $B$ into

$$(4.38) \qquad\qquad B = L_B + L_B^T,$$

where $L_B$ is a strictly lower triangular non-negative matrix, and as $B$ is a block Jacobi matrix, its associated successive overrelaxation matrix is given by

$$(4.39) \qquad \mathcal{L}_{B,\omega} = (I - \omega L_B)^{-1}\{\omega L_B^T + (1 - \omega)I\}.$$

It is interesting to note that if $\tilde{A} \equiv I - B$ where $B \in S$, then the matrix $\tilde{A}$ is real, symmetric, and positive definite, and the *point* successive overrelaxation matrix associated with $\tilde{A}$ is just $\mathcal{L}_{B,\omega}$ of (4.39). But from Ostrowski's Theorem 3.6, we necessarily conclude that $\rho(\mathcal{L}_{B,\omega}) < 1$ for *all* $0 < \omega < 2$.

We now remark that the set $S$ of matrices $B$ of Definition 4.3 contains not only matrices such as

$$(4.40) \qquad\qquad B_1 = \begin{bmatrix} 0 & \alpha \\ \alpha & 0 \end{bmatrix}, \qquad 0 < \alpha < 1,$$

† Due to Kahan (1958) and Varga (1959b).

which are consistently ordered cyclic matrices of index 2, but also contains matrices such as

$$(4.41) \qquad B_2 = \begin{bmatrix} 0 & \alpha & 0 & \alpha \\ \alpha & 0 & \alpha & 0 \\ 0 & \alpha & 0 & \alpha \\ \alpha & 0 & \alpha & 0 \end{bmatrix}, \qquad 0 < \alpha < \tfrac{1}{2},$$

which are cyclic of index 2, but inconsistently ordered (see Exercise 4 of Sec. 4.1), as well as matrices such as

$$(4.42) \qquad B_3 = \begin{bmatrix} 0 & \alpha & \alpha \\ \alpha & 0 & \alpha \\ \alpha & \alpha & 0 \end{bmatrix}, \qquad 0 < \alpha < \tfrac{1}{2},$$

which are not cyclic of index 2.

As in the proof of Lemma 3.2 of Sec. 3.3, we also associate with $B$ the matrix

$$(4.43) \qquad M_B(\sigma) = \sigma L_B + L_B^T, \qquad \sigma \geq 0.$$

For $\sigma > 0$, $M_B(\sigma)$ is a non-negative irreducible matrix, and its spectral radius $m_B(\sigma) \equiv \rho(M_B(\sigma))$ is, from Lemma 3.2, a strictly increasing function of $\sigma$.

Specifically, for the matrix $B_1$ of (4.40), its associated function $m_{B_1}(\sigma)$ is readily verified to be

$$(4.44) \qquad m_{B_1}(\sigma) = \rho(B_1)\sigma^{1/2}, \qquad \sigma > 0.$$

Actually, *any* consistently ordered cyclic matrix $B$ of index 2 of the set $S$ has its associated function $m_B(\sigma)$ given by (4.44) as we now show. Since

$$M_B(\sigma) = \sigma^{1/2}\{\sigma^{1/2}L_B + \sigma^{-1/2}L_B^T\}, \qquad \sigma > 0,$$

it follows from Definition 4.2 that all the eigenvalues of $M_B(\sigma)$ are the same as the eigenvalues of the matrix

$$\sigma^{1/2}(L_B + L_B^T) = \sigma^{1/2}B,$$

so that

$$m_B(\sigma) \equiv \rho(M_B(\sigma)) = \rho(\sigma^{1/2}B) = \sigma^{1/2}\rho(B).$$

Regarding this behavior as ideal, we measure departures from this by means of the function

$$(4.45) \qquad h_B(\ln \sigma) \equiv \frac{m_B(\sigma)}{\rho(B)\sigma^{1/2}}, \qquad \sigma > 0,$$

defined for *all* $B \in S$. The point of the following discussion is to obtain upper and lower bounds on the function $h_B$ ($\ln \sigma$) for $\sigma > 0$ which in turn will give rise to upper and lower bounds for $\rho(\mathcal{L}_{B,\omega})$.

**Lemma 4.1.** *Let* $B \in S$. *If* $h_B$ ($\ln \sigma$) *is defined as in* (4.45), *then* $h_B(\alpha) = h_B(-\alpha)$ *for all real* $\alpha$, *and* $h_B(0) = 1$.

*Proof.* For $\sigma > 0$, there exists an eigenvector $\mathbf{x} > 0$ with

$$M_B(\sigma)\mathbf{x} = m_B(\sigma)\mathbf{x}.$$

From (4.43), we can write for $\sigma > 0$

$$M_B(\sigma) = \sigma\left(L_B + \frac{1}{\sigma}L_B^T\right) = \sigma M_B^T\left(\frac{1}{\sigma}\right).$$

Thus,

$$M_B^T\left(\frac{1}{\sigma}\right)\mathbf{x} = \frac{m_B(\sigma)}{\sigma}\mathbf{x}.$$

Since $M_B(\sigma)$ and $M_B^T(\sigma)$ have the same eigenvalues, then

$$(4.46) \qquad \sigma m_B\left(\frac{1}{\sigma}\right) = m_B(\sigma), \qquad \sigma > 0.$$

From the definition of $h_B$ ($\ln \sigma$) in (4.45), we see that $h_B(\alpha) = h_B(-\alpha)$, where $\alpha \equiv \ln \sigma$. For $\sigma = 1$, $m_B(1) = \rho(B)$ by definition, so that $h_B(0) = 1$, which completes the proof.

**Lemma 4.2.** *If* $B \in S$, *let* $A_B(\alpha) \equiv e^{\alpha}L_B + \gamma I + e^{-\alpha}L_B^T$, *where* $\alpha$ *and* $\gamma$ *are non-negative constants. For* $0 \leq \alpha_1 \leq \alpha_2$,

$$\rho(A_B(\alpha_1)) \leq \rho(A_B(\alpha_2)).$$

*Moreover, if* $\rho(A_B(\hat{\alpha}_1)) < \rho(A_B(\hat{\alpha}_2))$ *for some particular values* $0 \leq \hat{\alpha}_1 < \hat{\alpha}_2$, *then it is true for all* $0 \leq \alpha_1 < \alpha_2$.

*Proof.* If $C = L_B + \gamma I + L_B^T = (c_{i,j})$, then the matrix $C$ is non-negative and irreducible by the hypotheses of the lemma. Assume now that $\nu_m \neq 0$ is any product of entries of the matrix $C$ corresponding to a closed path of length $m$:

$$(4.47) \qquad \nu_m = \prod_{r=0}^{m-1} c_{i_r, i_{r+1}}, \qquad \text{where } c_{i_r, i_{r+1}} > 0, \qquad i_0 = i_m.$$

As the terms above and below the main diagonal for the corresponding product $\nu_m(\alpha)$ for $A_B(\alpha)$ are simply modified respectively by the multiplicative factors $e^{-\alpha}$ and $e^{\alpha}$, it is clear that $\nu_m(\alpha) = e^{q_m \alpha} \nu_m$, where $q_m$ is an integer. From the symmetry of the matrix $C$,

$$(4.48) \qquad \hat{\nu}_m = \prod_{r=0}^{m-1} c_{i_{r+1}, i_r} = \nu_m$$

is also a similar product of entries of the matrix $C$, which in turn gives $\hat{\nu}_m(\alpha) = e^{-q_m \alpha} \nu_m$ as a corresponding product for $A_B(\alpha)$. Since $\nu_m(\alpha)$ and $\hat{\nu}_m(\alpha)$ are both contained in the $i_0$th diagonal entry of $A_B^m(\alpha)$, it follows that the trace of $A_B^m(\alpha)$ is composed of terms of the form $2\nu_m \cosh(q_m \alpha)$. Using the monotonicity of $\cosh(x)$, we obtain for $0 \leq \alpha_1 \leq \alpha_2$,

$$(4.49) \qquad \text{tr}\,(A_B^m(\alpha_1)) \leq \text{tr}\,(A_B^m(\alpha_2)), \qquad m \geq 1.$$

If the matrix $C$ is primitive, then $A_B(\alpha)$ is also primitive for all real $\alpha \geq 0$. From Exercise 6 of Sec. 2.2, we thus have

$$(4.50) \qquad [\text{tr}\,(A_B^m(\alpha))]^{1/m} \sim \rho(A_B(\alpha)), \qquad m \to \infty.$$

Combining the results of (4.49) and (4.50), we conclude that

$$\rho(A_B(\alpha_1)) \leq \rho(A_B(\alpha_2)),$$

with the additional assumption that the matrix $C$ is primitive. But if $C$ is not primitive, $C + \beta I$, $\beta > 0$ certainly is, and as

$$\rho\{A_B(\alpha) + \beta I\} = \rho(A_B(\alpha)) + \beta,$$

the first result of this lemma is proved. The second part of this lemma, which we omit, is based on the same argument (see Exercise 4 of this section).

This lemma serves as a basis for

**Theorem 4.6.** *If $B \in S$, then either $h_B(\alpha) \equiv 1$ for all real $\alpha$, or $h_B(\alpha)$ is strictly increasing for $\alpha \geq 0$. Moreover, for any $\alpha \neq 0$,*

$$(4.51) \qquad 1 \leq h_B(\alpha) < \cosh\left(\frac{\alpha}{2}\right).$$

*Proof.* For $\sigma > 0$, consider the matrix

$$(4.52) \qquad P_B(\sigma) \equiv \frac{M_B(\sigma)}{\rho(B)\sigma^{1/2}} = \frac{1}{\rho(B)}\{\sigma^{1/2}L_B + \sigma^{-1/2}L_B^T\}.$$

By definition, $\rho(P_B(\sigma)) = h_B(\ln \sigma)$, and upon applying Lemma 4.2 to $P_B(\sigma)$, we conclude that either $h_B(\alpha) \equiv 1$ for all real $\alpha$, or $h_B(\alpha)$ is a strictly increasing function for $\alpha \geq 0$. To prove the second part of this theorem, we write $P_B(\sigma)$ in the form

$$(4.53) \qquad P_B(e^{\alpha}) = \cosh\left(\frac{\alpha}{2}\right)V_B + \sinh\left(\frac{\alpha}{2}\right)W_B,$$

where

$$(4.54) \qquad V_B \equiv \frac{1}{\rho(B)}(L_B + L_B^T); \qquad W_B \equiv \frac{1}{\rho(B)}(L_B - L_B^T),$$

so that $V_B$ and $W_B$ are respectively real symmetric and real skew-symmetric matrices. As $P_B(e^{\alpha})$ is a non-negative irreducible matrix, let $\mathbf{x} > \mathbf{0}$ be the eigenvector corresponding to the eigenvalue $h_B(\alpha)$. If $\mathbf{x}$ is normalized so that $\|\mathbf{x}\| = 1$, then

$$(4.55) \qquad h_B(\alpha) = \mathbf{x}^T P_B(e^{\alpha})\mathbf{x} = \cosh\left(\frac{\alpha}{2}\right)\mathbf{x}^T V_B\mathbf{x},$$

since $W_B$ is a real skew-symmetric matrix.† As the quadratic form $\mathbf{x}^T V_B\mathbf{x}$ can be equivalently written as the Rayleigh quotient $\mathbf{x}^T V_B\mathbf{x}/\mathbf{x}^T\mathbf{x}$, this is bounded above‡ by the largest real eigenvalue of $V_B$, as $V_B$ is a real symmetric matrix. But as $V_B$ is also a non-negative irreducible matrix, its largest real eigenvalue is equal to its spectral radius, which from (4.54) is obviously unity. Combining with the first part of this theorem, we then have

$$1 \leq h_B(\alpha) \leq \cosh\left(\frac{\alpha}{2}\right),$$

for all real $\alpha$. To complete the proof, let $\alpha \neq 0$, and suppose that

$$\mathbf{x}^T V_B\mathbf{x} = \rho(V_B) = 1.$$

From the symmetry of $V_B$, this implies that $\mathbf{x}$ is also an eigenvector of $V_B$, which from (4.53) further implies that $\mathbf{x}$ is an eigenvector of the real skew-symmetric matrix $W_B$. But as a real skew-symmetric matrix has only eigenvalues of the form $i\sigma$, where $\sigma$ is real, it remains to show that $W_B\mathbf{x}$ is not the null vector. By the irreducibility of $B$, there exists at least one

† For any real vector $\mathbf{y}$, $\mathbf{y}^T M\mathbf{y} = 0$ when $M$ is a real skew-symmetric matrix.
‡ See, for example, the proof of Theorem 1.3.

positive entry in the first row of $L_B^T$, and by direct multiplication, the first component of the column vector $W_B\mathbf{x}$ is a negative real number. This contradiction proves that $\mathbf{x}^T V_B\mathbf{x} < 1$ for $\alpha \neq 0$, and thus

$$1 \leq h_B(\alpha) < \cosh\left(\frac{\alpha}{2}\right)$$

for all $\alpha \neq 0$, completing the proof.

Before finding bounds for the spectral radius of $\mathcal{L}_{B,\omega}$, we bring out the relationship between the matrix property $h_B(\alpha) \equiv 1$ and the previously introduced concepts of a consistently ordered cyclic matrix of index 2.

**Theorem 4.7.** *Let $B \in S$. Then, $h_B(\alpha) \equiv 1$ if and only if $B$ is a consistently ordered cyclic matrix of index 2.*

In motivating the construction of the function $h_B(\alpha)$, we have already established that if $B$ is a consistently ordered cyclic matrix of index 2, then $h_B(\alpha) \equiv 1$. The converse depends on a combinatorial argument like that of Lemma 4.2, and is omitted. (See Exercise 9.)

The importance of the result of Theorem 4.7 is that it gives rise to a graph-theoretic interpretation of the concept of a consistently ordered cyclic matrix of index 2. From Lemma 4.2, $h_B(\alpha) \equiv 1$ for all real $\alpha$ is equivalent to the statement that every nonzero product

$$\nu_m(\alpha) = e^{q_m\alpha}\nu_m$$

of the matrix $P_B(e^\alpha)$ is such that $q_m = 0$ for all $m \geq 1$. Let $\hat{G}[B]$ be the *directed graph of type 2*† for the matrix $B \equiv (b_{i,j})$, so constructed that if $b_{i,j} \neq 0$, then the path from the node $P_i$ to the node $P_j$ is denoted by a double-arrowed path (see Figure 9) *only* if $j > i$; otherwise, a single-arrowed path is used as in our previous graphs. The former paths can be called *major paths*; the other paths can be called *minor paths*. Then, the matrix $B$

**Figure 9**

is consistently ordered only if every closed path of its directed graph $\hat{G}[B]$ has an equal number of major and minor paths. To illustrate this, the point Jacobi matrix $B$ of (1.7) has the directed graph of type 2

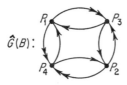

$\hat{G}(B)$:

**Figure 10**

† This terminology is taken from Harary (1955).

If the indices 2 and 3 are permuted we denote the corresponding point Jacobi matrix by $B_\phi$ which, in reality, turns out to be the matrix $B_2$ of (4.41) with $\alpha = \frac{1}{4}$. Its associated directed graph is

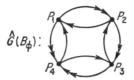

**Figure 11**

As there is a closed path of cycle of $\hat{G}[B_\phi]$ with an *unequal* number of major and minor paths, we see graphically that the matrix $B_\phi$ is inconsistently ordered. This graph-theoretic interpretation of consistent orderings will be useful to us in Chapter 6, where we derive difference equations from differential equations.

We now come to the major problem of finding bounds for $\rho(\mathcal{L}_{B,\omega})$ for any matrix $B \in S$. Since

$$m_B(\sigma) = \rho(B)\sigma^{1/2}h_B (\ln \sigma),$$

where $1 \leq h_B(\alpha) < \cosh (\alpha/2)$ for all real $\alpha \neq 0$, these inequalities for $h_B(\alpha)$ allow us to conclude the following upper and lower bounds for $m_B(\sigma)$:

$$(4.56) \qquad \rho(B)\sigma^{1/2} \leq m_B(\sigma) < \rho(B)\left(\frac{\sigma + 1}{2}\right), \qquad 0 < \sigma, \quad \sigma \neq 1,$$

which we have pictured in Figure 12. Note that these curves are tangent to one another at $\sigma = 1$.

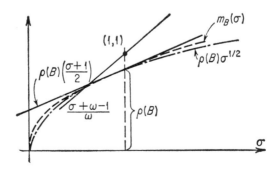

**Figure 12**

Let $\lambda$ be an arbitrary eigenvalue of $\mathcal{L}_{B,\omega}$ with

$$(4.57) \qquad \mathcal{L}_{B,\omega}\mathbf{x} = \lambda\mathbf{x}, \qquad \mathbf{x} \neq 0.$$

Directly from the definition of $\mathcal{L}_{B,\omega}$, this is true if and only if

$$(4.58) \qquad (\lambda L_B + L_B^T)\mathbf{x} = \left(\frac{\lambda + \omega - 1}{\omega}\right)\mathbf{x}, \quad \omega \neq 0,$$

which expresses $((\lambda + \omega - 1)/\omega)$ as an eigenvalue of $\lambda L_B + L_B^T$. Since

$$\left|\frac{\lambda + \omega - 1}{\omega}\right| \leq \rho(\lambda L_B + L_B^T) \leq \rho(|\lambda| L_B + L_B^T) = m_B(|\lambda|),$$

then from (4.56), we obtain

$$(4.59) \qquad \left|\frac{\lambda + \omega - 1}{\omega}\right| < \rho(B)\left(\frac{|\lambda| + 1}{2}\right), \quad \omega \neq 0, \quad |\lambda| \neq 1.$$

On the other hand, assume that $\lambda > 0$ and $\mathbf{x} > 0$ in (4.57). Then, as $B$ is non-negative and irreducible, it follows from (4.58) that $m_B(\lambda) = (\lambda + \omega - 1)/\omega$, and thus, from (4.56),

$$(4.60) \qquad \frac{\lambda + \omega - 1}{\omega} = m_B(\lambda) \geq \lambda^{1/2}\rho(B), \quad \omega \neq 0.$$

For $\lambda \neq 1$, moreover, Theorem 4.7 shows that equality is valid throughout (4.60) *only* if $B$ is a consistently ordered cyclic matrix of index 2.

For the special case $\omega = 1$, the proof of the Stein-Rosenberg Theorem 3.3 in the irreducible case shows that $\mathcal{L}_{B,1}$ does indeed have a positive eigenvalue and a positive eigenvector, and setting $\omega = 1$ in (4.59) and (4.60) directly gives the result of

**Theorem 4.8.** *If $B \in S$, then*

$$(4.61) \qquad \rho^2(B) \leq \rho(\mathcal{L}_{B,1}) < \frac{\rho(B)}{2 - \rho(B)},$$

*with equality possible only if $B$ is a consistently ordered cyclic matrix of index 2.*

Note that this theorem gives a sharpened form of statement 2 of the Stein-Rosenberg Theorem 3.3 for the set of matrices $S$.

With Figure 12 and the discussion of Sec. 4.3, it is geometrically evident that the straight line $g_\omega(\sigma) = (\sigma + \omega - 1)/\omega$ intersects the curves $m_B(\sigma)$ and $\rho(B)((\sigma + 1)/2)$ for *all* $\omega$ with $0 < \omega \leq \omega_b$, where $\omega_b$ is defined in (4.26'), and we can conclude, using a result of Exercise 4 of Sec. 3.6, that the inequalities of (4.59) and (4.60) can be applied to *every* $\omega$ in this interval. But for $0 < \omega \leq 1$, these values of $\omega$ are uninteresting from a practical point of view, since we have shown in Theorem 3.16 that the choice $\omega = 1$ minimizes $\rho(\mathcal{L}_{B,\omega})$ in this range.

For the more interesting case of $\omega = \omega_b > 1$, upper and lower bounds for $\rho(\mathcal{L}_{\omega,B})$ can be again deduced in the manner described for the specific case $\omega = 1$. First, let $\lambda = \rho(\mathcal{L}_{B,\omega_b})e^{i\theta}$. For each choice of $\theta$, the inequality of (4.59) gives an upper bound for $\rho(\mathcal{L}_{B,\omega_b})$, and it can be verified (Exercise 11) that the least such bound occurs for $\theta = 0$. But with $\theta = 0$, (4.59) reduces to

$$\rho(\mathcal{L}_{B,\omega_b}) < \left( \frac{\omega_b \rho(B) + 2(1 - \omega_b)}{2 - \omega_b \rho(B)} \right),$$

which can be equivalently written, using the expressions for $\omega_b$ in (4.26′), simply as

$$\rho(\mathcal{L}_{B,\omega_b}) < \sqrt{\omega_b - 1}.$$

Similarly, (4.60) for the case $\omega = \omega_b$ gives the lower bound $\rho(\mathcal{L}_{B,\omega_b}) \geq \omega_b - 1$, with equality only if $B$ is a consistently ordered cyclic matrix of index 2. Combining these inequalities, we have

**Theorem 4.9.** *Let $B \in S$. Then, if $\omega_b$ is defined as in* (4.26′),

$$(4.62) \qquad \omega_b - 1 \leq \rho(\mathcal{L}_{B,\omega_b}) < \sqrt{\omega_b - 1}$$

*with $\omega_b - 1 = \rho(\mathcal{L}_{B,\omega})$ if and only if $B$ is a consistently ordered cyclic matrix of index 2.*

Although a precise determination of the relaxation factor $\omega$ that minimizes $\rho(\mathcal{L}_{B,\omega})$ has not been obtained for all matrices $B \in S$, the importance of Theorem 4.9 lies in the fact that the use of the particular relaxation factor $\omega_b$ of (4.26′) in the successive overrelaxation iterative method derived from matrices $B \in S$ gives asymptotic convergence rates like those described in Sec. 4.3, and, thus, this development constitutes a theoretical extension of the successive overrelaxation theory for $p$-cyclic matrices.

Returning to the set $S$ of Definition 4.3, let the $n \times n$ matrix $B \in S$, and let $\phi$ be any *permutation* or *ordering* of the integers $1 \leq i \leq n$. If $\Lambda_\phi$ denotes the corresponding $n \times n$ permutation matrix, then $B_\phi \equiv \Lambda_\phi B \Lambda_\phi^{-1} = \Lambda_\phi B \Lambda_\phi^T$ is readily seen also to be an element of $S$. In other words, the set $S$ is invariant under similarity transformations by permutation matrices, and to each matrix $B \in S$ we can thus associate a subset $\{B_\phi\}$ of matrices of $S$, all having the same spectral radius. Again, as a consequence of (4.60), we have (Exercise 12)

**Theorem 4.10.** *Let $B \in S$, and let $\omega_b$ be defined as in (4.26'). Then, for all real $\omega$ and all orderings $\phi$,*

$$(4.63) \qquad \min_{\phi} \left\{ \min_{\omega} \rho(\mathfrak{L}_{B\phi,\omega}) \right\} \geq \omega_b - 1,$$

*with equality if and only if $B$ is a cyclic matrix of index 2.*

The importance of this result is that it shows that consistent orderings, when they exist, are *optimal*. Thus, in keeping with our goal of *maximizing* the asymptotic rate of convergence $\mathfrak{L}_{B,\omega}$, we have also determined which orderings $\phi$ give the greatest asymptotic rate of convergence, relative to the set $S$.

## EXERCISES

1. For the matrix $B_2$ of (4.41), show that

$$h_{B_2}(\alpha) = \frac{1}{\sqrt{2}} \left\{ 1 + \cosh\left(\frac{\alpha}{2}\right) \right\}^{1/2}.$$

Similarly, find $h_{B_3}(\alpha)$, where $B_3$ is defined in (4.42).

2. From the derivation of (4.59), it is clear that

$$\left| \frac{\lambda + \omega - 1}{\omega} \right| \leq m_B(|\lambda|),$$

where $\omega \neq 0$, and $\lambda$ is any eigenvalue of $\mathfrak{L}_{B,\omega}$. With the explicit knowledge of $h_{B_2}(\alpha)$ and $h_{B_3}(\alpha)$ from the previous exercise, determine sharper upper bounds than $\sqrt{\omega_b - 1}$ for $\rho(\mathfrak{L}_{B_i,\omega_b})$, $i = 2, 3$.

3. Let $A$ be an $n \times n$ *Stieltjes matrix* (see Sec. 3.5), and arbitrarily partition $A$ in the form (4.2). If $D$ is the corresponding block diagonal matrix of (4.3), form the matrix

$$B \equiv I - D^{-1/2} A D^{-1/2}.$$

If $B$ is irreducible, show that $B \in S$.

4. Prove the second part of Lemma 4.2.

*5. Prove (Kahan (1958)) that for any $B \in S$, $\ln m_B$ is a *convex* function of $\ln \sigma$, i.e.

$$\ln m_B(e^{\alpha}) \leq \frac{\ln m_B(e^{\alpha+h}) + \ln m_B(e^{\alpha-h})}{2}$$

for any real $\alpha$, and $h > 0$. (*Hint*: Extend the result of Lemma 4.2).

**6.** Show by means of graph theory that the partitioned matrix

$$
A = \begin{bmatrix}
4 & -1 & -1 & 0 & 0 \\
-1 & 4 & 0 & 0 & 0 \\
-1 & 0 & 4 & 0 & -1 \\
0 & 0 & 0 & 4 & -1 \\
0 & 0 & -1 & -1 & 4
\end{bmatrix}
$$

is a consistently ordered 2-cyclic matrix. Similarly, if the same matrix $A$ is partitioned so that its diagonal submatrices are all $1 \times 1$ matrices, show that $A$ is a consistently ordered 2-cyclic matrix.

**7.** Show by means of graph theory that the partitioned matrix

$$
A = \begin{bmatrix}
4 & -1 & 0 & 0 & 0 & 0 \\
-1 & 4 & -1 & 0 & -1 & 0 \\
0 & -1 & 4 & -1 & 0 & 0 \\
0 & 0 & -1 & 4 & -1 & 0 \\
0 & -1 & 0 & -1 & 4 & -1 \\
0 & 0 & 0 & 0 & -1 & 4
\end{bmatrix}
$$

is a consistently ordered 2-cyclic matrix. On the other hand, if $A$ is partitioned so that its diagonal submatrices are all $1 \times 1$ matrices, show that $A$ is an *inconsistently* ordered 2-cyclic matrix.

**8.** Let the $2n \times 2n$ matrix $B$ be given by

$$
B = \frac{\rho}{2n-1} \begin{bmatrix}
0 & 1 & 1 & \cdots & 1 \\
1 & 0 & 1 & \cdots & 1 \\
1 & 1 & 0 & \cdots & 1 \\
\vdots & & & \diagdown & \vdots \\
1 & 1 & 1 & \cdots & 0
\end{bmatrix},
$$

where $0 < \rho = \rho(B) < 1$. Note that $B \in S$. Show that

$$m_B(\sigma) = \frac{\rho}{2n-1} \left\{ \frac{\sigma - \sigma^{(1/2n)}}{\sigma^{(1/2n)} - 1} \right\},$$

and conclude that

$$h_B(\alpha) = \frac{1}{2n-1} \left\{ 2 \cosh \left[ \left( \frac{n-1}{2n} \right) \alpha \right] + 2 \cosh \left[ \left( \frac{n-2}{2n} \right) \alpha \right] \right.$$

$$\left. + \cdots + 2 \cosh \left[ \frac{\alpha}{2n} \right] + 1 \right\}.$$

*9. For Theorem 4.7, show that if $B \in S$ and $h_B(\alpha) \equiv 1$ for all $\alpha \neq 0$, then $B$ is a consistently ordered cyclic matrix of index 2.

10. Let $B \in S$. Prove that for $0 < \omega \leq 1$,

$$\frac{2(1-\omega) + \omega^2 \rho^2 + \omega \rho \sqrt{\omega^2 \rho^2 - 4(\omega - 1)}}{2} \leq \rho(\mathcal{L}_{B,\omega})$$

$$< \frac{2(1-\omega) + \omega \rho(B)}{2 - \omega \rho(B)},$$

with equality only if $B$ is a consistently ordered cyclic matrix of index 2.

11. With $\omega_b$ defined as in (4.26'), let $\lambda = re^{i\theta}$ in (4.57). Show that (4.59) implies that $r < q(\cos \theta)$, where $q(1) \leq q(\cos \theta)$.

12. Using (4.60) and the results of Sec. 4.3, prove Theorem 4.10.

## 4.5. Asymptotic Rates of Convergence

With $R_\infty(\mathcal{L}_{B,\omega}) = -\ln \rho(\mathcal{L}_{B,\omega})$, we now consider in part the effect of orderings on the asymptotic rate of convergence. First, we confine our attention to the Gauss-Seidel iterative method $\omega = 1$. From Theorem 4.8 we deduce

**Theorem 4.11.** *If $B \in S$, then for all orderings $\phi$,*

(4.64)
$$1 \geq \frac{R_\infty(\mathcal{L}_{B_\phi,1})}{2R_\infty(B)} > \frac{1}{2} + \frac{\ln (2 - \rho(B))}{-2 \ln \rho(B)}.$$

*Thus,*

(4.65)
$$\lim_{\rho(B) \to 1-} \frac{R_\infty(\mathcal{L}_{B_\phi,1})}{2R_\infty(B)} = 1.$$

*Proof.* The inequalities of (4.64) follow directly from (4.61). Applying L'Hospital's rule, we have

$$\lim_{\rho(B)\to 1-} \frac{\ln{(2 - \rho(B))}}{-2 \ln{\rho(B)}} = \frac{1}{2}.$$

from which (4.65) follows.

The result of Theorem 4.11 contains as a special case a proof of a conjecture by Shortley and Weller (1938) that different orderings of unknowns in the numerical solution of the Dirichlet problem have asymptotically no effect on rates of convergence of the Gauss-Seidel iterative method.

As we have no precise formula of the value of $\omega$ which maximizes $R_\infty(\mathcal{L}_{B,\omega})$ when $B$ is an arbitrary element of $S$, we consider then the effect of orderings on the asymptotic rate of convergence for the special case when $\omega = \omega_b$ of (4.26').

**Theorem 4.12.** *If $B \in S$,*

$$(4.66) \qquad 1 \geq \frac{R_\infty(\mathcal{L}_{B,\omega_b})}{-\ln{(\omega_b - 1)}} > \frac{1}{2},$$

*with equality on the left only if $B$ is a consistently ordered cyclic matrix of index 2.*

*Proof.* This follows directly from the inequalities of Theorem 4.9.

The importance of the result of (4.66) lies in the fact that use of the relaxation factor $\omega_b$ of (4.26') for *any* matrix $B \in S$, produces, except for a factor of 2, asymptotic convergence rates $R_\infty(\mathcal{L}_{B,\omega_b})$ like those determined for consistently ordered 2-cyclic matrices of Sec. 4.3.

Some final remarks about the selection of the relaxation factor $\omega$ which maximizes $R_\infty(\mathcal{L}_{B,\omega})$ can be made for arbitrary elements of the set $S$. If $h_B(\alpha) \not\equiv 1$ for all real $\alpha$, the results of Sec. 4.4 and Figure 12 would seem to indicate that the value of $\omega$ which optimizes $R_\infty(\mathcal{L}_{B,\omega})$ would necessarily be *greater* than $\omega_b$ of (4.26'). Indeed, numerical experiments of Young (1955) as well as theoretical results of Garabedian (1956) corroborate this, but the following counterexample of Kahan (1958) shows that this is not true in general. Let

$$(4.67) \qquad B = \frac{\rho(B)}{2} \begin{bmatrix} 0 & 1 & 1 \\ 1 & 0 & 1 \\ 1 & 1 & 0 \end{bmatrix}, \qquad 0 < \rho(B) < 1.$$

It is clear that $B \in S$, and that $h_B(\alpha) \not\equiv 1$. Nevertheless, the relaxation factor which minimizes $\rho(\mathcal{L}_{B,\omega})$ can be shown to be $\omega_b$ of (4.26'). Note from Theorem 4.10 that $\rho(\mathcal{L}_{B,\omega_b}) > \omega_b - 1$.

## EXERCISES

**1.** Determine the functions $m_B(\sigma)$ and $h_B(\alpha)$ for the matrix $B$ of (4.67). Also determine $\rho(\mathcal{L}_{B,\omega_b})$ where $\omega_b$ is defined in (4.26').

**2.** For $B \in S$, show that

$$\frac{\frac{1}{2}[-\ln(\omega_b - 1)]}{2(-2\ln\rho(B))^{1/2}} < \frac{R_\infty(\mathcal{L}_{B,\omega_b})}{2[R_\infty(\mathcal{L}_{B,1})]^{1/2}} < \frac{-\ln(\omega_b - 1)}{2 \cdot \left\{ \ln\left[ \frac{2 - \rho(B)}{\rho(B)} \right] \right\}^{1/2}}.$$

If $r_1(\rho(B))$ and $r_2(\rho(B))$ denote respectively these lower and upper bounds for

$$\frac{R_\infty(\mathcal{L}_{B,\omega_b})}{2[R_\infty(\mathcal{L}_{B,1})]^{1/2}},$$

show that

$$\lim_{\rho(B)\to 1-} r_1(\rho(B)) = \tfrac{1}{2}$$

$$\lim_{\rho(B)\to 1-} r_2(\rho(B)) = 1.$$

Thus, for $B \in S$ with $\rho(B)$ close to unity, one obtains a large increase in the asymptotic rate of convergence of the successive overrelaxation iterative method with $\omega = \omega_b$, as compared with the choice $\omega = 1$.

**3.** With the matrix $B$ of (4.67) and $\rho(B) = 0.999$, determine exactly the ratio

$$\frac{R_\infty(\mathcal{L}_{B,\omega_b})}{2[R_\infty(\mathcal{L}_{B,1})]^{1/2}},$$

using explicitly the results obtained from Exercise 1.

## BIBLIOGRAPHY AND DISCUSSION

**4.1.** As stated, the basis for the material of Sec. 4.1–4.3 is due to Young (1950). The generalization of Young's property $A$ to property $A^\pi$ is due to Arms, Gates, and Zondek (1956), whereas the generalization to $p$-cyclic matrices is due to Varga (1959a). Essentially the same definition of a $p$-cyclic matrix was given independently by Kjellberg (1961), although Kjellberg's formulation applies to bounded linear operators on Banach spaces as well.

With regard to Theorem 4.2, this type of determinantal equality has been generalized by Keller (1958) for matrices arising from elliptic partial differential equations. This in turn provides another generalization of the successive overrelaxation iterative method. See also Keller (1960a).

The terms *consistent* and *inconsistent orderings* are due to Young (1950); the eigenvalue formulation in Definition 4.2, however, is apparently new. In Exercise 8, the interesting property that *all* orderings of tri-diagonal block matrices are consistent orderings is due to Friedman (1957). For recent results on the number of consistent orderings, see Heller (1957).

**4.2.** Theorem 4.3 is a generalization by Varga (1959a) of original results of Young (1950) corresponding to the case $p = 2$. The functional equation (4.18) of Theorem 4.3, however, is not the only relationship known between the eigenvalues $\lambda$ of the successive overrelaxation matrix and the eigenvalues $\mu$ of the Jacobi matrix. Kjellberg (1961) obtains the equation

$$(\lambda + \omega - 1)^p = \lambda^k \omega^p \mu^p, \quad 1 \leq k \leq p - 1;$$

see Exercise 2. On the other hand, Keller (1958) obtains equations of the form

$$(\lambda + \alpha - 1)^2 = \frac{\alpha^2}{\beta} \mu^2 (\lambda + \beta - 1).$$

**4.3.** The determination of the optimum relaxation factor $\omega_b$ which minimized $\rho(\mathcal{L}_\omega)$ for the general $p$-cyclic case is due to Varga (1959a), which again generalizes the original results of Young (1950). For the simple case of $2 \times 2$ matrices, Ostrowski (1953) also determines an optimum relaxation factor, but he was apparently unaware of the earlier fundamental work of Young (1950).

As Exercise 2 indicates, one can use the functional equation

$$(\lambda + \omega - 1)^p = \lambda^{p-1} \omega^p \mu^p$$

to determine an optimum relaxation factor $\omega$ even when the eigenvalues of $B^p$ are not non-negative real numbers. For such generalizations, which rely on conformal mapping arguments, see Young (1954a) and Varga (1959a).

The geometrical determination for the case $p = 2$ of the optimum relaxation factor is suggested by results of Kahan (1958) and Varga (1959b).

**4.4.** Many of the results given in this section do not depend upon the symmetry and irreducible character of the matrices $B \in S$, and more general results can be found in Kahan (1958). The approach taken in this section, however, is due to the closely related, but independent papers of Kahan (1958) and Varga (1959b).

The graph-theoretic interpretation of a consistently ordered 2-cyclic matrix might be called new, although Young (1950) has given a similar discussion based on "circulation" for the numerical solution of the Dirichlet problem. See also Forsythe and Wasow (1960, p. 245), and Franklin (1959a).

The result (Varga (1959b)) of Theorem 4.10 contains the proof of a conjecture by Young (1950) that, among all orderings of a cyclic of index 2 matrix $B \in S$, only the consistent orderings give the greatest asymptotic rate of the convergence of the Gauss-Seidel iterative method. With respect to Theorem 4.9 Kahan (1958) first obtained these bounds for $\rho(\mathcal{L}_{B,\omega})$, but without a discussion of when equality is valid.

**4.5.** The particular result of Theorem 4.11, given by Varga (1959b), substantiates an empirical observation due to Shortley and Weller (1938). In the paper of Garabedian (1956), theoretical results are obtained for estimating the optimum relaxation factor in the numerical solution of the Dirichlet problem for small meshes for both the 5-point and 9-point approximations. In terms of the point successive overrelaxation scheme, the latter approximation gives rise to a primitive Jacobi iteration matrix, for which no ordering is consistent.

# CHAPTER 5

# SEMI-ITERATIVE METHODS

**5.1.** Semi-Iterative Methods and Chebyshev Polynomials

In our previous investigations concerning the acceleration of basic iterative methods, we would, in the final analysis, consider the convergence properties and rates of convergence of an iterative procedure of the form

$$(5.1) \qquad \mathbf{x}^{(m+1)} = M\mathbf{x}^{(m)} + \mathbf{g}, \qquad m \geq 0,$$

where $M$ is a fixed $n \times n$ matrix with $\rho(M) < 1$, and $\mathbf{x}^{(0)}$ is some given vector. Such an iterative procedure converges to the unique solution vector $\mathbf{x}$ of

$$(5.2) \qquad (I - M)\mathbf{x} = \mathbf{g}.$$

In particular, we defined

$$(5.3) \qquad \mathbf{\varepsilon}^{(m)} \equiv \mathbf{x}^{(m)} - \mathbf{x}, \qquad m \geq 0,$$

as the *error vector* for the $m$th iterate of (5.1) and, as we have previously seen, we could relate the error vector $\mathbf{\varepsilon}^{(m)}$ to the initial error vector $\mathbf{\varepsilon}^{(0)}$ through the equation

$$(5.4) \qquad \mathbf{\varepsilon}^{(m)} = M^m \mathbf{\varepsilon}^{(0)}, \qquad m \geq 0.$$

With the use of vector and matrix norms from Chapter 1,

$$(5.5) \qquad || \, \mathbf{\varepsilon}^{(m)} \, || = || \, M^m \mathbf{\varepsilon}^{(0)} \, || \leq || \, M^m \, || \cdot || \, \mathbf{\varepsilon}^{(0)} \, ||, \qquad m \geq 0,$$

and we defined

$$R(M^m) = \frac{-\ln \| M^m \|}{m}, \qquad m \geq 1,$$

as the *average rate of convergence* for $m$ iterations, provided that $\| M^m \| < 1$, and it was shown in Sec. 3.2 that

$$R(M^m) \leq R_\infty(M) = -\ln \rho(M), \qquad m \geq 0,$$

for any convergent matrix $M$. Reflecting on what has already been considered in previous chapters, it is evident that we have compared and optimized iterative methods *only* with respect to their asymptotic rates of convergence. The major aim of this chapter is now to introduce new iterative methods, and to make comparisons of various iterative methods with respect to *average* rates of convergence for every $m \geq 0$.

Suggested by the classical theory of summability of sequences, we consider now a more general iterative procedure defined by (5.1) and

$$(5.6) \qquad \mathbf{y}^{(m)} \equiv \sum_{j=0}^{m} \nu_j(m) \mathbf{x}^{(j)}, \qquad m \geq 0,$$

which we call a *semi-iterative method* with respect to the iterative method of (5.1). It is semi-iterative as we iterate first by means of (5.1), and then we algebraically combine these vector iterates by means of (5.6). As an example, letting

$$\nu_j(m) \equiv \frac{1}{m+1}, \qquad m \geq 0,$$

we have that the vectors $\mathbf{y}^{(m)}$ are arithmetic averages of the vectors $\mathbf{x}^{(j)}$, $0 \leq j \leq m$, which corresponds precisely to Cesàro (C, 1) summability.

Our object here is to prescribe the constants $\nu_j(m)$ of (5.6) in such a way that the vectors $\mathbf{y}^{(m)}$ tend rapidly to the unique solution $\mathbf{x}$ of (5.2), in some precise sense. First, we impose a natural restriction on the constants $\nu_j(m)$. If our initial vector approximation $\mathbf{x}^{(0)}$ were equal to the vector $\mathbf{x}$ of (5.2), it would follow from (5.4) and (5.3) that $\mathbf{x}^{(m)} = \mathbf{x}$ for all $m \geq 0$, and we require in this case that $\mathbf{y}^{(m)} = \mathbf{x}$ also for all $m \geq 0$. Thus,

$$(5.7) \qquad \sum_{j=0}^{m} \nu_j(m) = 1, \qquad m \geq 0.$$

Letting

$$(5.8) \qquad \tilde{\varepsilon}^{(m)} \equiv \mathbf{y}^{(m)} - \mathbf{x}, \qquad m \geq 0,$$

be the error vector associated with the vectors $\mathbf{y}^{(m)}$ of (5.6), we have from (5.8) and (5.7)

$$(5.9) \qquad \tilde{\boldsymbol{\varepsilon}}^{(m)} = \sum_{j=0}^{m} \nu_j(m)\mathbf{x}^{(j)} - \sum_{j=0}^{m} \nu_j(m)\mathbf{x} = \sum_{j=0}^{m} \nu_j(m)\boldsymbol{\varepsilon}^{(j)},$$

and thus

$$(5.10) \qquad \tilde{\boldsymbol{\varepsilon}}^{(m)} = \left(\sum_{j=0}^{m} \nu_j(m)M^j\right)\boldsymbol{\varepsilon}^{(0)},$$

using (5.4). If we define the polynomial $p_m(x)$ as

$$(5.11) \qquad p_m(x) = \sum_{j=0}^{m} \nu_j(m)x^j, \qquad m \geq 0,$$

we can write (5.10) in the form

$$(5.10') \qquad \tilde{\boldsymbol{\varepsilon}}^{(m)} = p_m(M)\boldsymbol{\varepsilon}^{(0)}, \qquad m \geq 0,$$

where $p_m(M)$ is a polynomial in the matrix $M$. The only restriction we have made thus far on the polynomials $p_m(x)$ is that $p_m(1) = 1$ from (5.7).

With our definitions of vector and matrix norms, it follows that

$$(5.12) \qquad \| \tilde{\boldsymbol{\varepsilon}}^{(m)} \| = \| p_m(M)\boldsymbol{\varepsilon}^{(0)} \| \leq \| p_m(M) \| \cdot \| \boldsymbol{\varepsilon}^{(0)} \|, \qquad m \geq 0.$$

If $\| p_m(M) \| < 1$, we analogously define

$$(5.13) \qquad R[p_m(M)] \equiv \frac{-\ln \| p_m(M) \|}{m}$$

as the *average rate of convergence* for $m$ iterations of the semi-iterative method of (5.6). Observe that by choosing $p_m(x) = x^m$, the vector $\mathbf{y}^{(m)}$ of (5.6) is just the vector $\mathbf{x}^{(m)}$, and moreover, the definition of the average rate of convergence of (5.13) for this particular semi-iterative method reduces exactly to the definition of the average rate of convergence in Sec. 3.2 for the iterative method of (5.1).

From (5.12), we see that making the quantities $\| p_m(M) \|$ smaller implies that the bounds for the norms of the errors $\tilde{\boldsymbol{\varepsilon}}^{(m)}$ for the semi-iterative method are proportionately smaller, and we are naturally led to the minimization problem

$$(5.14) \qquad \min_{p_m(1)=1} \| p_m(M) \|$$

for each $m \geq 0$. In a certain sense, this has a trivial solution for $m$ sufficiently large. Let $q_n(x) \equiv \det (xI - M)$ be the *characteristic polynomial*
for the $n \times n$ matrix $M$. From the Cayley-Hamilton Theorem[†], we then
know that $q_n(M) = O$. Defining

$$p_m(x) = \frac{x^{m-n}q_n(x)}{q_n(1)}, \qquad m \geq n,$$

where $q_n(1) \neq 0$ as $\rho(M) < 1$, it follows that $p_m(M)$ is the null matrix
for all $m \geq n$, and thus $\| p_m(M) \| = 0$ for $m \geq r$. In other words, if we
knew all the eigenvalues of $M$ a priori, a semi-iterative method could be
constructed so that at least $y^{(n)}$ would yield the desired solution of the
matrix problem (5.2). If the order $n$ of the matrix $M$ is very large, the
*a priori* determination of the eigenvalues or the characteristic polynomial of
$M$ is very difficult and time-consuming, and we thus consider the minimization problem of (5.14) with what we might call practical constraints.

We now assume that the matrix $M$ is Hermitian and that all polynomials $p_m(x)$ to be considered have real coefficients. Thus, from the
Corollary to Theorem 1.3,

$$(5.15) \qquad \| p_m(M) \| = \rho(p_m(M)) = \max_{1 \leq i \leq n} | p_m(\mu_i) |,$$

where $\mu_i$ is an eigenvalue of $M$. Since $M$ is Hermitian and convergent by
assumption, its eigenvalues $\mu_i$ lie in some real interval $-1 < a \leq \mu_i \leq b < 1$,
and clearly,

$$(5.16) \qquad \| p_m(M) \| = \max_{1 \leq i \leq n} | p_m(\mu_i) | \leq \max_{a \leq x \leq b} | p_m(x) |.$$

In place of the minimization problem of (5.14), we consider the new minimization problem

$$(5.17) \qquad \min_{p_m(1)=1} \left\{ \max_{-1 < a \leq x \leq b < 1} | p_m(x) | \right\},$$

where $p_m(x)$ is a real polynomial. The solution of this problem is classical,
and is given in terms of the Chebyshev polynomials $C_m(x)$, defined by

$$(5.18) \qquad C_m(x) \equiv \begin{cases} \cos (m \cos^{-1} x), & -1 \leq x \leq 1, m \geq 0, \\ \cosh (m \cosh^{-1} x), & x \geq 1, m \geq 0. \end{cases}$$

For $-1 \leq x \leq 1$, we can write $C_m(x) = \cos m\theta$, where $\cos \theta \equiv x$. This is

---

[†] For a proof of this well-known result, see, for example, Birkhoff and MacLane
(1953), p. 320.

useful in that the simple trigonometric identity

$$\cos [(m-1)\theta] + \cos [(m+1)\theta] = 2 \cos \theta \cos m\theta$$

immediately gives the well known three-term recurrence relation

$$(5.18') \qquad C_{m+1}(x) = 2xC_m(x) - C_{m-1}(x), \qquad m \geq 1,$$

where $C_0(x) = 1$ and $C_1(x) = x$. Moreover, the expression $C_m(x) = \cos m\theta$ shows that

$$\max_{-1 \leq x \leq 1} |C_m(x)| = 1 \quad \text{and} \quad C_m(t_k) = (-1)^k \quad \text{for} \quad t_k = \cos^{-1}\left(\frac{k\pi}{m}\right),$$

$$0 \leq k \leq m.$$

For the minimization problem of (5.17), consider the polynomial $\tilde{p}_m(x)$ defined by

$$(5.19) \qquad \tilde{p}_m(x) = \frac{C_m\left[\dfrac{2x - (b+a)}{b-a}\right]}{C_m\left[\dfrac{2 - (b+a)}{b-a}\right]}, \qquad m \geq 0.$$

Clearly, $\tilde{p}_m(x)$ is a real polynomial satisfying $\tilde{p}_m(1) = 1$, and as

$$y = \frac{2x - (b+a)}{b-a}$$

is a 1–1 mapping of $a \leq x \leq b$ onto $-1 \leq y \leq 1$, then

$$\max_{a \leq x \leq b} |\tilde{p}_m(x)| = \frac{\max\limits_{-1 \leq y \leq 1} |C_m(y)|}{C_m\left[\dfrac{2 - (b+a)}{b-a}\right]} = \frac{1}{C_m\left[\dfrac{2 - (b+a)}{b-a}\right]},$$

and evidently this maximum value is taken on with alternating sign in $m + 1$ distinct points $a = x_0 < x_1 < \cdots < x_m = b$. Suppose now that there exists a real polynomial $q_m(x)$ of degree $m$ with $q_m(1) = 1$ such that

$$\max_{a \leq x \leq b} |q_m(x)| < \max_{a \leq x \leq b} |\tilde{p}_m(x)|.$$

Form $R_m(x) = \tilde{p}_m(x) - q_m(x)$, which is a polynomial of degree at most $m$. Then, $\{R_m(x_j)\}_{j=0}^m$ is a set of $m + 1$ nonzero numbers which alternates in sign, and from the continuity of $R_m(x)$, there must exist $m$ values $s_j$, $1 \le j \le m$ with $a \le x_{j-1} < s_j < x_j \le b$ such that $R_m(s_j) = 0$. But as $R_m(1) = 0$, then $R_m(x)$ vanishes at $m + 1$ distinct points, and is thus identically zero. This establishes† that the polynomial $\tilde{p}_m(x)$ is *one* polynomial which solves (5.17). A slight extension‡ of the above argument, moreover, shows that $\tilde{p}_m(x)$ is the *only* polynomial for which this is true, and we have

**Theorem 5.1.** *For each* $m \ge 0$, *let* $S_m$ *be the set of all real polynomials* $p_m(x)$ *of degree* $m$ *satisfying* $p_m(1) = 1$. *Then, the polynomial* $\tilde{p}_m(x) \in S_m$ *of* (5.19) *is the unique polynomial which solves the minimization problem of* (5.17).

In contrast to the solution of the minimization problem of (5.14), which required a knowledge of *all* the eigenvalues of the matrix $M$, we see that the solution of the minimization problem of (5.17) requires only a knowledge of bounds for the largest and smallest eigenvalues of the Hermitian matrix $M$. We recall that as $M$ is also convergent, we must have that

$$-\rho(M) \le a \le b \le \rho(M) < 1,$$

and

$$\max \{ |a|, |b| \} = \rho(M).$$

We now concentrate on the particular case where we require as *a priori* knowledge only the spectral radius $\rho(M)$ of $M$. With $b \equiv \rho(M) \equiv -a$, the polynomials $\tilde{p}_m(x)$ of (5.19) also satisfy a three-term recurrence relation derived from (5.18′) and (5.19):

$$(5.20) \qquad C_{m+1}\left(\frac{1}{\rho}\right) \tilde{p}_{m+1}(x) = \frac{2x}{\rho} C_m\left(\frac{1}{\rho}\right) \tilde{p}_m(x) - C_{m-1}\left(\frac{1}{\rho}\right) \tilde{p}_{m-1}(x),$$

$$m \ge 1,$$

where $\rho \equiv \rho(M)$, $\tilde{p}_0(x) = 1$, and $\tilde{p}_1(x) = x$. Replacing the variable $x$ by the matrix $M$ and using (5.10′), we have

$$C_{m+1}\left(\frac{1}{\rho}\right) \bar{\varepsilon}^{(m+1)} = \frac{2C_m(1/\rho)}{\rho} M\bar{\varepsilon}^{(m)} - C_{m-1}\left(\frac{1}{\rho}\right) \bar{\varepsilon}^{(m-1)}, \qquad m \ge 1.$$

† The proof given here is due to Flanders and Shortley (1950). For more general results, including approximation by rational functions, see Achieser (1956).
‡ See Exercise 7.

After a small amount of algebra, using (5.18'), (5.8), and (5.2), we can write this equivalently as

$$(5.21) \qquad \mathbf{y}^{(m+1)} = \omega_{m+1}\{M\mathbf{y}^{(m)} + \mathbf{g} - \mathbf{y}^{(m-1)}\} + \mathbf{y}^{(m-1)}, \qquad m \geq 0,$$

where

$$(5.22) \qquad \omega_{m+1} = \frac{2C_m(1/\rho)}{\rho C_{m+1}(1/\rho)} = 1 + \frac{C_{m-1}(1/\rho)}{C_{m+1}(1/\rho)}, \qquad m \geq 1, \omega_1 = 1.$$

We shall call this the *Chebyshev semi-iterative method with respect to* (5.1). For $m = 0$, (5.21) reduces to $\mathbf{y}^{(1)} = M\mathbf{y}^{(0)} + \mathbf{g}$, but as $\mathbf{y}^{(0)} = \mathbf{x}^{(0)}$ from the restriction of (5.7), it follows that both the vectors $\mathbf{y}^{(0)}$ and $\mathbf{y}^{(1)}$ of the semi-iterative method are respectively equal to the vector iterates $\mathbf{x}^{(0)}$ and $\mathbf{x}^{(1)}$ of (5.1). But what is more important, the recursive relation of (5.21) shows that it is *unnecessary* to form the auxiliary vector iterates $\mathbf{x}^{(m)}$ of (5.1) in order to determine the vectors $\mathbf{y}^{(m)}$. Thus, whereas we had originally thought of forming the vectors $\mathbf{y}^{(m)}$ by means of (5.6), the relationship of (5.21) shows that we can directly determine the vectors $\mathbf{y}^{(m)}$ in a manner very similar to the determination of the vectors $\mathbf{x}^{(m)}$ of (5.1). We note, however, that whereas $\mathbf{x}^{(m+1)}$ depended explicitly only on $\mathbf{x}^{(m)}$ in (5.1), $\mathbf{y}^{(m+1)}$ depends on the two previous approximations $\mathbf{y}^{(m)}$ and $\mathbf{y}^{(m-1)}$. In terms of computing machine applications, the semi-iterative method of (5.21) requires an additional vector of storage over the iterative method of (5.1), an item which can be of considerable weight in the solution of large matrix problems.

Returning to the behavior of the matrix norms $\| \tilde{p}_m(M) \|$, we have from (5.15), by our assumption that the matrix $M$ is Hermitian and convergent,

$$\| \tilde{p}_m(M) \| = \max_{1 \leq i \leq n} | \tilde{p}_m(\mu_i) | = \max_{1 \leq i \leq n} \frac{| C_m(\mu_i/\rho) |}{C_m(1/\rho)}, \qquad m \geq 0.$$

We know that the eigenvalues $\mu_i$ of $M$ are real and satisfy

$$-\rho(M) \leq \mu_i \leq \rho(M),$$

with $| \mu_j | = \rho(M)$ for some $j$, and as $| C_m(\pm 1) | = 1$ from (5.18), we conclude that

$$(5.23) \qquad \| \tilde{p}_m(M) \| = \frac{1}{C_m(1/\rho)}, \qquad m \geq 0.$$

Thus, we see from (5.18) that the sequence of matrix norms $\| \tilde{p}_m(M) \|$ is *strictly decreasing* for all $m \geq 0$. It is interesting that we can express

(5.23) in a different way, which we will find to be of considerable use. First, from (5.18) we have with $\rho \equiv \rho(M)$

$$C_m \left(\frac{1}{\rho}\right) = \cosh (m\sigma),$$

where

$$\sigma = \cosh^{-1} \left(\frac{1}{\rho}\right)$$

is chosen to be positive. Thus,

$$C_m \left(\frac{1}{\rho}\right) = \frac{e^{m\sigma} + e^{-m\sigma}}{2} = e^{m\sigma} \left(\frac{1 + e^{-2m\sigma}}{2}\right).$$

Now,

$$\sigma = \cosh^{-1} \left(\frac{1}{\rho}\right) = \ln \left\{ \frac{1}{\rho} + \sqrt{\frac{1}{\rho^2} - 1} \right\},$$

so that

$$e^{-m\sigma} = \left(\frac{\rho}{1 + \sqrt{1 - \rho^2}}\right)^m.$$

Combining these identities, and using the definition of $\omega_b$ in (4.26′), we obtain

$$(5.24) \qquad || \tilde{p}_m(M) || = (\omega_b - 1)^{m/2} \left\{ \frac{2}{1 + (\omega_b - 1)^m} \right\}, \qquad m \geq 0.$$

Note this again implies that the matrix norms $|| \tilde{p}_m(M) ||$ are strictly decreasing. Thus, with the polynomials $\tilde{p}_m(M)$ of (5.19), our conclusion is that the norms of the error vectors $\tilde{\varepsilon}^{(m)}$ for the vector approximations $\mathbf{y}^{(m)}$ satisfy

$$(5.25) \qquad || \tilde{\varepsilon}^{(m)} || \leq \left(\frac{2(\omega_b - 1)^{m/2}}{1 + (\omega_b - 1)^m}\right) || \varepsilon^{(0)} ||, \qquad m \geq 0.$$

In terms of the average rate of convergence for $m$ steps for this semi-iterative method, we have from (5.24)

$$(5.26) \qquad R[\tilde{p}_m(M)] = -\frac{1}{2} \ln (\omega_b - 1) - \frac{\ln \left(\frac{2}{1 + (\omega_b - 1)^m}\right)}{m},$$

$$m \geq 0,$$

and thus

$$(5.26') \qquad \lim_{m \to \infty} R[\tilde{p}_m(M)] = -\tfrac{1}{2} \ln (\omega_b - 1) = \ln \left( \frac{1}{\sqrt{\omega_b - 1}} \right).$$

The identities used in rewriting the norms $|| \tilde{p}_m(M) ||$ of (5.23) in the form (5.24) can be used to show that the constants $\omega_m$ of (5.22) are also strictly decreasing for $m \geq 2$, and

$$(5.27) \qquad \lim_{m \to \infty} \omega_m = \omega_b,$$

where $\omega_b$ is defined in (4.26').

## EXERCISES

1. Verify (5.21).

2. Verify (5.27).

3. Let $M$ be an $n \times n$ convergent Hermitian matrix whose eigenvalues $\mu_i$ satisfy

$$-\rho(M) \leq a \leq \mu_i \leq b = p(M) < 1.$$

With $\tilde{p}_m(x)$ defined as in (5.19),

a. Deduce a three-term recurrence relation for the vectors $y^{(m)}$ analogous to (5.21).

b. if $\omega_m$ is the associated parameter analogous to $\omega_m$ of (5.21) derived from part a, find

$$g_1(a) \equiv \lim_{m \to \infty} \hat{\omega}_m.$$

c. If $R[\tilde{p}_m(M)]$ is the average rate of convergence for $m$ iterations of this method, find

$$g_2(a) \equiv \lim_{m \to \infty} R[\tilde{p}_m(M)].$$

4. Compute the norms $|| \tilde{p}_m(B) ||$, $m \geq 1$, for the $4 \times 4$ matrix $B$ of (1.7). How do these norms compare with the norms $|| \mathcal{L}_1^m || = \sqrt{5}/4^m$; $m \geq 1$, of Sec. 3.3., where $\mathcal{L}_1$ is the point successive overrelaxation matrix derived from $B$?

5. Let the $n \times n$ matrix $M$ of (5.1) be convergent and Hermitian, and consider any "generalized arithmetic mean" semi-iterative method (i.e., one for which each $\nu_j(m) > 0$ for all $0 \leq j \leq m$ and all $m \geq 1$, where

$$\sum_{j=0}^{m} \nu_j(m) = 1).$$

If $g_m(M)$ is the associated polynomial of $(5.10')$, and $M$ is such that $\rho(M)$ is itself an eigenvalue of $M$, show that

$$R[g_m(M)] < -\ln \rho(M) = R[M^m] \quad \text{for all } m \geq 1.$$

Thus, with these hypotheses, generalized arithmetic mean semi-iterative methods are always iteratively slower than the Jacobi iterative method.

**6.** In solving the matrix equation $Ax = k$, *Richardson's iterative method* is defined by

$$\mathbf{y}_{m+1} = \mathbf{y}_m + \alpha_m(-A\mathbf{y}_m + \mathbf{k}), \quad m \geq 0,$$

where the $\alpha_m$'s are real scalars.

**a.** Show that the error vectors $\varepsilon_m$ for this method satisfy

$$\varepsilon_m = q_m(A)\varepsilon_0, \quad m \geq 0,$$

where $q_m(x) \equiv \prod_{j=0}^{m-1} (1 - \alpha_j x).$

**b.** If $A$ is Hermitian and positive definite, with eigenvalues $0 < \lambda_1 \leq \lambda_2 \leq \cdots \leq \lambda_n$, then

$$\| q_m(A) \| \leq \max_{0 < \lambda_1 \leq x \leq \lambda_n} | q_m(x) |.$$

Also, by definition $q_m(0) = 1$. If $\tilde{S}_m$ is the set of all real polynomials $q_m(x)$ with $q_m(0) = 1$, then, as in Theorem 5.1, characterize completely

$$\min_{q_m \,\epsilon\, \tilde{S}_m} \left\{ \max_{0 < \lambda_1 \leq x \leq \lambda_n} | q_m(x) | \right\}.$$

(Young (1954b)).

**7.** Show that the polynomial $\tilde{p}_m(t)$ of $(5.19)$ is the *unique* polynomial of degree $m$ which solves $(5.17)$.

**8.** Show that the constant $\omega_i$ of $(5.22)$ can also be expressed as

$$\omega_{i+1} = \frac{1}{1 - \left(\dfrac{\rho^2 \omega_i}{4}\right)}, \quad i \geq 2.$$

## 5.2. Relationship of Semi-Iterative Methods to Successive Overrelaxation Iterative Methods

Our discussion of semi-iterative methods in Sec. 5.1 was based on the simple assumption that the $n \times n$ matrix $M$ of $(5.1)$ was Hermitian and convergent. In contrast to these assumptions, we recall that we made assumptions that the corresponding matrix $M$ was a consistently ordered weakly cyclic matrix of index $p$, such that $M^p$ had non-negative real eigenvalues, in obtaining the results of Sec. 4.3. Similarly, the results of Sec. 4.4 were obtained relative to the set of matrices of Definition 4.3. For these basic differences in assumptions, we would assume that the results of Sec. 5.1 on semi-iterative methods would be substantially different. With regard to asymptotic rates of convergence, this, however, is not the case.

We now make the slightly weaker assumption that the $n \times n$ complex matrix $M$ of (5.1) is convergent with real eigenvalues, and we consider the new matrix problem

$$\mathbf{x} = M\mathbf{y} + \mathbf{g},$$

(5.28)

$$\mathbf{y} = M\mathbf{x} + \mathbf{g},$$

where the column vectors $\mathbf{x}$ and $\mathbf{y}$, each with $n$ components, are sought. In matrix notation, this is

(5.29) $$\mathbf{z} = \hat{J}\mathbf{z} + \mathbf{h},$$

where

(5.30) $$\mathbf{z} = \begin{bmatrix} \mathbf{x} \\ \mathbf{y} \end{bmatrix}, \quad \hat{J} = \left[ \begin{array}{c|c} O & M \\ \hline M & O \end{array} \right], \quad \mathbf{h} = \begin{bmatrix} \mathbf{g} \\ \mathbf{g} \end{bmatrix}.$$

The $2n \times 2n$ matrix $\hat{J}$ is evidently weakly cyclic of index 2, and as

$$\hat{J}^2 = \left[ \begin{array}{c|c} M^2 & O \\ \hline O & M^2 \end{array} \right],$$

it follows that $\rho(\hat{J}) = \rho(M) < 1$. Thus, the matrix equation (5.29) possesses a unique solution vector $\mathbf{z}$, and we see that the vector components $\mathbf{x}$ and $\mathbf{y}$ of $\mathbf{z}$ are necessarily both equal to the vector solution of $\mathbf{x} = M\mathbf{x} + \mathbf{g}$.

With Definition 4.2 of Sec. 4.1, we immediately see that the matrix $\hat{J}$, as partitioned in (5.30), is a consistently ordered weakly cyclic matrix of index 2, and the knowledge that $\hat{J}$ is convergent with real eigenvalues allows us to apply rigorously the results of Sec. 4.3. Specifically, the successive overrelaxation iterative method applied to the matrix problem of (5.28) is

$$\mathbf{x}^{(m+1)} = \omega\{M\mathbf{y}^{(m)} + \mathbf{g} - \mathbf{x}^{(m)}\} + \mathbf{x}^{(m)},$$

(5.31)

$$\mathbf{y}^{(m+1)} = \omega\{M\mathbf{x}^{(m+1)} + \mathbf{g} - \mathbf{y}^{(m)}\} + \mathbf{y}^{(m)}, \quad m \geq 0,$$

and if $\hat{\mathcal{L}}_\omega$ is the associated $2n \times 2n$ successive overrelaxation iteration matrix, then we have directly from Theorem 4.4

**Theorem 5.2.** *Let $M$ be an $n \times n$ convergent matrix with real eigenvalues. If $\hat{\mathcal{L}}_\omega$ is the $2n \times 2n$ successive overrelaxation iteration matrix derived from the matrix $\hat{J}$ of (5.30), then*

$$(5.32) \qquad R_\infty(\hat{\mathcal{L}}_\omega) < R_\infty(\hat{\mathcal{L}}_{\omega_b}) = -\ln(\omega_b - 1)$$

*for all $\omega \neq \omega_b$, and*

$$(5.33) \qquad \omega_b = \frac{2}{1 + \sqrt{1 - \rho^2(M)}}.$$

Our first conclusion from Theorem 5.2 is that the theoretical results of Sec. 4.3 are more generally applicable than we may have thought. This application of the results of Sec. 4.3, however, was made at the expense of dealing with matrices of *twice* the order of those of (5.2). From a practical point of view, the iterative method of (5.31) is best carried out in the following manner. As we recall, the unique solution of (5.29) was such that $\mathbf{x} = \mathbf{y}$, where $\mathbf{x}$ is the unique vector solution of (5.2). Thus, as both sets of vector iterates $\{\mathbf{x}^{(m)}\}_{m=0}^\infty$ and $\{\mathbf{y}^{(m)}\}_{m=0}^\infty$ of (5.31) converge (for $0 < \omega < 2$) to the same limit vector $\mathbf{x}$ of (5.2), we define the new set of vector iterates $\{\boldsymbol{\zeta}^{(m)}\}_{m=0}^\infty$ by means of

$$(5.34) \qquad \boldsymbol{\zeta}^{(2l)} \equiv \mathbf{x}^{(l)}, \qquad \boldsymbol{\zeta}^{(2l+1)} \equiv \mathbf{y}^{(l)}, \qquad l \geq 0.$$

In terms of the vectors $\boldsymbol{\zeta}^{(m)}$, we can write (5.31) in the simpler form

$$(5.35) \qquad \boldsymbol{\zeta}^{(m+1)} = \omega\{M\boldsymbol{\zeta}^{(m)} + \mathbf{g} - \boldsymbol{\zeta}^{(m-1)}\} + \boldsymbol{\zeta}^{(m-1)}, \qquad m \geq 1,$$

where $\boldsymbol{\zeta}^{(0)} = \mathbf{x}^{(0)}$, $\boldsymbol{\zeta}^{(1)} = \mathbf{y}^{(0)}$, are the initial vector approximations of $\mathbf{x}$ of (5.2). But this is *identical* in form with the Chebyshev semi-iterative method of (5.21), the only exception being that the relaxation factor $\omega$ in (5.35) is held fixed throughout the course of iteration, whereas $\omega$ is varied in a precise way (5.22) in the Chebyshev semi-iterative method. Thus, actual computing machine applications of (5.35) and the Chebyshev semi-iterative method are identical in terms of vector and matrix storage as well as in terms of arithmetic calculations. In terms of the direct application of the successive overrelaxation iterative method to the original matrix equation (5.2), however, it is clear that the use of (5.35), or (5.21), requires an additional vector iterate of storage over that of the successive overrelaxation iterative method, and this can be important in practical problems.

Our second conclusion, derived from Theorem 5.2, is that for very large numbers of iterations, there is very little difference between (5.35) with $\omega = \omega_b$ and (5.21), since we know that the sequence of constants $\omega_m$ of (5.22) are monotonically decreasing to $\omega_b$ for $m \geq 2$. In fact, Golub (1959) has pointed out that with a finite number of significant figures on a given computing machine, the Chebyshev semi-iterative method of (5.21) degenerates *precisely* into the successive overrelaxation iterative method of (5.35).

Because of the strong similarity of these two iterative methods, we now look carefully at their average and asymptotic rates of convergence, under the assumption that the matrix $M$ of (5.2) is Hermitian and convergent. We first denote the error vectors for the vector iterates of the successive overrelaxation iterative method of (5.35) by

$$(5.36) \qquad \hat{\varepsilon}^{(m)} = \zeta^{(m)} - \mathbf{x}, \qquad m \geq 0,$$

where $\mathbf{x}$ is the unique solution of the matrix equation $\mathbf{x} = M\mathbf{x} + \mathbf{g}$. But, from (5.35) we obtain the following recurrence relation for the error vectors $\hat{\varepsilon}^{(m)}$:

$$(5.37) \qquad \hat{\varepsilon}^{(m+1)} = \omega M \hat{\varepsilon}^{(m)} + (1 - \omega)\hat{\varepsilon}^{(m-1)}, \qquad m \geq 1.$$

With $\hat{\varepsilon}^{(0)}$ and $\hat{\varepsilon}^{(1)}$ arbitrary initial error vectors, corresponding to the fact that both vectors $\mathbf{x}^{(0)}$ and $\mathbf{y}^{(0)}$ in (5.31) can be independent initial approximations of the unique vector $\mathbf{x}$, we can express the error vector $\hat{\varepsilon}^{(m+1)}$ as a sum of two polynomials in the matrix $M$ applied respectively to the initial error vectors $\hat{\varepsilon}^{(0)}$ and $\hat{\varepsilon}^{(1)}$. Specifically, we can verify by induction that

$$(5.38) \qquad \hat{\varepsilon}^{(m)} = \alpha_{m-1}(M)\hat{\varepsilon}^{(1)} + (1 - \omega)\alpha_{m-2}(M)\hat{\varepsilon}^{(0)}, \qquad m \geq 2,$$

where the polynomials $\alpha_m(M)$ are defined from $\alpha_0(M) = I$, $\alpha_1(M) = \omega M$, and

$$(5.39) \qquad \alpha_{m+1}(M) = \omega M \alpha_m(M) + (1 - \omega)\alpha_{m-1}(M), \qquad m \geq 1.$$

For $\omega \neq 0$, we see that $\alpha_m(M)$ is a polynomial of degree $m$ in the matrix $M$. We remark that the expression in (5.38) is just a natural generalization of expressions such as

$$\varepsilon^{(m)} = M^m \varepsilon^{(0)}, \qquad \tilde{\varepsilon}^{(m)} = p_m(M)\varepsilon^{(0)}.$$

For the Chebyshev semi-iterative method of (5.21), it is necessary only to have a single initial vector approximation of the vector $\mathbf{x}$, and with $b = -a = \rho(M)$, it turns out (from (5.10′) and (5.19)) that

$$\tilde{\varepsilon}^{(1)} = M\varepsilon^{(0)}.$$

In order to compare the Chebyshev semi-iterative method and the successive overrelaxation iterative method under similar hypotheses, we now assume that

$$\tilde{\varepsilon}^{(0)} = \hat{\varepsilon}^{(0)} = \varepsilon^{(0)}; \qquad \tilde{\varepsilon}^{(1)} = \hat{\varepsilon}^{(1)} = M\varepsilon^{(0)},$$

so that both iterative methods have their first two error vectors identical. Thus, we conclude from (5.38) that

$$(5.40) \qquad\qquad \hat{\varepsilon}^{(m)} = r_m(M)\,\varepsilon^{(0)}, \qquad m \geq 0,$$

where $r_0(M) = I$, $r_1(M) = M$, and

$$(5.41) \qquad r_m(M) = M\alpha_{m-1}(M) + (1-\omega)\alpha_{m-2}(M), \qquad m \geq 2.$$

For $\omega \neq 0$, $r_m(M)$ is a polynomial of degree $m$ in the matrix $M$. By virtue of our construction, $r_m(x)$ is a real polynomial, and as the matrix $M$ is Hermitian by assumption, we can write (from the Corollary to Theorem 1.3)

$$(5.42) \qquad\qquad \| \hat{\varepsilon}^{(m)} \| \leq \| r_m(M) \| \cdot \| \varepsilon^{(0)} \|, \qquad m \geq 0,$$

where

$$(5.43) \qquad\qquad \| r_m(M) \| = \rho\{r_m(M)\} = \max_{1 \leq i \leq n} | r_m(\mu_i) |;$$

again, the $\mu_i$'s are eigenvalues of the matrix $M$. Similarly, if $\tilde{\varepsilon}^{(m)}$ denotes the error vectors for the Chebyshev semi-iterative method, then

$$(5.44) \qquad\qquad \tilde{\varepsilon}^{(m)} = \tilde{p}_m(M)\,\varepsilon^{(0)}, \qquad m \geq 0,$$

and

$$(5.45) \qquad\qquad \| \tilde{\varepsilon}^{(m)} \| \leq \| \tilde{p}_m(M) \| \cdot \| \varepsilon^{(0)} \|, \qquad m \geq 0,$$

where $\| \tilde{p}_m(M) \|$ has already been explicitly evaluated in (5.24). To compare the matrix norms $\| r_m(M) \|$ and $\| \tilde{p}_m(M) \|$, we shall show that

$$\| r_m(M) \| = \max_{1 \leq i \leq n} | r_m(\mu_i) | = \max_{-\rho(M) \leq x \leq \rho(M)} | r_m(x) |,$$

and that $r_m(x)$ is an element of the set $S_m$ of normalized polynomials. Once this is established, the *uniqueness* of the polynomial $\tilde{p}_m(x)$ in solving the minimization problem of (5.17),

$$\min_{p_m(1)=1} \{ \max_{-\rho(M) \leq x \leq \rho(M)} | p_m(x) | \},$$

will give us the result that $\| \tilde{p}_m(M) \| < \| r_m(M) \|$, $m \geq 2$.

To evaluate the matrix norms $\| r_m(M) \|$, we replace the matrix $M$ in (5.39) by the real variable $\mu$, so that the expression (5.39) for $\alpha_{m+1}(M)$ can be interpreted as a linear difference equation for each fixed $\mu$. Such linear difference equations with constant coefficients can be solved by assuming solutions of the form $\phi^m(\mu)$. Indeed, it can be verified that

$$(5.46) \qquad \alpha_m(\mu) = \begin{cases} \dfrac{\phi_1^{m+1}(\mu) - \phi_2^{m+1}(\mu)}{\phi_1(\mu) - \phi_2(\mu)}; & \phi_1 \neq \phi_2, \\[4mm] (m+1)\phi_1^m(\mu); & \phi_1 = \phi_2, \end{cases} \qquad m \geq 0,$$

where $\phi_1(\mu)$ and $\phi_2(\mu)$ are the roots of

$$(5.47) \qquad \phi^2 - \omega\mu\phi + (\omega - 1) = 0.$$

But with $\lambda = \phi^2$, this takes the familiar form

$$(5.48) \qquad (\lambda + \omega - 1)^2 = \lambda\omega^2\mu^2,$$

which is the special case $p = 2$ of the functional equation (4.18) of Theorem 4.3. Since $M$ is Hermitian and convergent,

$$-\rho(M) \leq \mu \leq \rho(M),$$

and if we set

$$\omega = \omega_b = \frac{2}{1 + \sqrt{1 - \rho^2(M)}},$$

then we know from Sec. 4.3 that the solutions $\lambda$ of (5.48) are of modulus $\omega_b - 1$. From this it follows that $\lambda = (\omega_b - 1)e^{\pm 2i\theta}$, where $\cos\theta = \mu/\rho(M)$. With these restrictions on $\mu$ and $\omega$, we have that

$$(5.49) \qquad \phi(\mu) = (\omega_b - 1)^{1/2}e^{\pm i\theta}$$

are the roots of (5.47) and with this expression, we see from (5.46) that

$$(5.50) \qquad \alpha_m(\mu) = (\omega_b - 1)^{m/2}\left\{\frac{\sin\left[(m+1)\theta\right]}{\sin\theta}\right\},$$

$$m \geq 0, 0 \leq \theta \leq \pi.$$

Now, with the definition of the polynomial $r_m(\mu)$ in (5.41) with $\omega = \omega_b$, the above expression for $\alpha_m(\mu)$ in (5.50) and the fact that

$$(\omega_b - 1)^{1/2} = \frac{\rho(M)}{1 + \sqrt{1 - \rho^2(M)}}$$

from (4.26'), we obtain

$$(5.51) \qquad r_m(\mu) = (\omega_b - 1)^{m/2} \left\{ (1 + \sqrt{1 - \rho^2(M)}) \left( \frac{\cos \theta \sin m\, \theta}{\sin \theta} \right) \right.$$

$$\left. - \left( \frac{\sin (m - 1)\, \theta}{\sin \theta} \right) \right\}, \qquad m \geq 0,$$

and it is now easy to deduce from this (Exercise 2) that

$$(5.52) \qquad \max_{-\rho(M) \leq \mu \leq \rho(M)} |\, r_m(\mu)\,| = |\, r_m(\pm \rho(M))\,|$$

$$= (\omega_b - 1)^{m/2} \{ 1 + m\sqrt{1 - \rho^2(M)} \}.$$

But as either $\rho(M)$ or $-\rho(M)$ is necessarily an eigenvalue of $M$, then

$$\|\, r_m(M)\,\| = \max_{1 \leq i \leq n} |\, r_m(\mu_i)\,| = \max_{-\rho(M) \leq \mu \leq \rho(M)} |\, r_m(\mu)\,|,$$

and we have thus shown that

$$(5.53) \qquad \|\, r_m(M)\,\| = (\omega_b - 1)^{m/2} \{ 1 + m\sqrt{1 - \rho^2(M)} \}, \qquad m \geq 0.$$

To show now that $r_m(x)$ is an element of the set $S_m$, we know that, for $\omega \neq 0$, $r_m(x)$ is indeed a polynomial of degree $m$, and by virtue of its definition in (5.41), it follows inductively that $r_m(+1) = 1$ for all $m \geq 0$. Finally, it is easy to see that $r_m(x) \not\equiv \tilde{p}_m(x)$ for all $m \geq 2$, and using the result of Theorem 5.1 concerning the uniqueness of the polynomials $\tilde{p}_m(x)$ for each $m \geq 0$ in solving the minimization problem of (5.17), we have

**Theorem 5.3.** *Let $M$ be an $n \times n$ Hermitian and convergent matrix, and let $\|\, \tilde{p}_m(M)\,\|$ and $\|\, r_m(M)\,\|$ be respectively the matrix norms for the matrix operators for $m$ iterations of the Chebyshev semi-iterative method (5.21), and the successive overrelaxation method (5.35). If $\hat{\varepsilon}^{(1)} \equiv M\varepsilon^{(0)}$, then*

$$(5.54) \qquad \|\, \tilde{p}_m(M)\,\| < \|\, r_m(M)\,\|, \qquad m \geq 2,$$

*and thus,*

$$(5.55) \qquad R[\tilde{p}_m(M)] > R[r_m(M)], \qquad m \geq 2.$$

*Moreover,*

$$(5.55') \qquad \lim_{m \to \infty} R[\tilde{p}_m(M)] = \lim_{m \to \infty} R[r_m(M)] = \tfrac{1}{2} \{ -\ln (\omega_b - 1) \}.$$

We remark that (5.55') follows directly from (5.26') and (5.53). Note that assuming only that $M$ is Hermitian and convergent, one obtains asymptotic convergence rates for the Chebyshev semi-iterative method which are very close to those given in Theorem 4.12 for the more restricted set $S$ of matrices of Sec. 4.4.

Of course, other variants of the successive overrelaxation iterative method of (5.35) based on different assumptions concerning the initial errors $\hat{\varepsilon}^{(m)}$, $m = 0, 1$ (e.g., $\hat{\varepsilon}^{(1)} = -\varepsilon^{(0)}$), can be similarly compared (Golub-Varga (1961a)) with the Chebyshev semi-iterative method. The basic idea which is used in such comparisons is again the uniqueness of the polynomials $\tilde{p}_m(x)$ in solving the minimization problem (5.17).

Summarizing, the Chebyshev semi-iterative method of (5.21) can be viewed simply as a *variant* of the successive overrelaxation iterative method applied to the expanded matrix equations of (5.28), in which the relaxation factors vary with iteration. While this Chebyshev iterative method does not give improvements in the asymptotic rate of convergence (cf. (5.55')), it nevertheless gives improved *average* rates of convergence.

In the next section, we shall look at comparisons of average rates of convergence for different iterative methods, assuming that the matrix $M$ is a weakly cyclic Hermitian matrix.

## EXERCISES

1. Verify (5.46) and (5.47).

2. Directly verify the sequence of steps leading from Equations (5.50) through (5.52).

3. Show that the sequence of matrix norms $\| r_m(M) \|$ of (5.53) is strictly decreasing for $m \geq 0$.

4. By means of (5.53) and (5.24), directly verify that

$$\| \tilde{p}_m(M) \| < \| r_m(M) \|$$

for all $m \geq 2$.

5. Consider the variant in which we assume that $\varepsilon^{(1)} = -\varepsilon^{(0)}$. Show (Golub and Varga (1961a)) by means of a similar analysis that the corresponding matrix operator $s_m(M)$ for $m$ iterations of the successive overrelaxation iterative method is such that

$$\| s_m(M) \| = (\omega_b - 1)^{(m-1)/2}\{| m | + | m - 1 | \cdot \sqrt{\omega_b - 1}\}, \qquad m \geq 0.$$

Show also that this sequence of positive constants can actually be *initially increasing* for $\rho(M)$ sufficiently close to unity.

**5.3.** Comparison of Average Rates of Convergence: Weakly Cyclic Case

In this section, we assume that $M$ is an $n \times n$ Hermitian and convergent matrix which is of the special form

(5.56)
$$M = \left[ \begin{array}{c|c} O & F \\ \hline F^* & O \end{array} \right],$$

where the null diagonal submatrices of $M$ are square. In other words, $M$ is assumed to be in the normal form of a matrix which is weakly cyclic of index 2. In considering the successive overrelaxation iterative method for solving

$$(I - M)\mathbf{x} = \mathbf{g},$$

we partition the vectors $\mathbf{x}$ and $\mathbf{g}$ relative to the partitioning of (5.56), giving the pair of equations

(5.57)
$$\mathbf{x}_1 = F\mathbf{x}_2 + \mathbf{g}_1,$$

$$\mathbf{x}_2 = F^*\mathbf{x}_1 + \mathbf{g}_2.$$

The successive overrelaxation iterative method is then defined by

(5.58)
$$\mathbf{x}_1^{(m+1)} = \omega\{F\mathbf{x}_2^{(m)} + \mathbf{g}_1 - \mathbf{x}_1^{(m)}\} + \mathbf{x}_1^{(m)},$$

$$\mathbf{x}_2^{(m+1)} = \omega\{F^*\mathbf{x}_1^{(m+1)} + \mathbf{g}_2 - \mathbf{x}_2^{(m)}\} + \mathbf{x}_2^{(m)}, \qquad m \geq 0.$$

The assumptions we have made concerning the matrix $M$ are such that the results of Sec. 4.3 apply directly to (5.58). Indeed, we know that the constant

$$\omega_b = \frac{2}{1 + \sqrt{1 - \rho^2(M)}}$$

gives the fastest asymptotic rate of convergence of the successive overrelaxation iterative method of (5.58) and specifically,

$$R_\infty(\mathcal{L}_{\omega_b}) = -\ln (\omega_b - 1),$$

where $\mathcal{L}_{\omega_b}$ is the associated $n \times n$ successive overrelaxation iteration matrix.

The assumption of the cyclic nature of $M$ dispenses with the need for the artifice in Sec. 5.2 of doubling the size of the matrices in question so as to permit a rigorous application of the successive overrelaxation theory. On the other hand, if we now apply the Chebyshev semi-iterative method of Sec. 5.1 without modifications to the matrix equation $\mathbf{x} = M\mathbf{x} + \mathbf{g}$, we would obtain from (5.26′) an asymptotic rate of convergence

$$\lim_{m \to \infty} R[\tilde{p}_m(M)] = -\tfrac{1}{2} \ln (\omega_b - 1),$$

which is now *smaller* than the asymptotic rate of convergence of the successive overrelaxation iterative method by a factor of 2. Moreover, such a direct application of the Chebyshev semi-iterative method suffers from the practical problem of requiring additional vector storage.

To modify appropriately the Chebyshev semi-iterative method in this cyclic case, first consider the direct application of the Chebyshev semi-iterative method of (5.21) to the matrix equation $\mathbf{x} = M\mathbf{x} + \mathbf{g}$. With partitioning of the vector iterates, this becomes

(5.59)

$$\mathbf{x}_1^{(m+1)} = \omega_{m+1}\{F\mathbf{x}_2^{(m)} + \mathbf{g}_1 - \mathbf{x}_1^{(m-1)}\} + \mathbf{x}_1^{(m-1)},$$

$$\mathbf{x}_2^{(m+1)} = \omega_{m+1}\{F^*\mathbf{x}_1^{(m)} + \mathbf{g}_2 - \mathbf{x}_2^{(m-1)}\} + \mathbf{x}_2^{(m-1)}, \qquad m \geq 1,$$

where $\mathbf{x}_1^{(1)} = F\mathbf{x}_2^{(0)} + \mathbf{g}_1$ and $\mathbf{x}_2^{(1)} = F^*\mathbf{x}_1^{(0)} + \mathbf{g}_2$, and these equations determine the vector sequences $\{\mathbf{x}_1^{(m)}\}_{m=0}^{\infty}$ and $\{\mathbf{x}_2^{(m)}\}_{m=0}^{\infty}$. But, it is interesting that the proper subsequences $\{\mathbf{x}_1^{(2m+1)}\}_{m=1}^{\infty}$ and $\{\mathbf{x}_2^{(2m)}\}_{m=1}^{\infty}$ can be generated from

(5.60)

$$\mathbf{x}_1^{(2m+1)} = \omega_{2m+1}\{F\mathbf{x}_2^{(2m)} + \mathbf{g}_1 - \mathbf{x}_1^{(2m-1)}\} + \mathbf{x}_1^{(2m-1)}, \qquad m \geq 1,$$

$$\mathbf{x}_2^{(2m+2)} = \omega_{2m+2}\{F^*\mathbf{x}_1^{(2m+1)} + \mathbf{g}_2 - \mathbf{x}_2^{(2m)}\} + \mathbf{x}_2^{(2m)}, \qquad m \geq 0,$$

where again $\mathbf{x}_1^{(1)} = F\mathbf{x}_2^{(0)} + \mathbf{g}_1$. This process is indicated schematically in Figure 13 below. We shall call this particular iterative method the *cyclic*

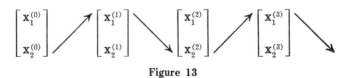

**Figure 13**

*Chebyshev semi-iterative method*, since it makes use of the cyclic nature of the matrix $M$. For practical purposes note that this iterative method differs from the application of the successive overrelaxation iterative method of (5.58) *only* in the use of a sequence $\omega_i$ of parameters, as opposed to a single fixed parameter such as $\omega_b$ in (5.58). Also, whereas the successive overrelaxation iterative method of (5.58) requires initial vector approximations of $\mathbf{x}_1$ and $\mathbf{x}_2$, this new semi-iterative method requires but a single vector approximation to $\mathbf{x}_2$. Moreover, this shows that there are no essential differences between these iterative methods as far as vector storage or arithmetic computations are concerned.

It is now intuitively clear from Figure 13 that the cyclic Chebyshev semi-iterative method, by virtue of its skipping half of the vector iterates of the Chebyshev semi-iterative method, must have an asymptotic rate of convergence equal to twice that of the Chebyshev semi-iterative method, namely $-\ln{(\omega_b - 1)}$, and although this is now the same as the asymptotic rate of convergence of the successive overrelaxation iterative method of (5.58), we are again interested in comparisons of their respective average rates of convergence for $m$ iterations for *all* $m \geq 0$. Such a comparison can indeed be carried out, but the essential theoretical steps used in this cyclic case to carry forth this comparison are somewhat different from those encountered in the previous section.

For the cyclic Chebyshev semi-iterative method, we partition the error vectors $\tilde{\boldsymbol{\varepsilon}}^{(m)}$ of the Chebyshev semi-iterative method relative to the cyclic form of the matrix $M$ in (5.56), and by virtue of the definition of the cyclic Chebyshev semi-iterative method in (5.60), the error vector after $m$ iterations of this method is

$$(5.61) \qquad \boldsymbol{\delta}^{(m)} \equiv \begin{pmatrix} \tilde{\boldsymbol{\varepsilon}}_1^{(2m-1)} \\ \tilde{\boldsymbol{\varepsilon}}_2^{(2m)} \end{pmatrix}, \qquad m \geq 1,$$

where $\tilde{\boldsymbol{\varepsilon}}_1^{(0)} = \boldsymbol{\varepsilon}_1^{(0)}$ and $\tilde{\boldsymbol{\varepsilon}}_2^{(0)} = \boldsymbol{\varepsilon}_2^{(0)}$ are the vector components of the initial error vector $\boldsymbol{\varepsilon}^{(0)}$. We now make two useful observations. First, it is readily seen from the recurrence relation (5.18′) for the Chebyshev polynomials that the polynomials $\tilde{p}_m(x)$ of (5.19) for $b = -a = \rho(M)$ of odd degree are odd, whereas those of even degree are even, so that we can represent the polynomials $\tilde{p}_m(x)$ by means of

$$(5.62) \qquad \tilde{p}_{2m}(x) = r_m(x^2); \qquad \tilde{p}_{2m+1}(x) = x \cdot s_m(x^2), \qquad m \geq 0.$$

Second, the weakly cyclic form of the matrix $M$ in (5.56) allows us to express the powers of $M$ simply as

$$M^{2m} = \left[ \begin{array}{c|c} (FF^*)^m & 0 \\ \hline 0 & (F^*F)^m \end{array} \right]; \qquad M^{2m+1} = \left[ \begin{array}{c|c} 0 & (F^*F)^m F \\ \hline (F^*F)^m F^* & 0 \end{array} \right],$$

$$m \geq 0.$$

Since $\tilde{\boldsymbol{\varepsilon}}^{(m)} = \tilde{p}_m(M)\,\boldsymbol{\varepsilon}^{(0)}$, these observations, with the definition of the error vector $\delta^{(m)}$ in (5.61), allow us to write

$$(5.63) \qquad \delta^{(m)} \equiv \tilde{P}_m(M) \cdot \varepsilon^{(0)}, \qquad m \geq 1,$$

where

$$(5.64) \qquad \tilde{P}_m(M) = \left[ \begin{array}{c|c} O & s_{m-1}(FF^*) \cdot F \\ \hline O & r_m(F^*F) \end{array} \right], \qquad m \geq 1.$$

In this particular form, direct computation of the matrix product

$$(\tilde{P}_m(M))^* \cdot \tilde{P}_m(M)$$

coupled with the definitions of (5.62) shows that

$$(5.65) \qquad \| \tilde{P}_m(M) \| = \{ \| \tilde{p}_{2m-1}(M) \|^2 + \| \tilde{p}_{2m}(M) \|^2 \}^{1/2}, \qquad m \geq 1,$$

where we have already evaluated $\| \tilde{p}_m(M) \|$ in (5.24). With (5.24), we conclude that this sequence of matrix norms must be *strictly decreasing*. On the other hand, it has been shown by Sheldon (1959) that

$$(5.66) \qquad \| \mathcal{L}_{\omega_b}^m \| = \left( \frac{2m}{\rho(M)} + \sqrt{\frac{4m^2}{\rho^2(M)} + 1} \right) \cdot (\omega_b - 1)^m, \qquad m \geq 0,$$

where $\mathcal{L}_{\omega_b}$ is the $n \times n$ successive overrelaxation matrix of (5.58). This shows that the norms $\| \mathcal{L}_{\omega_b}^m \|$ can be initially *increasing* for $\rho(M)$ sufficiently close to unity. It was Sheldon's idea (Sheldon (1959)) that the simple device of using $\omega = 1$ for the first complete iteration of (5.58) and using $\omega = \omega_b$ for all subsequent relaxation factors of the successive overrelaxation iterative method would improve upon the behavior of the matrix norms $\| \mathcal{L}_{\omega_b}^m \|$, and indeed he showed that

$$(5.67) \qquad \| \mathcal{L}_{\omega_b}^{m-1} \cdot \mathcal{L}_1 \| = \{ t_{2m-1}^2 + t_{2m}^2 \}^{1/2}, \qquad m \geq 1,$$

where

$$(5.67') \qquad t_m = (\omega_b - 1)^{(m-1)/2} \rho(M) \cdot (1 + (m-1)\sqrt{1 - \rho^2(M)}),$$

$$m \geq 1;$$

it can be readily verified that the norms $\| \mathcal{L}_{\omega_b}^{m-1} \cdot \mathcal{L}_1 \|$ are strictly decreasing. As the cyclic Chebyshev semi-iterative method of (5.60) also uses $\omega = 1$ as its first relaxation factor, direct comparisons of these various norms above is indicated. These comparisons, unlike those in Sec. 5.2, do not depend solely on the uniqueness of the polynomials $\tilde{p}_m(x)$ in Theorem 5.1; instead, they involve direct algebraic comparisons of these norms, which we omit. The major result, however, is

**Theorem 5.4.** *Let M be an $n \times n$ Hermitian and convergent weakly cyclic*

*matrix of the form* (5.56) *and let* $\| \tilde{P}_m(M) \|$, $\| \mathcal{L}_{\omega_b}^m \|$, *and* $\| \mathcal{L}_{\omega_b}^{m-1} \cdot \mathcal{L}_1 \|$ *be respectively the matrix norms for the matrix operators for m iterations of the the cyclic Chebyshev semi-iterative method of* (5.60), *the successive overrelaxation iterative method of* (5.58), *and its variant by Sheldon. Then,*

$$(5.68) \qquad \| \tilde{P}_m(M) \| < \| \mathcal{L}_{\omega_b}^m \|; \qquad \| \tilde{P}_m(M) \| < \| \mathcal{L}_{\omega_b}^{m-1} \cdot \mathcal{L}_1 \|,$$

$$m \geq 2,$$

*and thus*

$$(5.69) \qquad R[\tilde{P}_m(M)] > R_m(\mathcal{L}_{\omega_b}); \qquad R[\tilde{P}_m(M)] > R[\mathcal{L}_{\omega_b}^{m-1} \cdot \mathcal{L}_1],$$

$$m \geq 2.$$

*Moreover,*

$$(5.69') \qquad \lim_{m \to \infty} R[\tilde{P}_m(M)] = \lim_{m \to \infty} R(\mathcal{L}_{\omega_b}^m) = \lim_{m \to \infty} R[\mathcal{L}_{\omega_b}^{m-1} \cdot \mathcal{L}_1]$$

$$= -\ln (\omega_b - 1).$$

*Summarizing, the cyclic Chebyshev semi-iterative method of* (5.60) *can be viewed as a variant of the successive overrelaxation iterative method of* (5.58), *which gives improved average rates of convergence.*

Several comments are in order now. The comparison of the different iterative methods just completed rests firmly on the assumptions that the matrix $M$ is weakly cyclic of index 2 and in the precise form of (5.56), and that our basis for comparison is in terms of the spectral norms of the associated matrix operators. From our discussions in Chapter 4, the matrix $M$ of (5.56) is certainly consistently ordered, but it is only one of many consistent orderings. Restricting ourselves for the moment to the special set of matrices $S$ in Sec. 4.4, we recall that in terms of asymptotic rates of convergence, all consistent orderings are equally attractive. Nevertheless, numerical results† persistently indicate that there are considerable differences between various consistent orderings as far as norm behavior is concerned, and it must be stated that the particular consistent ordering of (5.56) is not always convenient to use in practical computations.

Again, from the point of view of practical computations, the vectors norms $\| x \|_1$ and $\| x \|_\infty$ are arithmetically far simpler to employ than the Euclidean vector norm $\| x \|$. Thus, the extension of the above results of comparisons of various iterative methods to different matrix norms would be most desirable.

† For example, see Forsythe and Wasow (1960), pp. 259–260 and Golub and Varga (1961b).

## EXERCISES

1. Using (5.62), verify that the norm of the matrix $\tilde{P}_m(M)$ of (5.64) is given by (5.65).

2. By direct computation from (5.65), show that

$$\lim_{m\to\infty} R[\tilde{P}_m(M)] = -\ln(\omega_b - 1).$$

3. From (5.65) and (5.66), verify that $\|\tilde{P}_m(M)\| < \|\mathcal{L}_{\omega_b}^m\|$ for all $m \geq 2$ and all $0 < \rho(M) < 1$.

4. With the $n \times n$ convergent matrix $M$ in the form (5.56), show that the Gauss-Seidel iteration matrix $\mathcal{L}_1$ obtained from the special case $\omega = 1$ in (5.58) is such that

$$\|\mathcal{L}_1^m\| = \rho^{2m-1}(M)\{1 + \rho^2(M)\}^{1/2}, \qquad m \geq 1,$$

which is a strictly decreasing sequence of positive numbers. [*Hint*: Compute the norms $\|\mathcal{L}_1^m\|$ by using (5.72').]

5. Prove that $\|\mathcal{L}_{\omega_b}^m\|/\|\tilde{P}_m(M)\| = \beta_m$ is strictly increasing with $m$ and that $\beta_m = 0(m)$ as $m \to \infty$. Does the ratio $\beta_m/m$ have a limit as $m \to \infty$?

6. From (5.67) and (5.67'), verify that the norms $\|\mathcal{L}_{\omega_b}^{m-1} \cdot \mathcal{L}_1\|$ are strictly decreasing with increasing $m \geq 1$.

## 5.4. Cyclic Reduction and Related Iterative Methods

In this section we shall tie together several theoretical observations and results which are closely related to the results of the previous sections of this chapter. Again, we assume that $M$ is an $n \times n$ Hermitian and convergent matrix of the weakly cyclic form

$$(5.70) \qquad M = \left[\begin{array}{c|c} O & F \\ \hline F^* & O \end{array}\right],$$

where the diagonal submatrices of $M$ are square and null. Applying the Gauss-Seidel iterative method to the solution of the matrix equation $\mathbf{x} = M\mathbf{x} + \mathbf{g}$, the vector iterates of this method are defined by

$$\mathbf{x}_1^{(m+1)} = F\mathbf{x}_2^{(m)} + \mathbf{g}_1,$$

$$(5.71)$$

$$\mathbf{x}_2^{(m+1)} = F^*\mathbf{x}_1^{(m+1)} + \mathbf{g}_2, \qquad m \geq 0,$$

or equivalently,

(5.72)
$$\mathbf{x}^{(m+1)} = \mathcal{L}_1 \mathbf{x}^{(m)} + \mathbf{k}, \qquad m \geq 0,$$

where

(5.72')
$$\mathcal{L}_1 = \left[ \begin{array}{c|c} O & F \\ \hline O & F^*F \end{array} \right].$$

As the nonzero eigenvalues $\lambda_i$ of $\mathcal{L}_1$ are just the squares of the nonzero eigenvalues of $M$, it follows that

(5.73)
$$0 \leq \lambda_i \leq \rho^2(M), \qquad 1 \leq i \leq n.$$

There is nothing to stop us from applying semi-iterative methods to the basic iterative method of (5.72), and in fact this has been done by several authors (Varga (1957a), Sheldon (1959, 1960), and Wachspress (1955)). Theoretically, the determination of the associated spectral norms $\| p_m(\mathcal{L}_1) \|$ follows the pattern given in the previous section.

What interests us here is a closed related iterative method. We know by Frobenius' Theorem 2.6 that the $k$th power of a matrix which is weakly cyclic of index $k$ is *completely reducible*. From (5.70), it is obvious by direct multiplication that the form of the matrix

(5.74)
$$M^2 = \left[ \begin{array}{c|c} FF^* & O \\ \hline O & F^*F \end{array} \right]$$

is predicted by the special case $k = 2$ of that theorem. Moreover, the matrix equation $\mathbf{x} = M\mathbf{x} + \mathbf{g}$ is equivalent to the solution of the *uncoupled* equations

(5.75)
$$\mathbf{x}_1 = FF^*\mathbf{x}_1 + (F\mathbf{g}_2 + \mathbf{g}_1),$$
$$\mathbf{x}_2 = F^*F\mathbf{x}_2 + (F^*\mathbf{g}_1 + \mathbf{g}_2).$$

We shall refer to the equations of (5.75) as the *cyclic reduction* of the matrix equation $\mathbf{x} = M\mathbf{x} + \mathbf{g}$. Since the vectors $F\mathbf{g}_2 + \mathbf{g}_1$ and $F^*\mathbf{g}_1 + \mathbf{g}_2$ can be formed once at the beginning of an iterative method, we now focus our attention on the solution of the reduced matrix problem

(5.76)
$$\mathbf{x}_2 = F^*F\mathbf{x}_2 + \mathbf{h}, \qquad \mathbf{h} \equiv F^*\mathbf{g}_1 + \mathbf{g}_2,$$

where $\mathbf{h}$ is given. Clearly, $F^*F$ is an $r \times r$ Hermitian and non-negative definite convergent matrix, whose eigenvalues $\lambda_i$ satisfy $0 \leq \lambda_i \leq \rho^2(M)$. Thus, the solution $\mathbf{x}_2$ of (5.76) is unique, and having found the vector

$x_2$ (or an excellent approximation to it), we can simply form the vector $x_1$ from

$$x_1 = Fx_2 + g_1,$$

which is a matrix multiplication and vector addition. Note that in this way it is *not* necessary to solve the corresponding system of equations for $x_1$ from (5.75). As $F^*F$ is Hermitian and convergent, we now directly apply the Chebyshev semi-iterative method of Sec. 5.1 to the solution of (5.76). In terms of the matrix polynomials $\tilde{p}_m(F^*F)$ of (5.19), the bounds for the eigenvalue spectrum of $F^*F$ are such that we set $a = 0$, $b = \rho^2(M)$, and in this case

(5.77)
$$\tilde{p}_m(F^*F) = \frac{C_m\left(\dfrac{2F^*F}{\rho^2(M)} - I\right)}{C_m\left(\dfrac{2}{\rho^2(M)} - 1\right)}, \qquad m \geq 0.$$

Again, if $\check{\varepsilon}^{(m)}$ are the error vectors of the iterates $x_2^{(m)}$ which are approximations to the solution (5.76), then

(5.78)
$$\varepsilon_2^{(m)} = \tilde{p}_m(F^*F)\varepsilon_2^{(0)}, \qquad m \geq 0,$$

and thus

(5.78′)
$$\| \check{\varepsilon}^{(m)} \| \leq \| \tilde{p}_m(F^*F) \| \cdot \| \varepsilon_2^{(0)} \|,$$

where an easy extension of the results of Sec. 5.1 shows that

(5.79)
$$\| \tilde{p}_m(F^*F) \| = \frac{1}{C_m\left(\dfrac{2}{\rho^2(M)} - 1\right)}$$

$$= (\omega_b - 1)^m \left(\frac{2}{1 + (\omega_b - 1)^{2m}}\right), \qquad m \geq 0.$$

In comparing these spectral norms with our previous results, we must keep in mind that the error vectors under consideration in this iterative method have *fewer* components than those considered previously. The comparable error vectors for the cyclic Chebyshev semi-iterative method of Sec. 5.3 are seen from (5.61) to be precisely the partitioned vectors

$$\delta_2^{(m)} = \tilde{\varepsilon}_2^{(2m)}, \qquad m \geq 0,$$

and from (5.63) and (5.64), we find that

(5.80)
$$\delta_2^{(m)} = \tilde{p}_{2m}(M)\varepsilon_2^{(0)}, \qquad m \geq 0.$$

But, as we have evaluated the spectral norm of the $\bar{p}_m(M)$ in (5.24), we observe that

$$|| \bar{p}_{2m}(M) || = (\omega_b - 1)^m \left\{ \frac{2}{1 + (\omega_b - 1)^{2m}} \right\}, \quad m \geq 0,$$

which is thus *identical* with the result of (5.79). In other words, the Chebyshev semi-iterative method applied to the cyclically reduced equation (5.76) gives *precisely* the same sequence of spectral norms as does the cyclic Chebyshev semi-iterative method applied to the partitioned error vectors of (5.80). Thus, a comparison of corresponding average rates of convergence favors neither of the methods. If these two methods are under consideration for actual computations, the final decision as to which method is preferable may come from practical considerations such as ease of arithmetic computation and vector storage. The iterative method based on the cyclic reduction of the matrix $M$ does have the advantage that a vector with fewer components is sought, although the matrix $F^*F$ will not in general be as sparse as $M$ if the product $F^*F$ is formed explicitly. But, from (5.72') the direct application of the Gauss-Seidel matrix $\mathcal{L}_1$ in (5.71) *implicitly* forms the product $F^*F$.

Finally, we make use of a result (Theorem 2.6) about the primitive character of the diagonal submatrices of the completely reduced powers of a cyclic matrix. Let us assume that the matrix $M$ of (5.70) is in addition *non-negative* and *irreducible*. Then, we know that $F^*F$ is a *primitive* square matrix. In fact, as no row or column of $M$ can be identically zero since $M$ is irreducible, it is easy to verify from the symmetry of $M$ that all the diagonal entries of $M^2$ are positive. However, this gives rise to an improved basic iterative method by means of the regular splitting theory. Let

$$C \equiv I_r - F^*F.$$

With $D_1 = I_r$, $E_1 = F^*F$, we see that $C = D_1 - E_1$ is a regular splitting of $C$. If $D_2$ is the positive diagonal matrix formed from the diagonal entries of $C$, and $E_2$ is defined from

$$C = D_2 - E_2,$$

this is also a regular splitting of $C$, where $O \leq E_2 \leq E_1$, equality excluded. As the matrix $C^{-1} > O$, we then conclude from Theorem 3.15 that

$$0 < \rho(D_2^{-1}E_2) < \rho(D_1^{-1}E_1) < 1.$$

The point of this observation is that semi-iterative methods can now be applied to the matrix $M \equiv D_2^{-1}E_2$ with corresponding improvements in rates of convergence. As an illustration, consider the simple $4 \times 4$ matrix

$M$ below, which was first described in Sec. 1.2. Squaring the matrix $M$ gives

$$M = \frac{1}{4}\left[\begin{array}{cc|cc} 0 & 0 & 1 & 1 \\ 0 & 0 & 1 & 1 \\ \hline 1 & 1 & 0 & 0 \\ 1 & 1 & 0 & 0 \end{array}\right]; \qquad M^2 = \frac{1}{16}\left[\begin{array}{cc|cc} 2 & 2 & 0 & 0 \\ 2 & 2 & 0 & 0 \\ \hline 0 & 0 & 2 & 2 \\ 0 & 0 & 2 & 2 \end{array}\right].$$

The matrix $F^*F$ is such that $\rho(F^*F) = \frac{1}{4}$ by inspection. On the other hand,

$$I_2 - F^*F = \frac{1}{8}\left[\begin{array}{cc} 7 & -1 \\ -1 & 7 \end{array}\right] = \frac{7}{8}I_2 - \frac{1}{8}\left[\begin{array}{cc} 0 & 1 \\ 1 & 0 \end{array}\right].$$

In this case, $\rho(D_2^{-1}E_2) = \frac{1}{7}$, and the ratio of the asymptotic convergence rates of $D_2^{-1}E_2$ and $F^*F$ is 1.404, indicating an improvement.

### EXERCISES

**1.** Verify that

$$\| \tilde{p}_m(F^*F) \| = (\omega_b - 1)^m \left\{ \frac{2}{1 + (\omega_b - 1)^{2m}} \right\}, \qquad m \geq 0.$$

**2.** Let

$$M = \frac{\alpha}{2}\left[\begin{array}{cc|cc} 0 & 0 & 1 & 1 \\ 0 & 0 & 1 & 1 \\ \hline 1 & 1 & 0 & 0 \\ 1 & 1 & 0 & 0 \end{array}\right], \qquad 0 < \alpha < 1.$$

From the matrix $M^2$, which determines $F^*F$, evaluate the norms $\| p_m(F^*F) \|$ of (5.79) for each $m \geq 0$. If $D$ denotes the positive diagonal matrix formed from the diagonal entries of $I - F^*F$, set $I - F^*F \equiv D - E$, which is a regular splitting of $I - F^*F$. Apply the Chebyshev semi-iterative method of Sec. 5.1 to the iteration matrix $D^{-1}E$, and compute the norms $\| \tilde{p}_m(D^{-1}E) \|$ for each $m \geq 0$ from (5.24). In particular, how do the asymptotic rates of convergence of these iterative methods compare as a function of $\alpha$?

**3.** With the particular polynomials $\tilde{p}_m(t)$ of (5.19) with $a = 0$, $b = \rho^2(M)$, determine the norms

$$\| \tilde{p}_m(\mathcal{L}_1) \|$$

for all $m \geq 0$, where the matrix $\mathcal{L}_1$ is given in (5.72').

## Bibliography and Discussion

**5.1.** The descriptive term *semi-iterative method* for the procedure of (5.6) was introduced by Varga (1957a). But the idea of considering polynomials in the matrix $M$ of (5.1) to accelerate convergence goes back some years. For example, Forsythe (1953) calls (5.6) *linear acceleration*. Von Neumann, also, had earlier considered such an approach. See Blair, et al (1959).

Closely related to this is the formation of polynomials in the matrix $A$, where one seeks to solve the matrix problem $A\mathbf{x} = \mathbf{k}$. See Exercise 6. Originally considered by Richardson (1910), the optimization of this procedure by means of Chebyshev polynomials to solve systems of linear equations has been discussed by many authors and is sometimes called *Richardson's method*. See Young (1954b), (1956a), Shortley (1953), Lanczos (1952), (1956), (1958), Milne (1953, p. 187), and Bellar (1961). Matrix Chebyshev polynomials have also been used in eigenvalue problems. See Flanders and Shortley (1950), Lanczos (1950), (1956), and Stiefel (1956). For other orthogonal polynomials and norms and their applications to matrix theory, see Stiefel (1958). Richardson's method, as developed by Young (1954b), (1956a), does not make use of the three-term recurrence relationship (5.18') for Chebyshev polynomials. Compared with the Chebyshev semi-iterative method, Richardson's method has the advantage of requiring less machine storage. On the other hand, it has the disadvantage of being numerically unstable.

The origin of Theorem 5.1 goes back to Markoff (1892), but the proof we have given was formulated by Flanders and Shortley (1950). Achieser (1956) considers the more general problem of approximation by rational functions.

The particular form (5.21) defining the Chebyshev semi-iterative method is also closely related to what is called the *second-order Richardson iterative method*. See Frankel (1950), Riley (1954), and Wachspress, Stone, and Lee (1958). In terms of norms and average rates of convergence, it is proved in Golub and Varga (1961a) that the Chebyshev semi-iterative method is superior to this latter iterative method.

The determination of the matrix norms of (5.24) was formulated by Golub (1959), and Golub and Varga (1961a).

**5.2.** The idea of essentially doubling the order of a matrix problem in order to obtain a weakly cyclic (of index 2) iteration matrix goes back to Lanczos (1956, p. 190) and Aitken (1950). Their basic purpose, however, was to obtain Hermitian matrices. The remaining material of this section is taken from Golub and Varga (1961a).

**5.3.** The material covered in this section is basically taken from Golub and Varga (1961b). See also Golub (1959) and Sheldon (1959), (1960). The distinction that has been made in Sec. 5.2 and Sec. 5.3 between the weakly cyclic case and the general Hermitian case is theoretically important. In earlier papers, such as Varga (1957a) and Young (1956a), this distinction was not made, and the results comparing asymptotic rates of convergence of the successive overrelaxation iteration method and the Chebyshev semi-iterative method of (5.21) naturally favored the successive overrelaxation iterative method.

Numerical results comparing the cyclic Chebyshev semi-iterative method and the successive overrelaxation iterative method with optimum relaxation factor are given in Golub and Varga (1961b). For recent numerical results for the Chebyshev semi-iterative method of (5.21), see W. Frank (1960).

**5.4.** The application of Chebyshev polynomials to accelerate the Gauss-Seidel iterative method has been considered by Wachspress (1955), Varga (1957a), and Sheldon (1959), (1960). For the application of semi-iterative methods to the successive overrelaxation matrix $\mathcal{L}_\omega$ for $\omega > 1$, see Varga (1957a).

The idea of directly squaring an iteration matrix to produce faster iterative methods is certainly not new. For example, it is mentioned by Geiringer (1949), who in turn refers to Hotelling (1943). Schröder (1954) however, was apparently the first one to couple the weakly cyclic nature of the matrix $M$ of (5.70) with the complete reducibility of $M^2$ for Laplace-type elliptic difference equations. For $M$ symmetric and irreducible, he also pointed out that $M^2$ had positive diagonal entries, and that this could possibly lead to improved convergence rates for point iterative methods. But if $M$ is also non-negative, our development shows, using the regular splitting theory, that such improvements *must* exist. Extensions of the cyclic reduction method to block methods and numerical results have been given recently by Hageman (1962).

The result that the Chebyshev semi-iterative method, applied to the cyclically reduced equation (5.76), gives precisely the same sequence of spectral norms as does the cyclic Chebyshev semi-iterative method and is apparently new.

# CHAPTER 6

# DERIVATION AND SOLUTION OF ELLIPTIC DIFFERENCE EQUATIONS

## 6.1. A Simple Two-Point Boundary-Value Problem

The basic aims in this chapter are to derive finite difference approximations for certain elliptic differential equations, to study the properties of the associated matrix equations, and to describe rigorous up-to-date methods for the solution of such matrix equations. In certain cases, the derived properties of the associated matrix equations are sufficiently powerful to allow us to prove (Theorem 6.2) rather easily that these finite difference approximations become more accurate with finer mesh spacings, but results in this area, especially in higher dimensions, require detailed developments which we omit.† Rather, in this chapter we concentrate on useful methods for deriving finite difference approximations as well as the treatment of boundary-value problems with internal interfaces, which are of interest in reactor physics.

To best understand different methods for deriving finite difference approximations, we look closely in this section at the following second-order self-adjoint ordinary differential equation,

$$(6.1) \qquad -\frac{d^2y(x)}{dx^2} + \sigma(x)y(x) = f(x), \qquad a < x < b,$$

subject to the two-point boundary conditions

$$(6.2) \qquad\qquad y(a) = \alpha; \qquad y(b) = \beta.$$

† For such developments, see Collatz (1960), p. 348, Forsythe and Wasow (1960), p. 283, and Kantorovich and Krylov (1958), p. 231.

Here, $\alpha$ and $\beta$ are given real constants, and $f(x)$ and $\sigma(x)$ are given real continuous functions on $a \le x \le b$, with

$$\sigma(x) \ge 0.$$

For simplicity, we place a *uniform* mesh of size

$$h = \frac{(b - a)}{N + 1}$$

on the interval $a \le x \le b$, and we denote the mesh points of the discrete problem by

$$x_j = a + j \frac{(b - a)}{N + 1} = a + jh, \qquad 0 \le j \le N + 1,$$

as illustrated in Figure 14.

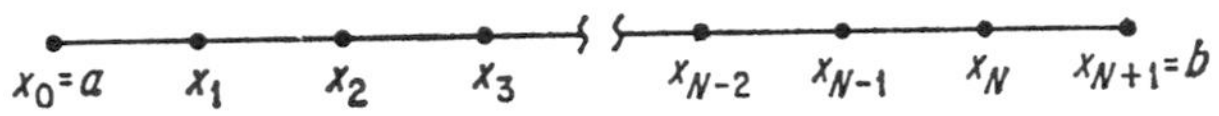

Figure 14

Perhaps the best-known method for deriving finite difference approximations to (6.1) is based on finite Taylor's series expansions of the solution $y(x)$ of (6.1). Specifically, let us assume that the (unique) solution† $y(x)$ of (6.1)–(6.2) is of class $C^4$ in $a \le x \le b$, i.e., $y^{(4)} \equiv d^4y/dx^4$ exists and is continuous in this interval. Denoting $y(x_i)$ by $y_i$, the finite Taylor expansion of $y_{i\pm1}$ is

$$y_{i\pm1} = y_i \pm hy_i^{(1)} + \frac{h^2}{2!} y_i^{(2)} \pm \frac{h^3}{3!} y_i^{(3)} + \frac{h^4}{4!} y^{(4)}(x_i + \theta_i^{\pm}h),$$

$$0 < |\theta_i^{\pm}| < 1,$$

from which it follows‡ that

$$-y_i^{(2)} = \frac{2y_i - y_{i-1} - y_{i+1}}{h^2} + \frac{h^2}{12} y^{(4)}(x_i + \theta_i h); \qquad 0 \le |\theta_i| < 1.$$

Using (6.1), we then obtain

$$(6.3) \qquad \frac{2y_i - y_{i+1} - y_{i-1}}{h^2} + \sigma_i y_i = f_i - \frac{h^2}{12} y^{(4)}(x_i + \theta_i h),$$

$$1 \le i \le N.$$

---

† For a proof of existence and uniqueness, see, for example, Henrici (1962), p. 347.
‡ We have made use of the continuity of $y^{(4)}(x)$ by writing

$$y^{(4)}(x_i + \theta_i^+h) + y^{(4)}(x_i + \theta_i^-h) = 2y^{(4)}(x_i + \theta_i h).$$

Note that as $y_0 = \alpha$ and $y_{N+1} = \beta$, we have $N$ equations for the $N$ unknowns $y_i$. In matrix notation, we can write this equivalently in the form

$$(6.4) \qquad\qquad A\mathbf{y} = \mathbf{k} + \mathbf{\tau}(y),$$

where $A$ is a real $N \times N$ matrix, and $\mathbf{y}$, $\mathbf{k}$, and $\mathbf{\tau}$ are column vectors, given specifically by

$$(6.5) \qquad A = \frac{1}{h^2} \begin{bmatrix} 2 + \sigma_1 h^2 & -1 & & & \mathbf{O} \\ -1 & 2 + \sigma_2 h^2 & -1 & & \\ & & \ddots & & -1 \\ \mathbf{O} & & & -1 & 2 + \sigma_N h^2 \end{bmatrix},$$

and

$$(6.5') \qquad \mathbf{y} = \begin{bmatrix} y_1 \\ y_2 \\ \cdot \\ \cdot \\ \cdot \\ y_N \end{bmatrix}; \quad \mathbf{k} = \begin{bmatrix} f_1 + \dfrac{\alpha}{h^2} \\ f_2 \\ \cdot \\ \cdot \\ f_N + \dfrac{\beta}{h^2} \end{bmatrix}; \quad \mathbf{\tau} = \frac{-h^2}{12} \begin{bmatrix} y^{(4)}(x_1 + \theta_1 h) \\ y^{(4)}(x_2 + \theta_2 h) \\ \cdot \\ \cdot \\ \cdot \\ y^{(4)}(x_N + \theta_N h) \end{bmatrix}.$$

Basic properties of the matrix $A$ are easily recognized from (6.5). The matrix $A$ is real, symmetric, and tridiagonal, and as $\sigma(x) \geq 0$ by assumption, then $A$ is, from Definition 1.7, *diagonally dominant* with positive diagonal entries. Furthermore, the directed graph of $A$ (Figure 15)

$G(A):$

**Figure 15**

is clearly strongly connected, implying that $A$ is irreducible, and as there is strict diagonal dominance in the first and last rows of $A$, we can state more precisely that $A$ is *irreducibly diagonally dominant*. But from the Corollary of Theorem 1.8, $A$ is then positive definite. As the off-diagonal entries of $A$ are nonpositive, $A$ is an irreducible *Stieltjes matrix*,† so that $A^{-1} > 0$.

† See Definition 3.4 and Corollary 3 of Sec. 3.5.

The point Jacobi matrix $B = (b_{i,j})$ derived from $A$ is given explicitly by

(6.6)

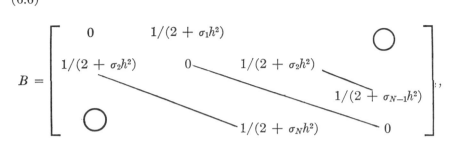

$$B = \begin{bmatrix} 0 & 1/(2 + \sigma_1 h^2) & & & \\ 1/(2 + \sigma_2 h^2) & 0 & 1/(2 + \sigma_2 h^2) & & \\ & & & & 1/(2 + \sigma_{N-1} h^2) \\ & & 1/(2 + \sigma_N h^2) & & 0 \end{bmatrix},$$

which is clearly non-negative, irreducible, and tridiagonal, with row sums satisfying $\sum_{j=1}^{N} b_{i,j} \leq 1$, $1 \leq i \leq N$, with strict inequality for $i = 1$ and $i = N$. But then (Theorem 1.7), $B$ is necessarily *convergent*.

Other properties of the matrix $B$ are also evident. The results of Sec. 4.1 can now be directly applied, since $B$ is a special case of the matrix of (1.8), which was shown to be weakly cyclic of index 2 and consistently ordered. It is more instructive, however, to establish these properties by means of graph theory. The directed graph of $B$ is shown in Figure 16,

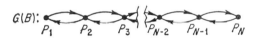

$$G(B):$$

$$P_1 \quad P_2 \quad P_3 \quad P_{N-2} \quad P_{N-1} \quad P_N$$

Figure 16

and as the greatest common divisor of the lengths of the closed paths of $G(B)$ is two, then $B$ is *cyclic of index* 2. Continuing, the directed graph $\hat{G}(B)$ of type 2 (see page 121) is illustrated in Figure 17 below,

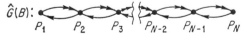

$$\hat{G}(B):$$

$$P_1 \quad P_2 \quad P_3 \quad P_{N-2} \quad P_{N-1} \quad P_N$$

Figure 17

and as every closed path of $\hat{G}(B)$ contains the same number of major and minor paths, then we see graphically that $B$ is *consistently ordered*. Moreover (Exercise 5 of Sec. 6.2), $B$ has real eigenvalues. Combining these results, we thus have proved

**Theorem 6.1.** *Let $A$ be the $N \times N$ matrix of* (6.5), *derived from the*

*boundary-value problem of* (6.1)–(6.2), *where* $\sigma(x)$ *of* (6.1) *is non-negative.
Then,*

1. *A is real and symmetric, with positive diagonal entries and nonpositive
   off-diagonal entries. Moreover, A is irreducibly diagonally dominant,
   so that A is positive definite.*
2. *A is a tridiagonal irreducible Stieltjes matrix, and thus $A^{-1} > 0$.*
3. *If B is the $N \times N$ point Jacobi matrix derived from A, then B is a
   non-negative, irreducible, consistently ordered cyclic matrix of index 2
   with real eigenvalues, and $\rho(B) < 1$.*

Although we have confined our attention to the simple two-point boundary
value problem in (6.1)–(6.2), the detailed results derived in Theorem 6.1
do essentially extend (Theorems 6.3 and 6.4) to more general boundary-
value problems in one and more spatial dimensions.

We now define the solution vector **z** of

$$(6.7) \qquad\qquad A\mathbf{z} = \mathbf{k}$$

as our *discrete approximation* to the solution $y(x)$ of (6.1)–(6.2). Note
that this approximation is obtained simply by neglecting the vector $\boldsymbol{\tau}(y)$ in
(6.4). The particular result of Theorem 6.1 that establishes that the matrix
$A^{-1}$ is positive is interesting, not only because it connects with our developments of Chapter 2, but because it plays a fundamental role in the proof
below of *convergence*, due to Gerschgorin (1930), of the discrete approximation **z** to the solution $y(x)$ as $h \to 0$. By convergence, we mean that the
difference between the discrete approximation $z_i$ and $y(x_i)$ can be made
arbitrarily small for all $0 \le i \le N + 1$ simply by choosing the mesh spacing
$h$ sufficiently small, as in (6.8').

   **Theorem 6.2.** *Let $y(x)$ be the solution of the two-point boundary-value
problem of* (6.1)–(6.2), *where $\sigma(x)$ is non-negative, and let its discrete ap-
proximation* **z** *be defined by* (6.7). *If $|y^{(4)}(x)| \le M_4$ for all $a \le x \le b$,
then*

$$(6.8) \qquad |y_i - z_i| \le \frac{M_4 h^2}{24} [(x_i - a)(b - x_i)],$$

$$x_i = a + ih = a + \frac{i(b-a)}{N+1}, \qquad 0 \le i \le N + 1.$$

*Thus,*

$$(6.8') \qquad \max_{1 \le i \le N} |y_i - z_i| \equiv \|\mathbf{y} - \mathbf{z}\|_\infty \le \frac{M_4 h^2 (b-a)^2}{96}.$$

*Proof.* It necessarily follows from (6.7) and (6.4) that

$$\mathbf{y} - \mathbf{z} = A^{-1}\boldsymbol{\tau}(y).$$

Recalling the notation (Definition 2.1) that $|\mathbf{x}|$ is the non-negative vector with components $|x_i|$, we then deduce that

$$|\mathbf{y} - \mathbf{z}| = |A^{-1}\boldsymbol{\tau}(y)| \leq A^{-1}|\boldsymbol{\tau}(y)|,$$

since $A^{-1} > O$ from Theorem 6.1. From (6.5'), the hypotheses show that $|\tau_i(y)| \leq M_4 h^2/12$. Next, if $A_0$ is the $N \times N$ matrix of (6.5) with $\sigma \equiv 0$, then $A_0^{-1}$ is also a positive matrix from Theorem 6.1, which dominates (Exercise 9 of Sec. 3.5) $A^{-1}$:

$$O < A^{-1} \leq A_0^{-1}.$$

Combining these results gives

$$|\mathbf{y} - \mathbf{z}| \leq \frac{M_4 h^2}{12} A_0^{-1}\boldsymbol{\xi},$$

where $\boldsymbol{\xi}$ is the vector whose components are all unity. In other words, if $\mathbf{w}$ is the unique solution of $A_0 \mathbf{w} = \boldsymbol{\xi}$, then

$$|y_i - z_i| \leq \frac{M_4 h^2}{12} w_i, \qquad 1 \leq i \leq N.$$

But it is a simple matter to verify that $w(x) \equiv (x - a)(b - x)/2$, $a \leq x \leq b$, when evaluated at the mesh points $x_i$, is the solution of $A_0\mathbf{w} = \boldsymbol{\xi}$, which proves (6.8). To complete the proof, observe that

$$\max_{a \leq x \leq b} w(x) = w\left(\frac{b + a}{2}\right) = \frac{(b - a)^2}{8},$$

which proves (6.8').

It is interesting to point out that (6.8') can also be established by directly computing an upper bound for $\| A^{-1} \|_\infty$ (see Exercise 4).

Thus far we have used Taylor's series as a means of generating finite difference approximations of (6.1)–(6.2). We now describe other methods for deriving such finite difference approximations. We observe that by integrating (6.1) from $\eta_1$ to $\eta_2$, the solution $y(x)$ of (6.1)–(6.2) necessarily satisfies the equation

$$(6.9) \qquad \frac{-dy(\eta_2)}{dx} + \frac{dy(\eta_1)}{dx} + \int_{\eta_1}^{\eta_2} \sigma(x)y(x)\,dx = \int_{\eta_1}^{\eta_2} f(x)\,dx,$$

$$a \leq \eta_1 \leq \eta_2 \leq b.$$

In particular, choose $\eta_2 = x_{i+1/2} \equiv x_i + h/2$ and $\eta_1 = x_{i-1/2} \equiv x_i - h/2$ and approximate (by central differences) the terms of (6.9) by

(6.10)
$$\frac{dy(x_{i+1/2})}{dx} \doteq \frac{y_{i+1} - y_i}{h}$$

and

(6.10')
$$\int_{x_{i-1/2}}^{x_{i+1/2}} g(x)\,dx \doteq g_i \cdot (x_{i+1/2} - x_{i-1/2}) = g_i h.$$

In this way, we are led to

$$\frac{2y_i - y_{i+1} - y_{i-1}}{h} + h\sigma_i y_i = hf_i + h\tilde{\tau}_i(y), \qquad 1 \le i \le N,$$

which, upon dividing by $h$, simply reduces to

$$A\mathbf{y} = \mathbf{k} + \tilde{\boldsymbol{\tau}}(y),$$

using (6.5) and (6.5'). In other words, the matrix equation (6.7) can also be deduced from discrete approximations to (6.9). We also mention that since the matrix $A$ and the vectors $\mathbf{k}$ and $\mathbf{y}$ are necessarily the same as those in (6.4), we conclude that the corresponding vectors $\tilde{\boldsymbol{\tau}}(y)$ above and $\boldsymbol{\tau}(y)$ of (6.4) must be identical.

The third method for deriving difference approximations to (6.1) is based on a *variational formulation* of the solution of (6.1)–(6.2). Specifically, let $w(x)$ be any function whose derivative is piecewise continuous† in $a \le x \le b$, satisfying the boundary conditions $w(a) = \alpha$, $w(b) = \beta$, and consider the functional of $w(x)$:

(6.11)
$$F(w) \equiv \frac{1}{2} \int_a^b \left\{ \left(\frac{dw(x)}{dx}\right)^2 + \sigma(x)\,(w(x))^2 - 2w(x)f(x) \right\} dx.$$

If we write $w(x) \equiv y(x) + \epsilon(x)$, where $y(x)$ is the solution of (6.1)–(6.2), then simply substituting this sum in (6.11) yields

$$F(w) = F(y) + \int_a^b \left\{ \frac{dy}{dx}\frac{d\epsilon}{dx} + \sigma y\epsilon - \epsilon f \right\} dx + \frac{1}{2} \int_a^b \left\{ \left(\frac{d\epsilon}{dx}\right)^2 + \sigma(\epsilon)^2 \right\} dx.$$

By integrating the first term of the first integral by parts, we can then write the first integral as

$$\epsilon(b)\frac{dy(b)}{dx} - \epsilon(a)\frac{dy(a)}{dx} + \int_a^b \epsilon(x) \left\{ -\frac{d^2y}{dx^2} + \sigma y - f \right\} dx;$$

---

† The function $f(x)$ is *piecewise continuous* in $a \le x \le b$ if it is continuous at all but a finite number of points in this interval, and possesses right- and left-hand limits at all interior points of the interval and appropriate one-sided limits at the end points.

but the integral above surely vanishes from (6.1). Since $w(x)$ and $y(x)$ satisfy the same boundary conditions of (6.2), then $\epsilon(a) = \epsilon(b) = 0$, and we finally conclude that

$$(6.12) \qquad F(w) = F(y) + \frac{1}{2} \int_a^b \left\{ \left(\frac{d\epsilon(x)}{dx}\right)^2 + \sigma(x)(\epsilon(x))^2 \right\} dx \geq F(y).$$

Thus, the solution $y(x)$ of (6.1)–(6.2) *minimizes* $F(w)$ for all sufficiently smooth functions $w(x)$ satisfying the boundary conditions $w(a) = \alpha$, $w(b) = \beta$.

There are two different ways of utilizing the preceding results for the functional $F(w)$. The first is directly to approximate numerically the functional $F(w)$, which we express as the sum

$$(6.11') \qquad F(w) = \frac{1}{2} \sum_{i=0}^N \int_{x_i}^{x_{i+1}} \left\{ \left(\frac{dw(x)}{dx}\right)^2 + \sigma(x)(w(x))^2 - 2w(x)f(x) \right\} dx.$$

Making central difference approximations to the integrals of the form

$$\int_{x_i}^{x_{i+1}} \left(\frac{dw(x)}{dx}\right)^2 dx \doteq \left(\frac{w_{i+1} - w_i}{h}\right)^2 \cdot (x_{i+1} - x_i) = \frac{(w_{i+1} - w_i)^2}{h}$$

and

$$\int_{x_i}^{x_{i+1}} g(x)\, dx \doteq \left(\frac{g_{i+1} + g_i}{2}\right) \cdot (x_{i+1} - x_i) = \left(\frac{g_{i+1} + g_i}{2}\right) h$$

leads to the discrete approximation $\tilde{F}(\mathbf{w})$, defined as

$$(6.13) \qquad \tilde{F}(\mathbf{w}) \equiv \frac{1}{2} \sum_{i=0}^N \left\{ \frac{(w_{i+1} - w_i)^2}{h} + \left(\frac{\sigma_{i+1} w_{i+1}^2 + \sigma_i w_i^2}{2}\right) h \right.$$

$$\left. - (w_{i+1} f_{i+1} + w_i f_i) h \right\}.$$

Using the explicit form of the matrix $A$ and the vector $\mathbf{k}$ in (6.5) and (6.5'), it can be verified (Exercise 7) that we can express $2\tilde{F}(\mathbf{w})/h$ as the *quadratic form*:

$$(6.14) \qquad \frac{2\tilde{F}(\mathbf{w})}{h} = \mathbf{w}^T A \mathbf{w} - 2\mathbf{w}^T \mathbf{k} + \frac{\beta^2}{h^2} + \frac{\alpha^2}{h^2} + \frac{1}{2}\{\sigma(1)\beta^2 + \sigma(0)\alpha^2\}$$

$$- (\beta f(1) + \alpha f(0)),$$

where $\mathbf{w}$ is a vector with arbitrary components $w_1, \cdots, w_N$. If $\mathbf{w} \equiv \mathbf{z} + \boldsymbol{\varepsilon}$

where $A\mathbf{z} = \mathbf{k}$, then it follows from the real symmetric and positive definite nature of the matrix $A$ that

$$(6.15) \qquad \frac{2\tilde{F}(\mathbf{z} + \boldsymbol{\varepsilon})}{h} = \frac{2\tilde{F}(\mathbf{z})}{h} + \boldsymbol{\varepsilon}^T A \boldsymbol{\varepsilon} \geq \frac{2\tilde{F}(\mathbf{z})}{h},$$

which is the discrete analogue of the minimization property of (6.12). From this, we see that minimizing the discrete functional $\tilde{F}(\mathbf{w})$ by means of

$$\frac{\partial \tilde{F}(\mathbf{w})}{\partial w_j} = 0, \qquad 1 \leq j \leq N,$$

leads us again precisely to the solution of the matrix equation

$$A\mathbf{z} = \mathbf{k},$$

whose properties have already been discussed.

It is interesting to mention at this point that our finite difference approximation $A\mathbf{z} = \mathbf{k}$ of the two-point boundary value problem of (6.1)–(6.2) bears a close relationship with the *Ritz Method*[†] applied to the functional $F(w)$ of (6.11). Specifically, let us now restrict our class of admissible functions to *continuous piecewise linear* functions $w(x)$, where $w(a) = \alpha$ and $w(b) = \beta$. With $w_1, \cdots, w_N$ as arbitrary real numbers, define

$$w(x) = w_j + \left(\frac{w_{j+1} - w_j}{h}\right)(x - x_j), \qquad x_j \leq x \leq x_{j+1}, \qquad 0 \leq j \leq N,$$

as shown in Figure 18.

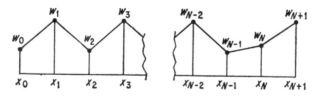

**Figure 18**

In this case, the functional $F(w) = F(w_1, w_2, \cdots, w_N)$ of (6.11) is a quadratic form in the parameters $w_1, \cdots, w_N$, which can be minimized as a function of these parameters by solving

$$\frac{\partial F(w)}{\partial w_i} = 0, \qquad 1 \leq i \leq N.$$

[†] See Collatz (1960), pp. 202–208, or Courant-Hilbert (1953), p. 175.

This results (Exercise 8) in the system of linear equations

$$(6.16) \quad \left(\frac{w_i - w_{i+1}}{h}\right) + \left(\frac{w_i - w_{i-1}}{h}\right) + \int_{x_{i-1}}^{x_{i+1}} \sigma(x)w(x)t_i(x)\,dx$$

$$= \int_{x_{i-1}}^{x_{i+1}} f(x)t_i(x)\,dx, \quad 1 \le i \le N,$$

where $t_i(x)$ is the continuous piecewise linear function defined by

$$t_i(x_j) = \begin{cases} 1, & j = i \\ 0, & j \ne i \end{cases}, \quad 0 \le j \le N + 1.$$

We express the totality of linear equations as the matrix equation

$$(6.16') \qquad\qquad \mathcal{Q}\mathbf{w} = \tilde{\mathbf{k}}.$$

Of course, if $\sigma(x) \equiv 0$, we see from (6.16) that the matrix $(1/h)\mathcal{Q}$ is identical with the matrix $A$ of (6.5). In general, $\mathcal{Q}$ will differ from $A$, but all the essential properties possessed by $A$, such as being real, symmetric, positive definite, and tridiagonal, are also shared by $\mathcal{Q}$ (see Exercise 8). Because the entries of $\mathcal{Q}$ and the components of $\tilde{\mathbf{k}}$ depend on various integrals of $\sigma(x)$ and $f(x)$, it must be mentioned that the practical derivation of the matrix equation of (6.16') for such one-dimensional problems requires more in the way of initial calculations.

To summarize, the three different methods of deriving finite difference approximations of (6.1)–(6.2), based on Taylor's series, integration, and the variational formulation of this problem, can be used to produce (with the exception of the Ritz form of the variational method) *identical* matrix equations, and it would appear difficult to choose among these three methods of derivation. Yet, each has distinct advantages. The Taylor's series method seems indispensible in deducing the order of approximation of the discrete methods, and is applicable to differential equations in general. Although it is not apparent from the formulation above, the variational principle is especially attractive in that *nonessential* boundary conditions appear as a direct consequence of the minimum property of the functional equation.† Next, from (6.13) we see that the quadratic form $\mathbf{w}^T A \mathbf{w}$ of the discrete functional $\tilde{F}(\mathbf{w})$ in (6.14) can be expressed as a sum of squares, and this is often useful in establishing the positive definite character of the matrix of the discrete approximation. Unfortunately, the variational method is based on the self-adjoint property of the differential equation (6.1), and thus is not applicable to general differential equations. The integration technique, on the other hand, is applicable

---

† This is treated at some length in Collatz (1960), p. 205, Kantorovich and Krylov (1958), p. 244, and Forsythe and Wasow (1960), p. 182. See also Exercise 8 of Sec. 6.2.

in the general case, and gives particularly simple discretizations to problems with internal interfaces and nonuniform meshes, which we will discuss next.

To illustrate more clearly these different discretization methods, such as the Ritz form of the variational method, we have included numerical results for a sample one-dimensional problem in Appendix A. The reader will find it instructive to derive these different matrix discretizations.

## EXERCISES

**1.** If the solution $y(x)$ of (6.1)–(6.2) is of class $C^3$ on $a \leq x \leq b$, show that the corresponding vector $\tau(y)$ of (6.4) has components $\tau_i$ satisfying†

$$\tau_i = O(h), \quad 1 \leq i \leq N, \quad h \rightarrow 0.$$

Similarly, if $y(x)$ is only of class $C^2$, show that $\tau_i = o(1)$, $1 \leq i \leq N$, $h \rightarrow 0$.

**2.** If the solution $y(x)$ of (6.1)–(6.2) is of class $C^3$ on $0 \leq x \leq 1$, show that the following analogue of (6.8) is valid:

$$| y_i - z_i | \leq \frac{M_3 h}{6} [x_i(1 - x_i)], \quad x_i = \frac{i}{N+1}, \quad 0 \leq i \leq N + 1,$$

so that

$$\| \mathbf{y} - \mathbf{z} \|_\infty \leq \frac{M_3 h}{24}, \quad M_3 \equiv \max_{0 \leq x \leq 1} | y^{(3)}(x) |.$$

Similarly, if $y(x)$ is of class $C^2$, show that $\| \mathbf{y} - \mathbf{z} \|_\infty = o(1)$, $h \rightarrow 0$.

**3.** If $\sigma(x) \equiv 0$ in (6.1), verify that the inverse of the $N \times N$ matrix $A$ of (6.5) is given explicitly by $A_0^{-1} = (r_{i,j})$ where

$$r_{i,j} = r_{j,i} \quad \text{and} \quad r_{j,i} = \frac{i(N + 1 - j)h^2}{N+1}, \quad i \leq j$$

(Todd (1950) and Rutherford (1952)).

Using the results of Exercise 2, Sec. 1.3, show then that

$$\| A_0^{-1} \|_1 = \| A_0^{-1} \|_\infty = \begin{cases} \dfrac{(m + 1)^3 h^2}{(2m + 2)} & \text{if } N = 2m + 1 \text{ is odd.} \\[3ex] \dfrac{m(m + 1)h^2}{2} & \text{if } N = 2m \text{ is even,} \end{cases}$$

and conclude that $\| A_0^{-1} \|_1 = \| A_0^{-1} \|_\infty \leq (b - a)^2/8$ for all $h > 0$.

† We say that $f(x) = O(g(x))$ as $x \rightarrow a$ if $|f(x)| \leq Mg(x)$ as $x \rightarrow a$. Similarly, $f(x) = o(g(x))$ as $x \rightarrow a$ if $\lim_{x \rightarrow a} f(x)/g(x) = 0$.

4. If $\sigma(x)$ is *any* non-negative continuous function in (6.1), using the results of Exercise 9, Sec. 3.5, show that

$$|| A^{-1} ||_1 = || A^{-1} ||_\infty \leq (b - a)^2/8,$$

where the $N \times N$ matrix $A$ is defined in (6.5). Starting with

$$\mathbf{y} - \mathbf{z} = A^{-1}\boldsymbol{\tau}(\mathbf{y}),$$

directly establish the result (6.8') of Theorem 6.2, using norm inequalities.

5. If $\sigma(x) \equiv \sigma \geq 0$, show directly from (6.5) that

$$|| A^{-1} || = \frac{h^2}{\sigma h^2 + 4 \sin^2 \left[ \pi h/2(b - a) \right]} \leq \frac{h^2}{4 \sin^2 \left[ \pi h/2(b - a) \right]}$$

and establish that $|| A^{-1} || \sim (b - a)^2/\{\pi^2 + \sigma(b - a)^2\}$ as $h \to 0$.

6. If the vector $\mathbf{z}$ has components $z_i$, $1 \leq i \leq N$, it is convenient to define the (normalized) vector norms

$$|| \mathbf{z} ||_{(p)} \equiv \left( \frac{1}{N} \sum_{i=1}^{N} | z_i |^p \right)^{1/p}, \qquad p \geq 1.$$

Assume that the solution $y(x)$ of (6.1)–(6.2) is of class $C^4$ on $0 \leq x \leq 1$. With the results of Exercises 4 and 5, show that

$$|| \mathbf{y} - \mathbf{z} ||_{(2)} \leq \frac{M_4 h^4}{48 \, (\sin \pi h/2)^2} = O(h^2), \qquad h \to 0,$$

and

$$|| \mathbf{y} - \mathbf{z} ||_{(1)} \leq \frac{M_4 h^2}{96}, \qquad M_4 \equiv \max_{0 \leq x \leq 1} | y^{(4)}(x) |.$$

How do these results compare with bounds for $|| \mathbf{y} - \mathbf{z} ||_{(2)}$ and $|| \mathbf{y} - \mathbf{z} ||_{(1)}$ computed directly from (6.8)?

7. From the definition of the discrete functional $\tilde{F}(w)$ in (6.13), verify the expressions of (6.14) and (6.15) using (6.5) and (6.5').

*8. Assuming $w(x)$ to be piecewise linear as in Figure 18, verify that $\partial F/\partial w_i = 0$ leads to the expression of (6.16). Next, show that the matrix $\mathcal{A}$ of (6.16') is real, symmetric, positive definite, and tridiagonal for all $h > 0$. Moreover, show that $\mathcal{A}$ is an irreducible Stieltjes for all $h$ sufficiently small. Is $\mathcal{A}$ always irreducible for $h > 0$? If the diagonal entries of $\mathcal{A}$ are all $2/h^2$ when $\sigma(x) \equiv 0$, and $y(x)$, the solution of (6.1)–(6.2), is of class $C^4$, show that $\mathcal{A}\mathbf{y} = \tilde{\mathbf{k}} + \boldsymbol{\tau}(\mathbf{y})$, where each $\tau_i = O(h^2)$, $h \to 0$.

9. Consider the differential equation of (6.1) with $\sigma \equiv 0$, $f(x) \equiv 2$, subject to the boundary conditions $y(0) = 3$, $y(1) = 2$, whose solution is clearly

$y(x) = 3 - x^2$. Show that $y_i = y(x_i) = 3 - x_i^2$ is also a solution of the system of linear equations

$$\frac{2y_i - (y_{i+1} + y_{i-1})}{h^2} = f_i = 2, \qquad 1 \le i \le N,$$

where $y_0 = 3$, $y_{N+1} = 2$, and $h = 1/(N + 1)$. Now, let the vector $\mathbf{w}$ with components $w_i$ be the solution of the perturbed system of linear equations

$$\frac{2w_i - (w_{i+1} + w_{i-1})}{h^2} = 2 + 2\left(1 + \cos\left(\frac{\pi}{N+1}\right)\right)(-1)^i \sin\left(\frac{\pi i}{N+1}\right),$$

$$1 \le i \le N,$$

where $w_0 = 3$ and $w_{N+1} = 2$. Whereas the corresponding vector $\boldsymbol{\tau}$ of (6.4) has components given by

$$\tau_i = 2\left(1 + \cos\left(\frac{\pi}{N+1}\right)\right)(-1)^i \sin\left(\frac{\pi i}{N+1}\right) = O(1) \quad \text{as} \quad h \to 0,$$

nevertheless show that $\| \mathbf{y} - \mathbf{w} \|_\infty = h^2$ for all $h > 0$. (*Hint*: What are the eigenvectors of the matrix $A$ of (6.5) in this case?)

**10.** For the special case

$$-\frac{d^2y}{dx^2} + y(x) = -x, \qquad 0 < x < 1$$

where $y(0) = 0$, $y(1) = 1$, and $h = \frac{1}{4}$, determine explicitly the $3 \times 3$ matrix $\mathcal{A}$ and the associated matrix equation of (6.16'). (This can be checked in Appendix A.)

## 6.2. General Second-Order Ordinary Differential Equations

In this section, we derive finite difference approximations of general second-order ordinary differential equations. Specifically, consider

(6.17)   $$\frac{-d}{dx}\left\{p(x)\frac{dy(x)}{dx}\right\} + q(x)\frac{dy(x)}{dx} + \sigma(x)y(x) = f(x),$$

$$a < x < b,$$

subject to the general two-point boundary conditions

(6.18)   $$\alpha_1 y(a) - \beta_1 y'(a) = \gamma_1; \quad \alpha_2 y(b) + \beta_2 y'(b) = \gamma_2.$$

It is well known that if $p'(x)$, $q(x)$, $\sigma(x)$, and $f(x)$ are continuous in $a \le x \le b$, then it is possible to reduce† (6.17) by means of a change of

† See, for example, Kantorovich and Krylov (1958), p. 218.

independent and dependent variables to the simple form of (6.1). Numerically, this reduction is often not at all convenient to use. Moreover, in many problems arising from reactor physics and engineering, $p$, $q$, $\sigma$, and $f$ are only *piecewise continuous*, and actually possess a finite number of discontinuities in the interval of interest $a \leq x \leq b$. Physically, such problems arise from steady-state diffusion problems for heterogeneous materials, and the points of discontinuity represent *interfaces* between successive homogeneous compositions. For such problems, we assume that $y(x)$ and $p(x) \, dy/dx$ are both continuous in $a \leq x \leq b$. Since $p(x)$ itself is only piecewise continuous, this latter assumption implies that

$$(6.19) \qquad p(t+) \frac{dy(t+)}{dx} = p(t-) \frac{dy(t-)}{dx}, \quad a < t < b$$

at all interior points $t$ of $a \leq x \leq b$, where in general $g(t+)$ and $g(t-)$ denote respectively the right- and left-hand limits of $g(x)$ at $x = t$. Finally, we assume that the given quantities of (6.17) and (6.18) satisfy

$$(6.20) \qquad p(x) > 0, \quad \sigma(x) > 0, \quad a \leq x \leq b,$$

and

$$(6.21) \qquad \alpha_i \geq 0, \quad \beta_i \geq 0, \quad \alpha_i + \beta_i > 0, \quad i = 1, 2.$$

Because of the similarity of results for the different methods of derivation, our discrete approximations to (6.17)–(6.18) will be formulated, for brevity, only in analogy with the integration method of (6.9). Choose now *any* distinct set of points $a = x_0 < x_1 < \cdots < x_{N+1} = b$, subject only to the requirement that the discontinuities of the given functions $p$, $q$, $\sigma$, and $f$ be a subset of these points. We shall call these points $x_i$ the *mesh points* of the discrete problem. Note that the mesh spacings $h_i \equiv x_{i+1} - x_i$ need *not* be constant in this discussion. We integrate (6.17) over the interval

$$x_i \leq x \leq x_i + \frac{h_i}{2} \equiv x_{i+1/2},$$

which precisely gives

$$(6.22) \qquad -p_{i+1/2} \frac{dy_{i+1/2}}{dx} + p(x_i+) \frac{dy(x_i+)}{dx} + \int_{x_i}^{x_{i+1/2}} q(x) \frac{dy}{dx} \, dx$$

$$+ \int_{x_i}^{x_{i+1/2}} \sigma(x) y(x) \, dx = \int_{x_i}^{x_{i+1/2}} f(x) \, dx.$$

If $x_i$ is not a boundary mesh point, we can also integrate (6.17) over the interval $x_{i-1/2} \leq x \leq x_i$, which similarly gives

(6.22′)  $\quad -p(x_i-) \dfrac{dy(x_i-)}{dx} + p_{i-1/2} \dfrac{dy_{i-1/2}}{dx} + \displaystyle\int_{x_{i-1/2}}^{x_i} q(x) \dfrac{dy}{dx} \, dx$

$$+ \int_{x_{i-1/2}}^{x_i} \sigma(x) y(x) \, dx = \int_{x_{i-1/2}}^{x_i} f(x) \, dx.$$

Now, adding these two equations and employing the continuity condition of (6.19) results in the exact expression

(6.22″)  $\quad -p_{i+1/2} \dfrac{dy_{i+1/2}}{dx} + p_{i-1/2} \dfrac{dy_{i-1/2}}{dx} + \displaystyle\int_{x_{i-1/2}}^{x_{i+1/2}} q(x) \dfrac{dy}{dx} \, dx$

$$+ \int_{x_{i-1/2}}^{x_{i+1/2}} \sigma(x) y(x) \, dx = \int_{x_{i-1/2}}^{x_{i+1/2}} f(x) \, dx.$$

The derivative terms can be conveniently approximated by central differences. The integral terms, on the other hand, can be approximated in a variety of ways,† but basically, for simplicity, we use approximations like those of (6.10′), of the form

$$\int_{x_{i-1/2}}^{x_{i+1/2}} g(x) \, dx = \int_{x_{i-1/2}}^{x_i} g(x) \, dx + \int_{x_i}^{x_{i+1/2}} g(x) \, dx \doteq g_i^- \cdot \left(\frac{h_{i-1}}{2}\right)$$

$$+ g_i^+ \cdot \left(\frac{h_i}{2}\right).$$

Making these approximations in (6.22″) gives us

(6.23)  $\quad -p_{i+1/2} \left(\dfrac{y_{i+1} - y_i}{h_i}\right) + p_{i-1/2} \left(\dfrac{y_i - y_{i-1}}{h_{i-1}}\right) + q_i^+ \left(\dfrac{y_{i+1} - y_i}{2}\right)$

$$+ q_i^- \left(\frac{y_i - y_{i-1}}{2}\right) + y_i \left(\frac{\sigma_i^- h_{i-1} + \sigma_i^+ h_i}{2}\right)$$

$$= \left(\frac{f_i^- h_{i-1} + f_i^+ h_i}{2}\right) + \left(\frac{h_{i-1} + h_i}{2}\right) \tau_i,$$

† For example, knowing $\sigma(x)$ permits us to approximate

$$\int_{x_{i-1/2}}^{x_{i+1/2}} \sigma(x) y(x) \, dx$$

by

$$y_i \int_{x_{i-1/2}}^{x_{i+1/2}} \sigma(x) \, dx.$$

This is considered by Tikhonov and Samarskii (1956). See also Exercise 9.

which, upon omitting the final term, will be the linear equation for our discrete approximation to (6.17) for the mesh point $x_i$. The quantity $\tau_i$ again can be estimated in terms of higher derivatives of $y(x)$ and the given functions. (See Exercises 1 and 2.)

For boundary conditions, if $\beta_1 = 0$ in (6.18), then $y_0 = \gamma_1/\alpha_1$ is a specified value. On the other hand, if $\beta_1 > 0$, then $y_0$ is unknown. In this case, directly substituting the boundary condition of (6.18) in (6.22) for the case $i = 0$ and making similar discrete approximations result in

$$(6.23') \qquad -p_{1/2}\left(\frac{y_1 - y_0}{h_0}\right) + p_0^+\left(\frac{\alpha_1 y_0 - \gamma_1}{\beta_1}\right) + q_0^+\left(\frac{y_1 - y_0}{2}\right) + \frac{\sigma_0^+ y_0 h_0}{2}$$

$$= \frac{f_0^+ h_0}{2} + \frac{h_0}{2}\tau_0.$$

The analogous treatment of the right-hand boundary is evident. Thus, these various expressions can be written in the form

$$(6.24) \qquad D_i y_i - L_i y_{i-1} - U_i y_{i+1} = k_i + \left(\frac{h_{i-1} + h_i}{2}\right)\tau_i, \quad 1 \le i \le N,$$

where

$$L_i = \frac{p_{i-1/2}}{h_{i-1}} + \frac{q_i^-}{2}; \qquad U_i = \frac{p_{i+1/2}}{h_i} - \frac{q_i^+}{2},$$

$$(6.24') \qquad D_i = L_i + U_i + \frac{\sigma_i^- h_{i-1} + \sigma_i^+ h_i}{2},$$

$$k_i = \frac{f_i^- h_{i-1} + f_i^+ h_i}{2}, \qquad 1 \le i \le N.$$

Again, if $\beta_1 > 0$ in (6.18), then

$$L_0 = 0, \qquad U_0 = \frac{p_{1/2}}{h_0} - \frac{q_0^+}{2},$$

$$(6.24'') \qquad D_0 = L_0 + U_0 + \frac{p_0^+ \alpha_1}{\beta_1} + \frac{\sigma_0^+ h_0}{2},$$

$$k_0 = \frac{f_0^+ h_0}{2} + \frac{p_0^+ \gamma_1}{\beta_1}.$$

Summarizing then, we have derived a system of equations which, in matrix notation, is

$$(6.25) \qquad \hat{A}\mathbf{y} = \mathbf{k} + \hat{\tau}(y).$$

Note that $\hat{A}$ is an $m \times m$ matrix, where $m = N, N + 1$, or $N + 2$, depending on the boundary conditions.

As in the case of the matrix $A$ of (6.5), we can also quickly deduce properties of the matrix $\hat{A}$ from (6.24). Clearly, $\hat{A}$ is a real tridiagonal matrix and from the assumptions of (6.20) and (6.21), we see that the diagonal entries of $\hat{A}$ are positive, while the off-diagonal entries of $\hat{A}$ are nonpositive, *providing that the mesh spacings $h_i$ are all sufficiently small*. Assuming this to be true, then $\hat{A}$ is also irreducibly diagonally dominant, and thus (Corollary 1 of Theorem 3.11) $\hat{A}$ is an irreducible $M$-matrix with $(\hat{A})^{-1} > O$. As in Theorem 6.1, we conclude (using Theorem 3.10) that the point Jacobi matrix $\hat{B}$ associated with $\hat{A}$ is non-negative and convergent. It is also clear that the use of graph theory, as in Sec. 6.1, similarly establishes that $\hat{B}$ is a consistently ordered cyclic matrix of index 2. Moreover (Exercise 5), $\hat{B}$ has real eigenvalues. Thus, we have proved

**Theorem 6.3.** *Let $\hat{A}$ be the $m \times m$ matrix of (6.25), derived from the boundary-value problem of (6.17)–(6.18), subject to the conditions of (6.19)–(6.21). Then, for all mesh spacings sufficiently small,*

1. *$\hat{A}$ is real, with positive diagonal entries and nonpositive off-diagonal entries. Moreover, $\hat{A}$ is irreducibly diagonally dominant.*
2. *$\hat{A}$ is a tridiagonal irreducible $M$-matrix, and $(\hat{A})^{-1} > O$.*
3. *If $\hat{B}$ is the $m \times m$ point Jacobi matrix derived from $\hat{A}$, then $\hat{B}$ is a non-negative, irreducible, consistently ordered cyclic matrix of index 2 with real eigenvalues, and $\rho(\hat{B}) < 1$.*

One interesting observation we can make from the above derivation is that boundary conditions are used *in conjunction* with the differential equation, which seems desirable in general. The same is true of applications of the variational method introduced in Sec. 6.1. Another fact worth noting is that if $q(x) \equiv 0$ and $p'(x)$ and $\sigma(x)$ are continuous, then the differential equation of (6.17) is *self-adjoint*, which suggests that the associated matrix $\hat{A}$ should be symmetric in this case, even for *nonuniform* mesh spacings. However, more generally, with $p(x)$ and $\sigma(x)$ only piecewise continuous, we can deduce the symmetry of $\hat{A}$ directly from (6.24′) when $q \equiv 0$, which we state as the

**Corollary.** *Let $q(x) \equiv 0$ in (6.17). Then, for all mesh spacings, the matrix $\hat{A}$ of (6.25) is real, symmetric, and positive definite, and all the conclusions of Theorem 6.3 are valid.*

In contrast to this Corollary, it is interesting to point out that the direct use of Taylor's series for deriving finite difference approximations to (6.17) does *not* in general produce symmetric matrices for general mesh spacings for the case $q \equiv 0$. (See Exercise 1.)

Because of the existence of discontinuous coefficients in (6.17) (arising from physical interfaces), the *a priori* estimation of the error vector

$\tilde{\tau}(y)$ of (6.25) in terms of higher derivatives of $y(x)$ is now quite tenuous. For this reason, we shall not pursue extensions of Gerschgorin's Theorem 6.2.

If our goal were now to solve by iterative methods the matrix equation

$$(6.26) \qquad\qquad \hat{A}\mathbf{z} = \mathbf{k},$$

obtained from (6.25) by setting $\tilde{\tau}(y) = \mathbf{0}$, we would have many convergent methods from which to choose. Assuming sufficiently small mesh spacings, the properties of Theorem 6.3 tell us, for example, that the point Jacobi iterative method is convergent. Likewise, when $q(x) \equiv 0$, from the Corollary we know (Theorem 3.6) that the point successive overrelaxation method is convergent for any $\omega$ satisfying $0 < \omega < 2$. Practically, however, such tridiagonal matrix problems are never solved now by iterative methods, but instead are solved directly. As we will discuss in Sec. 6.4, this is useful in block iterative methods and also serves to introduce the alternating direction implicit methods of Chapter 7.

The finite difference methods we have introduced to solve two-point boundary-value problems such as (6.17)–(6.18) have been restricted to three-point approximations, that is, linear relationships among three neighboring mesh points. Clearly, more point formulas can also be derived (cf. Collatz (1960), p. 538) as finite difference approximations, which hold the possibility of greater accuracy for the same number of mesh points, but complications arise near end points. Moreover, these higher-point formulas are based on more stringent differentiability assumptions on the solution $y(x)$ of the differential equation, which are generally not fulfilled by solutions of differential problems with interfaces. However, for differential equations with smooth coefficients, repeated differentiation of the differential equations leads to the solution of *modified* three-point approximations, which can often yield amazingly accurate results with but a few mesh points. For such methods, see Exercise 7 and Appendix A.

### EXERCISES

1. Let $p(x) \in C^3$ and $y(x) \in C^4$ in $a \le x \le b$, and consider a general mesh spacing $a = x_0 < x_1 < x_2 < \cdots < x_{N+1} = b$, where $x_i = h_i + x_{i+1}$, and $p_{i+1/2} = p(x_i + h_i/2)$. If $\bar{h}_i = \max(h_i, h_{i-1})$, show by means of Taylor's series that for any $1 \le i \le N$,

$$-\frac{d}{dx}\left\{ p(x_i) \frac{dy(x_i)}{dx} \right\} = \frac{-p_{i+1/2}\left(\dfrac{y_{i+1} - y_i}{h_i}\right) + p_{i-1/2}\left(\dfrac{y_i - y_{i-1}}{h_{i-1}}\right)}{\dfrac{h_i + h_{i-1}}{2}}$$

$$+ O(\bar{h}_i)$$

if $h_i \neq h_{i-1}$, and $O(\bar{h}_i^2)$ otherwise. Next, for the case $q \equiv 0$ and $\alpha_i = 1$, $\beta_i = 0$ in (6.17)–(6.18), show that this use of Taylor's series in approximating (6.17) leads to a matrix equation similar to that of (6.25), but the associated matrix now is in general *nonsymmetric*.

2. Let $q(x) \in C^2$ and $y(x) \in C^2$, $a \leq x \leq b$. Using the notation of the previous exercise, show that

$$q(x_i) \frac{dy(x_i)}{dx} = q_i \left( \frac{y_{i+1} - y_i}{2h_i} \right) + q_i \left( \frac{y_i - y_{i-1}}{2h_{i-1}} \right) + O(\bar{h}_i),$$

if $h_i \neq h_{i-1}$, and $O(\bar{h}_i^2)$ otherwise.

3. For the differential equation of (6.1), assume that $\sigma(x) \geq 0$ for $0 \leq x \leq 1$, and that

$$\frac{dy(0)}{dx} = 0 \quad \text{and} \quad y(1) = \beta$$

are now boundary conditions. Using a uniform mesh $h = 1/(N+1)$, determine explicitly the matrix $\hat{A}$ of (6.25) of order $N+1$ in this case.

4. In place of (6.20), assume that $\sigma(x) \geq 0$ for $a \leq x \leq b$. Show that if $\min(\beta_1, \beta_2) = 0$, then all the conclusions of Theorem 6.3 and its Corollary remain valid.

5. Let $A = (a_{i,j})$ be any $n \times n$ tridiagonal matrix, i.e., $a_{i,j} = 0$ for $|i - j| > 1$, with real entries such that $a_{i,i-1} \cdot a_{i-1,i} > 0$ for all $2 \leq i \leq n$. Show that there exists a diagonal matrix $D$ with positive diagonal entries such that $DAD^{-1}$ is real and symmetric. Thus, conclude that for all $h$ sufficiently small, the matrix $\hat{A}$ of (6.25) is diagonally similar to a real symmetric and positive definite matrix, and thus has *positive* real eigenvalues. Similarly, conclude that the matrices $B$ of Theorem 6.1 and $\hat{B}$ of Theorem 6.3 (for all $h$ sufficiently small) have *real* eigenvalues.

6. Let $\hat{A}$ be the $N \times N$ matrix of (6.25), where we assume that $\beta_1 = \beta_2 = 0$ in (6.21) and that $q(x) \equiv 0$ in (6.17). If

$$p(x) \geq k > 0, \quad h_{\max} \equiv \max_{0 \leq i \leq N} \{h_i \equiv x_{i+1} - x_i\},$$

show that

$$\| (\hat{A})^{-1} \| \leq \frac{h_{\max}}{4k \sin^2 \left( \dfrac{\pi}{2(N+1)} \right)}.$$

(*Hint*: If $A_0$ is the matrix of (6.5) with $\sigma \equiv 0$, express $\hat{A} = (kh^2/h_{\max}) A_0 + R$, and show that $R$ is Hermitian and non-negative definite.)

7. For the two-point boundary-value problem of (6.1)–(6.2), assume that $\sigma(x) \equiv \sigma \geq 0$ and that $f(x)$ of (6.1) is of class $C^4$ on the interval $a \leq x \leq b$.

Using Taylor's series, derive the following improved three-point difference approximation of (6.1) by the repeated differentiations of (6.1):

$$\frac{\left(2 + \sigma h^2 + \frac{\sigma^2 h^4}{12}\right) y_i - (y_{i+1} + y_{-1})}{h^2} = \frac{1}{12} \{(12 + \sigma h^2)f_i + h^2 f_i^{(2)}\}$$

$$+ O(h^4) = \frac{1}{12} \{(10 + \sigma h^2)f_i + f_{i-1} + f_{i+1}\} + O(h^4).$$

More generally, if $f(x)$ is of class $C^{2n}$, $n \geq 2$, on the interval $a \leq x \leq b$, derive improved difference approximations of (6.1) which, similar to the above expression, couple only three successive unknowns $y_{i-1}$, $y_i$, and $y_{i+1}$. If the associated $N \times N$ tridiagonal matrix is $A_n \equiv (a_{i,j}^{(n)})$, show that

$$a_{i,i-1}^{(n)} \equiv -1/h^2 \equiv a_{i,i+1}^{(n)} \quad \text{and} \quad \lim_{n\to\infty} a_{i,i}^{(n)} = 2 \cosh (h\sqrt{\sigma})/h^2.$$

If the analogue of (6.7) for these finite difference equations is

$$A_n \mathbf{z}^{(n)} = \mathbf{k}^{(n)}, \qquad n \geq 1,$$

show, following the result (6.8) of Theorem 6.2, that

$$|y_i - z_i^{(n)}| \leq \frac{M_{2n+2}h^{2n}}{(2n+2)!} (x_i - a)(b - x_i), \quad 0 \leq i \leq N + 1,$$

and thus

$$\|\mathbf{y} - \mathbf{z}^{(n)}\|_\infty \leq \frac{M_{2n+2}h^{2n}(b-a)^2}{4(2n+2)!}, \quad M_{2n+2} \equiv \max_{a\leq x\leq b} |y^{(2n+2)}(x)|.$$

8. Assume that $q(x) \equiv 0$ and that $p'(x)$, $\sigma(x)$, and $f(x)$ of (6.17) are continuous in $a \leq x \leq b$. Moreover, assume that $p(x) > 0$ and $\sigma(x) > 0$ in $a \leq x \leq b$, and that $\alpha_i > 0$, $\beta_i > 0$ in (6.21). For all sufficiently smooth functions $w(x)$ in $a \leq x \leq b$, consider the functional

$$F(w) \equiv \frac{1}{2} \int_a^b \left\{p(x)\left(\frac{dw}{dx}\right)^2 + \sigma(x)(w(x))^2 - 2w(x)f(x)\right\} dx$$

$$+ \frac{p(a)}{2\beta_1} \{\alpha_1 w^2(a) - 2\gamma_1 w(a)\} + \frac{p(b)}{2\beta_2} \{\alpha_2 w^2(b) - 2\gamma_2 w(b)\}.$$

a. If $y(x)$ is the solution of (6.17)–(6.18) under these assumptions, show that $F(w) = F(y + \epsilon) \geq F(y)$ for all such functions $w(x)$.

b. For *any* set of distinct mesh points $x_0 = a < x_1 < \cdots < x_{N+1} = b$, approximate the integral of $F(w)$ (using the approximation methods of Sec. 6.1., leading to (6.13)), and define from this the discrete functional $\tilde{F}(\mathbf{w})$. Show that the set of linear equations

$$\frac{\partial \tilde{F}(\mathbf{w})}{\partial w_j} = 0, \qquad 0 \leq j \leq N + 1,$$

is *identical* with the matrix equation of (6.26).

9. With $q(x) \equiv 0$ and $p(x) > 0$ in (6.17), derive from (6.17) by means of integration the exact result

$$p_{i-1/2}\frac{dy_{i-1/2}}{dx} = \frac{1}{\displaystyle\int_{x_{i-1}}^{x_i} dr/p(r)}.$$

$$\left\{ y_i - y_{i-1} - \int_{x_{i-1}}^{x_i} \frac{dr}{p(r)} \int_{x_{i-1/2}}^{r} \sigma(x)y(x)\ dx + \int_{x_{i-1}}^{x_i} \frac{dr}{p(r)} \int_{x_{i-1/2}}^{r} f(x)\ dx \right\},$$

and use this in (6.22'') to obtain an *exact* expression which resembles (6.23) (Tikhonov and Samarskii (1956)).

10. Let $D$ be the $m \times m$ diagonal matrix obtained by setting all the off-diagonal entries of the matrix $\hat{A}$ of (6.25) to zero. For all mesh spacings sufficiently small, show that $\hat{A} \equiv D - C$ is a regular splitting (Sec. 3.6) of $\hat{A}$. From this, directly prove that $\rho(D^{-1}C) < 1$.

11. Using the notation of the previous exercise, let $\hat{A} = D - C$, where $\hat{A}$ is the matrix of (6.25), with $q(x) \equiv 0$ in (6.17). Show that the point Jacobi matrix $\hat{B}$ derived from $\hat{A}$ is such that $D^{+1/2}\hat{B}D^{-1/2}$ is an element of the set $S$ of Sec. 4.4.

12. Consider the problem of (6.17)–(6.18), where $p(x) \equiv 1$, $q(x) \equiv 0$, $\sigma(x) \geq 0$, and $\alpha_i = 1$, $\beta_i = 0$. If we have $h_i \equiv h(\epsilon)$ for all $i \neq j$, and $h_j = \epsilon > 0, 0 \leq i \leq N$, where $(N - 1)h(\epsilon) + \epsilon = b - a$, let $\hat{B}(\epsilon)$ be the associated point Jacobi matrix of Theorem 6.3. While Theorem 6.3 tells us that $\rho(\hat{B}(\epsilon)) < 1$, prove that for $N$ fixed,

$$\lim_{\epsilon \to 0} \rho(\hat{B}(\epsilon)) = 1.$$

## 6.3. Derivation of Finite Difference Approximations in Higher Dimensions

The different methods introduced in Sec. 6.1 for deriving finite difference approximations to ordinary differential equations can also be applied in deriving finite difference approximations to second-order self-adjoint elliptic partial differential equations in two or more spatial variables, and it would be quite accurate to say simply that higher-dimensional forms of Theorems 6.1 and 6.3 are valid. Yet, the associated geometric detail in higher-dimensional derivations requires us to look carefully at such extensions. Again, our main object here is to derive such finite difference approximations, and questions of convergence of discrete solutions to the solutions of the differential equations in higher dimensions are omitted. Consider now the second-order self-adjoint elliptic partial differential equation

$$(6.27) \qquad -(P(x, y)u_x)_x - (P(x, y)u_y)_y + \sigma(x, y)u(x, y) = f(x, y),$$

$$(x, y) \in R,$$

defined in an open, bounded, and connected set $R$ in the plane. For boundary conditions, we are given that

$$(6.28) \qquad \alpha(x, y)u + \beta(x, y)\,\frac{\partial u}{\partial n} = \gamma(x, y), \qquad (x, y) \in \Gamma,$$

where $\Gamma$, the boundary of $R$, is assumed to be sufficiently smooth. Here, $\partial u/\partial n$ refers to the *outward* pointing normal on $\Gamma$. For simplicity, we assume that the given functions $P$, $\sigma$, and $f$ of (6.27) are *continuous* in $\bar{R}$, the closure of $R$, and satisfy

$$(6.29) \qquad \left.\begin{array}{c} P(x, y) > 0 \\[2mm] \sigma(x, y) > 0 \end{array}\right\}, \qquad (x, y) \in \bar{R}.$$

Later, however, more general problems with internal interfaces will be considered. We also assume that the given functions $\alpha$, $\beta$, and $\gamma$ defined on the boundary $\Gamma$ of $R$ are piecewise continuous and satisfy

$$(6.30) \qquad \alpha(x, y) \geq 0, \quad \beta(x, y) \geq 0, \quad \alpha + \beta > 0, \quad (x, y) \in \Gamma.$$

As a concrete example, let $R$ be the quarter disk: $0 < x < 1, 0 < y < 1$, $0 < x^2 + y^2 < 1$, and consider the special case $P \equiv 1, \sigma \equiv 0.1$ of

$$(6.27') \qquad -u_{xx} - u_{yy} + 0.1u(x, y) = -4 + 0.1(x^2 + y^2),$$

$$(x, y) \in R,$$

with boundary conditions as shown in Figure 19(a).

Analogous to our discussion in Sec. 6.1 for the one-dimensional problem, the first step in deriving finite difference approximations to (6.27)–(6.28) is the definition of a rectangular spatial mesh. In the plane, draw a system of straight lines parallel to the coordinate axes:

$$x = x_0, \quad x = x_1, \cdots, x = x_n, \quad x_0 < x_1 < \cdots < x_n,$$

$$(6.31)$$

$$y = y_0, \quad y = y_1, \cdots, y = y_m, \quad y_0 < y_1 < \cdots < y_m,$$

as shown in Figure 19(b). The mesh spacings $h_i \equiv x_{i+1} - x_i$ and $k_j \equiv y_{j+1} - y_j$ need *not* be uniform in each coordinate direction. Denoting the intersection of these lines as *mesh points*, the boundary $\Gamma$ of $R$ is now approximated by the polygonal or discrete boundary $\Gamma_h$, formed by joining appropriate mesh points by line segments, as shown in Figure 19(b). We shall similarly call $R_h$ the corresponding open and connected point set

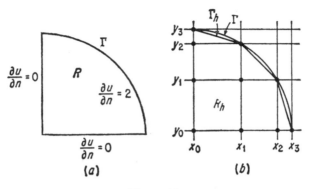

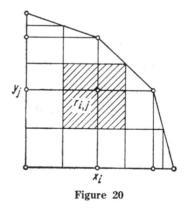

**Figure 19**

which approximates $R$. The actual mesh points of the discrete approximation are those mesh points which lie on or are interior to $\Gamma_h$. With $\Gamma_h$ and $R_h$, we now consider the numerical solution of (6.27)–(6.28), *relative* to $R_h$ and $\Gamma_h$. This implies that the boundary conditions specified for $\Gamma$ in (6.28) have also been possibly approximated in passing to $\Gamma_h$. For simplicity, assume that $\alpha$, $\beta$, and $\gamma$ are specified on $\Gamma_h$.

To derive finite difference approximations to (6.27)–(6.28), we again make use of the integration method introduced in Sec. 6.1. To this end, we associate to *each* mesh point $(x_i, y_j)$ of $R_h$ a closed *mesh region* $r_{i,j}$ defined as lying within $R_h$ and being bounded by the lines $x = x_i - h_{i-1}/2$, $x = x_i + h_i/2$, $y = y_j - k_{j-1}/2$, and $y = y_j + k_j/2$. For mesh points $(x_i, y_j)$ interior to $\Gamma_h$, these mesh regions are rectangular, whereas mesh regions for boundary points of $R_h$ are polygonal, although not always rectangular, as shown in Figure 20. For each mesh point $(x_i, y_j)$, for which $u(x_i, y_j) \equiv u_{i,j}$ is unknown, we now *integrate* (6.27) over the

**Figure 20**

corresponding mesh region $r_{i,j}$:

$$(6.32) \quad -\int\int_{r_{i,j}} \{(Pu_x)_x + (Pu_y)_y\}\, dx\, dy + \int\int_{r_{i,j}} \sigma u\, dx\, dy$$

$$= \int\int_{r_{i,j}} f\, dx\, dy.$$

By Green's Theorem,† we know that for any two differentiable functions $S(x, y)$ and $T(x, y)$ defined in $r_{i,j}$,

$$(6.33) \quad \int\int_{r_{i,j}} (S_x - T_y)\, dx\, dy = \int_{c_{i,j}} (T\, dx + S\, dy),$$

where $c_{i,j}$ is the boundary of $r_{i,j}$, and the line integral is taken in the positive sense. With $S = Pu_x$ and $T = Pu_y$, we can then write (6.32) as

$$(6.34) \quad -\int_{c_{i,j}} \{Pu_x\, dy - Pu_y\, dx\} + \int\int_{r_{i,j}} \sigma u\, dx\, dy = \int\int_{r_{i,j}} f\, dx\, dy.$$

If $g(x_i, y_j) \equiv g_{i,j}$, the double integrals above can be simply approximated by means of

$$(6.35) \quad \int\int_{r_{i,j}} g\, dx\, dy \doteq g_{i,j} \cdot a_{i,j},$$

where $a_{i,j}$ is the area of $r_{i,j}$. For the rectangular regions $r_{i,j}$ of Figure 21 (a),

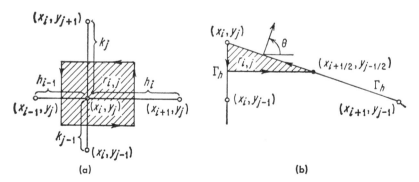

(a)　　　　　　　　(b)

**Figure 21**

this area is given by $a_{i,j} = (h_{i-1} + h_i) \cdot (k_{j-1} + k_j)/4$, but in any event, the approximations leading to $R_h$ and $\Gamma_h$ as in Figure 19 (b) are such that these areas $a_{i,j}$ are easily calculated. Referring again to Figure 21 (a),

† See, for example, Widder (1947), p. 191.

the line integral of (6.34) over the four sides of $r_{i,j}$ is approximated by means of central differences by

$$(6.36) \qquad - \int_{c_{i,j}} \{ P u_x \, dy - P u_y \, dx \}$$

$$\doteq \left( \frac{k_{j-1} + k_j}{2} \right) \left\{ P_{i+1/2,j} \left( \frac{u_{i,j} - u_{i+1,j}}{h_i} \right) + P_{i-1/2,j} \left( \frac{u_{i,j} - u_{i-1,j}}{h_{i-1}} \right) \right\}$$

$$+ \left( \frac{h_{i-1} + h_i}{2} \right) \left\{ P_{i,j+1/2} \left( \frac{u_{i,j} - u_{i,j+1}}{k_j} \right) + P_{i,j-1/2} \left( \frac{u_{i,j} - u_{i,j-1}}{k_{j-1}} \right) \right\},$$

where $P_{i+1/2,j} \equiv P(x_i + h_i/2, \, y_j)$, etc. For regions $r_{i,j}$, possibly non-rectangular as shown in Figure 21(b), stemming from mesh points on $\Gamma_h$ where $u(x_i, y_j)$ is unknown ($\beta_{i,j} > 0$ in (6.28)), the approximations of (6.36) can be used to treat those portions of the line integral of (6.34) that do not coincide with $\Gamma_h$; but on $\Gamma_h$ we suppose now that the normal to $\Gamma_h$, as shown in Figure 21(b), makes an angle $\theta$ with the positive $x$ axis. Thus, the portion of the curve $\Gamma_h$ of interest now can be described by the parameterization

$$x = x_{i+1/2} - t \sin \theta, \qquad y = y_{j-1/2} + t \cos \theta$$

and thus on $\Gamma_h$,

$$\frac{\partial u}{\partial n} = u_x \cos \theta + u_y \sin \theta.$$

The portion of the line integral of (6.34) of interest can be written as

$$- \int_0^l (P u_x \, dy - P u_y \, dx)$$

$$= - \int_0^l (P u_x \cos \theta + P u_y \sin \theta) \, dt$$

$$= - \int_0^l P \frac{\partial u}{\partial n} \, dt = - \int_0^l P \left( \frac{\gamma(t) - \alpha(t) u(t)}{\beta(t)} \right) dt$$

$$\doteq - P_{i,j} \cdot \left( \frac{\gamma_{i,j} - \alpha_{i,j} u_{i,j}}{\beta_{i,j}} \right) l, \qquad l = \tfrac{1}{2} \sqrt{h_i^2 + k_{j-1}^2},$$

where $l$ is the length of the path of integration. Note that we have directly used the boundary condition of (6.28) together with the differential equation of (6.27) in this derivation.

Summarizing, for *each* mesh point $(x_i, y_j)$ where $u_{i,j}$ is unknown, we have derived a finite difference expression of the form

(6.37)
$$D_{i,j}u_{i,j} - L_{i,j}u_{i-1,j} - R_{i,j}u_{i+1,j} - T_{i,j}u_{i,j+1} - B_{i,j}u_{i,j-1}$$

$$= s_{i,j} + \left(\frac{h_{i-1} + h_i}{2}\right)\left(\frac{k_{j-1} + k_j}{2}\right)\tau_{i,j},$$

where for mesh points $(x_i, y_j)$ of $R_h$, the quantities $D_{i,j}$, $L_{i,j}$, etc. can be deduced from (6.35) and (6.36), and are explicitly given by

$$L_{i,j} = P_{i-1/2,j}\left(\frac{k_{j-1} + k_j}{2h_{i-1}}\right), \quad R_{i,j} = P_{i+1/2,j}\left(\frac{k_{j-1} + k_j}{2h_i}\right),$$

$$T_{i,j} = P_{i,j+1/2}\left(\frac{h_{i-1} + h_i}{2k_j}\right), \quad B_{i,j} = P_{i,j-1/2}\left(\frac{h_{i-1} + h_i}{2k_{j-1}}\right),$$

(6.38)

$$D_{i,j} = L_{i,j} + R_{i,j} + T_{i,j} + B_{i,j} + \sigma_{i,j}\left(\frac{h_{i-1} + h_i}{2}\right)\left(\frac{k_{j-1} + k_j}{2}\right),$$

$$s_{i,j} = f_{i,j}\left(\frac{h_{i-1} + h_i}{2}\right)\left(\frac{k_{j-1} + k_j}{2}\right).$$

Expressions for these coefficients can be obtained similarly for mesh points $(x_i, y_j)$ on $\Gamma_h$ where $u_{i,j}$ is unknown. Because each mesh point $(x_i, y_j)$ is coupled to at most four other adjacent mesh points, it is called a *five-point formula*. We write this system of linear equations as

(6.39)
$$A\mathbf{u} = \mathbf{k} + \tilde{\tau}(u).$$

The properties of the matrix $A$ are again of interest to us. The derivation leading to (6.39) shows that $A$ is a real matrix, and as $P$ and $\sigma$ are positive in $\bar{R}$ in (6.29), then $A$ has positive diagonal entries and non-positive off-diagonal entries. In fact, it is evident from (6.38) not only that $A$ is *symmetric*, but as $\sigma > 0$ then $A$ is *strictly diagonally dominant*. For irreducibility properties of $A$, consider now the *point Jacobi matrix* $B$ derived from $A$. Again, as in Sec. 6.1, (6.38) shows us that $B$ is evidently a non-negative matrix with row sums less than unity, so that $B$ is convergent. For the discrete mesh of Figure 20, the *directed graph* of the point Jacobi matrix is given in Figure 22(a). Not only do we see from this that $B$ (and hence $A$!) is irreducible, but we also see that the greatest common divisor of the lengths of the closed paths of the directed graph $G(B)$ is two, which proves (Theorem 2.9 of Sec. 2.4) that $B$ is *cyclic of index* 2. Next, we number the mesh points where $u_{i,j}$ is unknown in Figure 22(b) from left

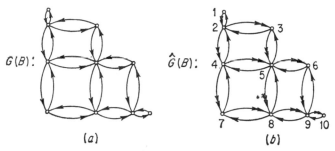

$$G(B): \qquad \hat{G}(B):$$

(a)                        (b)

**Figure 22**

to right, top to bottom, calling this the *natural ordering* of the mesh points. With this ordering, consider the *directed graph* $\hat{G}(B)$ *of type* 2 of Figure 22(b). Since each closed path has the same number of major and minor paths, then from the results of Sec. 4.4, $B$ is *consistently ordered*. Thus, almost all the conclusions drawn in Theorem 6.1 for the one-dimensional case remain valid in two dimensions. Although it is evident that the matrices $A$ and $B$ are no longer tridiagonal, there is a partitioning of $A$ which is *block tridiagonal*. In Figure 22(b), we partition the matrix $A$ by placing all mesh points of successive horizontal mesh line segments into successive sets, and since the particular derivation given here is based on the *five-point approximation*, then $A$ can be written as

(6.40)
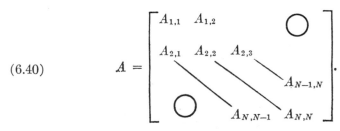

$$A = \begin{bmatrix} A_{1,1} & A_{1,2} & & & \bigcirc \\ A_{2,1} & A_{2,2} & A_{2,3} & & \\ & \ddots & \ddots & \ddots & \\ & & & & A_{N-1,N} \\ \bigcirc & & & A_{N,N-1} & A_{N,N} \end{bmatrix}.$$

Here, $N$ refers to the number of horizontal lines of the mesh $R_h$. Referring specifically to the matrix $A$ derived for the mesh configuration of Figure 22(b), the submatrices $A_{1,1}$, $A_{2,2}$, $A_{3,3}$, and $A_{4,4}$ are respectively $1 \times 1$, $2 \times 2$, $3 \times 3$, and $4 \times 4$ matrices. Evidently, each diagonal submatrix $A_{j,j}$ is tridiagonal, while the off-diagonal matrices have at most one nonzero entry per row. Combining these results gives us

**Theorem 6.4.** *Let $A$ be the $n \times n$ matrix of (6.39), derived from the boundary-value problem of (6.27)–(6.28), with the assumptions of (6.29)–(6.30). Then,*

1. *$A$ is real and symmetric, with positive diagonal entries and nonpositive off-diagonal entries. Moreover, $A$ is irreducibly diagonally dominant, so that $A$ is positive definite.*

2. *There is a partitioning* (6.40) *of A such that A is a block tridiagonal irreducible Stieltjes matrix, and thus* $A^{-1} > 0$.

3. *If B is the* $n \times n$ *point Jacobi matrix derived from A, then B is a non-negative, irreducible cyclic matrix of index 2 with real eigenvalues, and* $\rho(B) < 1$. *Moreover, for the natural ordering of the mesh points, B is consistently ordered.*

4. *If $\tilde{B}$ is the block Jacobi matrix obtained from the partitioned matrix A of* (6.40), *then $\tilde{B}$ is a non-negative, consistently ordered cyclic matrix of index 2 with* $\rho(\tilde{B}) < 1$.

There are several items in the derivation just presented that are worthy of comment. First, the irreducibility of the matrix $A$, established by graph theory, is essentially a consequence of the *connectedness* of the region $R$ of (6.27), since for all sufficiently fine mesh spacings $R$ is connected if and only if $A$ is irreducible. Note that we have used the mesh points of $\bar{R}_h$ precisely as the nodes of the directed graphs of Figure 22, which further emphasizes this geometrical relationship. Finally, to conclude that $A$ is *irreducibly diagonally dominant*, we see that it is sufficient to have but one $\sigma_{i,j} > 0$, or one boundary mesh point $(x_i, y_j)$, where $u_{i,j}$ is specified adjacent to a mesh point where $u$ is unknown, so that the condition $\sigma > 0$ in (6.29) is overly restrictive.

One might ask *why* the general self-adjoint partial differential equation

$$(6.27') \qquad -(P(x, y)u_x)_x - (Q(x, y)u_y)_y + \sigma(x, y)u(x, y) = f(x, y),$$

$$(x, y) \in R,$$

was not taken in place of (6.27), where $Q(x, y) > 0$ in $\bar{R}$. Briefly, it can be verified (Exercise 6) that all the steps of the previous derivation similarly follow, *except* when $\beta > 0$ on $\Gamma_h$, and the normal to $\Gamma_h$ makes an angle $\theta$ to the $x$-axis, where $\sin 2\theta \neq 0$. In this case, the boundary condition of (6.28) is not the "natural" boundary condition, and (6.28) should now include a tangential derivative as well.

To show the similarity of the results of the previous derivation with the two-dimensional use of the variational principle, consider again the differential equation of (6.27) with, for simplicity, boundary conditions $\alpha \equiv 1$ and $\beta \equiv 0$ in (6.28), so that $u(x, y) = \gamma(x, y)$ is specified on $\Gamma$. By analogy with (6.11), define

$$(6.41) \quad F(w) \equiv$$

$$\frac{1}{2} \iint_{\bar{R}} \{P(x, y)(w_x)^2 + P(x, y)(w_y)^2 + \sigma(w)^2 - 2wf\} \, dx \, dy,$$

where $w(x, y)$ is any sufficiently smooth function defined in $\bar{R}$, satisfying the boundary conditions $w(x, y) = \gamma(x, y)$, $(x, y) \in \Gamma$. If we write

$$w(x, y) = u(x, y) + \epsilon(x, y),$$

where $u(x, y)$ is the unique solution of (6.27)–(6.28), then the two-dimensional analogue of integrating by parts in Sec. 6.1 is now Green's theorem (6.33), from which we verify† that

$$(6.42) \qquad F(w) = F(u) + \frac{1}{2} \int\!\!\!\int_{\bar{R}} \{P(\epsilon_x)^2 + P(\epsilon_y)^2 + \sigma(\epsilon)^2\} \, dx \, dy$$

$$\geq F(u),$$

so that $F(u)$ again *minimizes* $F(w)$. This relationship can also be used in deriving finite difference approximations to (6.27)–(6.28). In fact, we merely state that an approximation of $F(w)$, analogous to $\tilde{F}(w)$ in Sec. 6.1, can be made which, when minimized, is *identical* to the matrix equation of (6.39).

It is also of some interest to mention that the Ritz form of the variational method introduced in Sec. 6.1 can be applied in two dimensions by again restricting attention to the class of piecewise linear functions. Geometrically, the region $\bar{R}_h$ is first decomposed into triangular sub-regions, as shown in Figure 23, the mesh points of unknowns remaining the same. For any choice of values $w_{i,j}$ at mesh points $(x_i, y_j)$ of $\bar{R}_h$, where $w(x, y)$ is not specified, we can then define a function $w(x, y)$ in $\bar{R}_h$ so that $w(x, y)$ is *continuous* in $\bar{R}_h$, and *linear* in each triangular sub-region of $R_h$. Referring to the shaded triangular subregion of Figure 23,

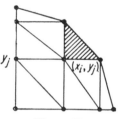

**Figure 23**

$$w(x, y) \equiv w_{i,j} + \left(\frac{w_{i,j+1} - w_{i,j}}{k_j}\right)(y - y_j) + \left(\frac{w_{i+1,j} - w_{i,j}}{h_i}\right)(x - x_i).$$

In this case, $F(w) = F(w_{i,j})$ is a quadratic form in the unknowns $w_{i,j}$, which can again be minimized by setting

$$\frac{\partial F(w)}{\partial w_{i,j}} = 0.$$

The resulting system of linear equations

$$\mathfrak{a}\mathbf{w} = \mathbf{k}$$

† See Exercise 7.

is such that $\mathcal{Q}$ shares all the essential properties with the matrix $A$ of Theorem 6.4. Moreover, $\mathcal{Q}$ has at most five nonzero entries per row, and generates a *five-point* formula very much like that of the matrix $A$ of (6.39) when $\sigma \equiv 0$ in (6.27).

Before continuing with more general problems, let us consider the special case $P \equiv 1$ in (6.27) for the uniform mesh spacings $h_i = h$, $k_j = k$. In this case, we can write the difference equation of (6.38), upon dividing by $kh$, as

$$(6.38') \qquad \left(\frac{2u_{i,j} - u_{i+1,j} - u_{i-1,j}}{h^2}\right) + \left(\frac{2u_{i,j} - u_{i,j+1} - u_{i,j-1}}{k^2}\right)$$
$$+ \sigma_{i,j} u_{i,j} = f_{i,j} + \tau_{i,j}.$$

However, this is what we would obtain from Taylor's series considerations! In fact, if $u_{xxxx}$ and $u_{yyyy}$ exist and are both bounded in $\bar{R}$ by $M_4$, then by means of Taylor's series developments of $u(x, y)$, it can be verified that

$$\tau_{i,j} = -\frac{h^2}{12}\{u_{xxxx}(x_i + \theta_1 h, y_i) + u_{yyyy}(x_i, y_i + \theta_2 h)\},$$

so that

$$|\tau_{i,j}| \le \frac{M_4 h^2}{6}.$$

The point here is that extensions of Gerschgorin's Theorem 6.2 make use of this bound.

We have paid particular attention to the self-adjoint problem of (6.27), but several extensions of our derivation are useful in practice. First, consider in place of (6.27) and (6.28)

$$(6.27'') \qquad -\{Pu_x\}_x - \{Pu_y\}_y + \sigma u = f, \qquad (x, y) \in R,$$

with, for simplicity,

$$(6.28'') \qquad u(x, y) = \gamma(x, y), \qquad (x, y) \in \Gamma,$$

where the coefficients $P$, $\sigma$, and $f$ are only *piecewise continuous* in $\bar{R}$, resulting from physical internal interfaces $\gamma$, as shown in Figure 24, where we assume that these internal interfaces are composed of horizontal and vertical line segments that do not terminate in $R$. Analogous to the continuity condition (6.19), we ask that $u(x, y)$ be continuous in $\bar{R}$, and that

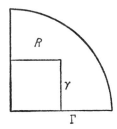

**Figure 24**

$$(6.43) \qquad P(x, y)\frac{\partial u(x, y)}{\partial n} \quad \text{be continuous across any segment of } \gamma.$$

The derivation we have given based on integration can be extended, and the essential idea is to integrate over the four separate parts of $r_{i,j}$ in Figure 21(a). Because of (6.43), all interior line integrals vanish, and we can *exactly* use the approximation of (6.36). In (6.35), however, the value $g_{i,j}$ is merely replaced by the *average* of $g(x, y)$ in the region $r_{i,j}$. Summarizing, the results of Theorem 6.4 apply equally to such problems with physical interfaces.

The extension of these derivations of finite difference approximations to second-order elliptic partial differential equations in three dimensions is now almost obvious for seven-point formulas. The essential steps are to replace $R$ and $\Gamma$ by $R_h$ and $\Gamma_h$ so that $\bar{R}_h$ is the convex hull of a finite set of mesh points in three dimensions. Then, integrations are carried over three-dimensional regions $r_{i,j}$.

In two-dimensional calculations, we have seen that the region $R_h$ approximates $R$, and it might appear that simply choosing mesh points sufficiently close *near the boundary* would give better geometrical approximations of $R$. Unfortunately, rapid changes in mesh spacings have a disastrous effect on the spectral radius of the associated point Jacobi matrix, as shown in Exercise 2. In certain physical problems with interfaces, this cannot always be avoided within machine limitations. In the next section, however, we shall see that part of this decrease in rate of convergence can be offset by the use of block iterative methods.

Finally, one key idea we have used in the derivation of finite difference approximations is the integration over mesh regions. Although the mesh regions considered were generally rectangular, more general mesh regions can also be treated. Following MacNeal (1953), consider the region of Figure 25 as an example. Again, one can assign to each mesh point $(x_i, y_j)$, where $u_{i,j}$ is unknown, a mesh region $r_{i,j}$. If $u_{i,j}$ is to be coupled to $u_{k,l}$, then

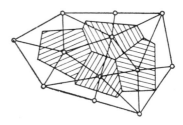

**Figure 25**

there is a boundary edge of $r_{i,j}$ that bisects the line segment joining $(x_i, y_j)$ to $(x_k, y_l)$. In this way, mesh regions more general than rectangles can be composed, and integrating the differential equation over each mesh region leads to a matrix equation $A\mathbf{u} = \mathbf{k} + \boldsymbol{\tau}(u)$ for which $A$ has positive diagonal entries, nonpositive off-diagonal entries. Moreover, $A$ is real, symmetric, and diagonally dominant. For applications to triangular and hexagonal meshes, see Exercises 3 and 4.

## EXERCISES

1. Consider the solution of the Laplace equation $-\nabla^2 u = S$, in the unit square $0 < x, y < 1$, with the following boundary conditions:

$$u(t, 0) = u(0, t), \qquad 0 \le t \le 1,$$

$$\frac{\partial u}{\partial y}(t, 0) = -\frac{\partial}{\partial x} u(0, t), \qquad 0 \le t \le 1,$$

$$u(t, 1) = u(1, t) = 1, \qquad 0 \le t \le 1.$$

Such a problem is said to have *rotational symmetry*. With a uniform mesh $x_j = y_j = j/(N+1)$, $0 \le j \le N + 1$, for this region, derive a system of linear equations approximating this boundary-value problem of the form

$$A\mathbf{z} = \mathbf{k},$$

where $A$ is an $m \times m$ real, symmetric, and positive definite matrix with $m = N(N+1) + 1$, and each row of $A$ has at most five nonzero entries. Show that $A$ and its point Jacobi matrix (for some ordering) satisfy the results of Theorem 6.4. Show also that each unknown mesh point is *not* always coupled to its nearest four mesh points.

2. To show the effect of abrupt changes in mesh spacings on rates of convergence, consider the solution of the Dirichlet problem, i.e., $\nabla^2 u = 0$ in the unit square, with boundary values $u = 1$. With four interior mesh points as shown, and with the mesh spacings as indicated, where $2\alpha + \epsilon = 1$, $\epsilon > 0$, derive a system of linear equations of the form

$$A(\epsilon)\mathbf{u} = \mathbf{k},$$

where $A(\epsilon)$ is a $4 \times 4$ matrix. If $B(\epsilon)$ is the associated $4 \times 4$ point Jacobi matrix derived from $A(\epsilon)$, find $\rho[B(\epsilon)]$. Prove that

$$\lim_{\epsilon \to 0} \rho[B(\epsilon)] = 1.$$

Also, find $R_\infty[B(0.001)]/R_\infty[B(\tfrac{1}{3})]$.

**3.** Consider the regular hexagonal mesh in the figure. Derive, by means of integration, a four-point difference approximation to

$$-\nabla^2 u + \sigma u = f.$$

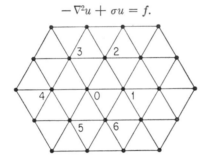

**4.** Consider the regular triangular mesh in the figure. Derive, by means of integration, a seven-point difference approximation to

$$-\nabla^2 u + \sigma u = f.$$

**5.** Consider the problem of finding five-point difference approximations to Laplace's equation in cylindrical coordinates,

$$\frac{\partial^2 u}{\partial r^2} + \frac{1}{r}\frac{\partial u}{\partial r} + \frac{\partial^2 u}{\partial z^2} = 0,$$

over a uniform mesh $\Delta z = \Delta r$. By means of Taylor's series and central differences, one obtains nonsymmetric difference equations. However, show that, by means of integration over wedge-shaped regions in three dimensions of the three-dimensional form of this differential equation,

$$\frac{\partial^2 u}{\partial r^2} + \frac{1}{r}\frac{\partial u}{\partial r} + \frac{1}{r^2}\frac{\partial^2 u}{\partial \theta^2} + \frac{\partial^2 u}{\partial z^2} = 0,$$

*symmetric* difference equations can be obtained where $u(r, z, \theta)$ is independent of $\theta$ (axially symmetric).

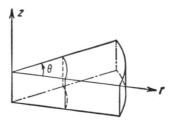

6. Assume that the boundary $\Gamma_h$ in two-dimensions is composed only of horizontal or vertical line segments, i.e., $\sin 2\theta = 0$, where $\theta$ is the angle between the normal to $\Gamma_h$ and the positive real axis. Show in this case that the integration procedure carried out applies equally well to the self-adjoint equation

$$-(Pu_x)_x - (Qu_y)_y + \sigma u(x, y) = f(x, y), \qquad (x, y) \in R_h,$$

with the assumptions of (6.28)–(6.30). What happens, however, when $\sin 2\theta \neq 0$, and $\beta > 0$ on $\Gamma_h$?

7. Verify the result of (6.42).

8. For the differential equation:

$$-u_{xx} - u_{yy} + 0.1u = -4 + 0.1(x^2 + y^2)$$

in the quarter circle $0 < x < 1, 0 < y < 1, 0 < x^2 + y^2 < 1$, with the boundary conditions of Figure 19(a), let $x_0 = y_0 = 0$, $x_1 = y_1 = 0.5$, $x_2 = y_2 = 0.866$, and $x_3 = y_3 = 1$. Derive the associated finite difference approximation $A\mathbf{z} = \mathbf{k}$, where $A$ is a $10 \times 10$ matrix. Also, determine its point Jacobi matrix.

9. For the previous problem derive the finite difference approximation $\mathcal{Q}\mathbf{w} = \mathbf{k}$ associated with the Ritz form of the variational method for the particular triangularization of Figure 23.

10. Consider the partial differential equation:

$$-\{A(x, y)u_x\}_x - \{C(x, y)u_y\}_y + D(x, y)u_x + E(x, y)u_y$$
$$+ F(x, y)u = S(x, y)$$

defined in an open and bounded connected set $R$, with the boundary conditions of (6.28) applying on the boundary $\Gamma$ of $R$. Assuming that the coefficients are continuous in $\bar{R}$, and satisfy

$$A(x, y) > 0, \quad C(x, y) > 0, \quad F(x, y) \geq 0, \quad (x, y) \in \bar{R},$$

derive a finite difference approximation of this boundary-value problem for a rectangular mesh, and obtain a two-dimensional analogue of Theorem 6.3 for the properties of the derived matrix.

11. For the partial differential equation (B.1) with boundary conditions (B.2) of Appendix B, derive for the mesh configuration of Figure 31 the associated five-point finite difference approximation $A\mathbf{z} = \mathbf{k}$ summarized in (B.5).

## 6.4. Factorization Techniques and Block Iterative Methods

The previous sections of this chapter were aimed at the derivation of finite difference approximations of certain differential equations, but questions pertaining to the numerical solution of these matrix equations were left unanswered. Given these matrix equations along with their derived properties, we shall now examine, both theoretically and practically, numerical methods for solving these matrix equations.

Let us first look at the matrix equation

$$(6.44) \qquad A\mathbf{z} = \mathbf{k},$$

where $A$ is a given $n \times n$ tridiagonal matrix, and $\mathbf{k}$ is a given column vector, of the form

$$(6.45) \qquad A = \begin{bmatrix} b_1 & c_1 & & \text{O} \\ a_2 & \searrow & \searrow & \\ & \searrow & \searrow & c_{n-1} \\ \text{O} & & a_n & b_n \end{bmatrix}; \qquad \mathbf{k} = \begin{bmatrix} k_1 \\ k_2 \\ \cdot \\ \cdot \\ \cdot \\ k_n \end{bmatrix}.$$

This matrix problem can be directly solved by the following simple algorithm. Defining

$$(6.46) \qquad w_1 = \frac{c_1}{b_1}; \qquad w_i = \frac{c_i}{b_i - a_i w_{i-1}}, \qquad 2 \leq i \leq n,$$

$$g_1 = \frac{k_1}{b_1}; \qquad g_i = \frac{k_i - a_i g_{i-1}}{b_i - a_i w_{i-1}}, \qquad 2 \leq i \leq n,$$

the components $z_i$ of the solution vector $z$ are then given recursively by

$$(6.46') \qquad z_n = g_n; \qquad z_i = g_i - w_i z_{i+1}, \quad 1 \leq i \leq n - 1.$$

With the calculation of the quantities $w_i$ and $g_i$, the above procedure is equivalent to the reduction by Gaussian elimination† of the matrix equation to a new matrix equation, the new matrix being upper-triangular with unit diagonal entries. Thus, $(6.46')$ is just the backward substitution determining the vector solution **z**. In terms of arithmetic calculations, the above method requires at most *five* multiplications and *three* additions per unknown $z_i$, assuming for simplicity that divisions are roughly as time-consuming on digital computers as multiplications. On the other hand, the point successive overrelaxation iterative method applied to (6.44) can be written as

$$(6.47) \qquad z_i^{(m+1)} = z_i^{(m)} + \omega \left\{ \left( \frac{-a_i}{b_i} \right) z_{i-1}^{(m+1)} + \left( \frac{-c_i}{b_i} \right) z_{i+1}^{(m)} + \left( \frac{k_i}{b_i} \right) - z_i^{(m)} \right\},$$

$$m \geq 0.$$

† See, for example, either Householder (1953), p. 69, or Faddeeva (1959), p. 79.

Assuming that the coefficients $a_i/b_i$, $c_i/b_i$, and $k_i/b_i$ have been computed once and stored, it is apparent for $\omega$ different from zero or unity that this iterative method requires three multiplications and four additions *per mesh point per iteration*. Therefore, the direct method of (6.46)–(6.46'), which compares in total arithmetic effort with just two complete iterations of the point successive overrelaxation method, is numerically more desirable in practical problems. The only possible shortcoming of the direct inversion would be instability with respect to growth of rounding errors, but when the matrix $A$ is diagonally dominant and tridiagonal, Wilkinson (1961) has shown that the direct method is extremely stable with respect to the growth of rounding errors.

The application of this direct method to the numerical solution of matrix equations arising from finite difference approximations to elliptic partial differential equations in two and higher dimensions can be made in several ways. Recall that for two-dimensional problems, these finite difference approximations produce matrix equations $A\mathbf{z} = \mathbf{k}$, where the matrix $A$ has the form (Theorem 6.4)

$$(6.48) \qquad A = \begin{bmatrix} B_1 & C_1 & & & \\ A_2 & B_2 & C_2 & & \\ & & \ddots & & \\ & & & & C_{N-1} \\ & & & A_N & B_N \end{bmatrix}.$$

Here, the square diagonal submatrices $B_j$ are of order $n_j$, where $n_j$ corresponds to the number of mesh points on the $j$th horizontal mesh line of the discrete problem. The direct inversion method of (6.46)–(6.46') can be immediately generalized so as to apply to $A\mathbf{z} = \mathbf{k}$. Indeed, if the vectors $\mathbf{z}$ and $\mathbf{k}$ are partitioned relative to the matrix $A$ of (6.48), then define

$$W_1 = B_1^{-1}C_1; \qquad W_i = (B_i - A_iW_{i-1})^{-1}C_i,$$
$$2 \leq i \leq N,$$

$$(6.49)$$

$$G_1 = B_1^{-1}K_1; \qquad G_i = (B_i - A_iW_{i-1})^{-1}(K_i - A_iG_{i-1}),$$
$$2 \leq i \leq N,$$

and the vector components $Z_i$ of the solution of $A\mathbf{z} = \mathbf{k}$ are given recursively by

$$(6.49') \qquad Z_N = G_N; \qquad Z_i = G_i - W_iZ_{i-1}, \quad 1 \leq i \leq N - 1.$$

Of course, the advantage again of this generalized method is that it is *direct.* But this method now is quite costly as far as storage on a digital computer is concerned. For example, let us consider the discrete approximation of the partial differential equation

$$(6.50) \qquad -(P(x, y)u_x)_x - (P(x, y)u_y)_y + \sigma(x, y)u(x, y) = f(x, y)$$

on a square mesh $R_h$ consisting of $N$ rows, each with $N$ mesh points. With the five-point approximation, the symmetry of the derived matrix $A$ shows that at most three coefficients are needed per mesh point to completely specify the matrix $A$, so that at most $3N^2$ coefficients suffice to specify $A$. If the functions $P$, and $\sigma$ are positive, then the method of derivation of Sec. 6.3 shows that the matrices $B_i$ of (6.48) are all irreducible $N \times N$ Stieltjes matrices, and thus (Corollary 3 of Sec. 3.5) each entry of $B_i^{-1}$ is positive. Because of this, each matrix $W_i$ of (6.49) would require roughly $\frac{1}{2}N^2$ coefficients of storage, even taking into account the symmetry of $W_i$. Hence, just the storage of all the $W_i$ would require roughly $\frac{1}{2}N^3$ coefficients of storage for the general case, and because of this, the application of this direct method to the *general* problem of approximating elliptic partial differential equations seems to be computationally feasible only for moderate-sized $(N^2 \leq 500)$ matrix problems. Nevertheless, in rather *special* cases, such as in the solution of Laplace's or Poisson's equation with Dirichlet boundary conditions in a rectangle with *uniform* mesh spacings, simplifications† of (6.49)–(6.49') are possible which result in substantial reductions both in the total computing effort and in coefficient storage. These simplifications stem from the observation that, in the special cases mentioned, the matrices $B_i$ of (6.48) are all of the form

$$B = \begin{bmatrix} 4 & -1 & & \bigcirc \\ -1 & 4 & -1 & \\ & \ddots & \ddots & \ddots & -1 \\ \bigcirc & & -1 & 4 \end{bmatrix},$$

and that the matrices $A_i$ and $C_i$ of (6.48) are just $N \times N$ identity matrices. It is important to add, however, that rounding errors are *not* as well behaved in this generalization as in the original method of (6.46)–(6.46'), even for the special cases mentioned.

The practical use of the direct method of (6.46)–(6.46') is that it has

† For such simplifications, with emphasis on the numerical solution of elliptic partial differential equations, see notably Karlquist (1952), Cornock (1954), Schechter (1960), Egerváry (1960), and von Holdt (1962).

been successfully applied in conjunction with iterative solutions to two-
(and three-) dimensional elliptic difference equations in the following
way. Specifically, the partitioning of the matrix in (6.48) came from
considering all mesh points of a particular horizontal line as a block.
It is then clear that, for five-point approximations to (6.50), the corresponding matrix $A$ of (6.48) is real, symmetric, and positive definite,
and that *each* matrix $B_i$ is an irreducible *tridiagonal* Stieltjes matrix whose
nonzero entries give the coupling of a mesh point and its immediate neigh-
bors on the same horizontal line. Similarly, the matrices $A_i$ and $C_i$ contain
the coupling coefficients of a mesh point respectively to the mesh point of
the line above and the mesh point of the line below, and thus the matrices
$A_i$ and $C_i$ have at most one nonzero entry per row. Since the matrices $B_i$
are nonsingular, we can form the block Jacobi matrix $M$ as

$$(6.51) \qquad M \equiv I - D^{-1}A,$$

where

$$(6.51') \qquad D \equiv \begin{bmatrix} B_1 & & \mathbf{O} \\ & B_2 & \\ \mathbf{O} & & \ddots \\ & & & B_N \end{bmatrix}$$

is the block diagonal matrix of $A$. More precisely, we see from (6.51) that

$$(6.52) \qquad M = \begin{bmatrix} 0 & B_1^{-1}C_1 & & & \mathbf{O} \\ B_2^{-1}A_2 & 0 & B_2^{-1}C_2 & & \\ & \ddots & \ddots & \ddots & \\ & & & & B_{N-1}^{-1}C_{N-1} \\ \mathbf{O} & & & B_N^{-1}A_N & 0 \end{bmatrix}.$$

Comparing this with the matrix $B_2$ of (4.8) of Sec. 4.1, we necessarily
conclude that $M$ is a *consistently ordered weakly cyclic matrix of index* 2.
Moreover (Exercise 5), $M$ has real eigenvalues. Thus, the block successive
overrelaxation iterative method, which now can be rigorously applied to
the solution of $Az = k$, takes the form

$$(6.53) \qquad Z_i^{(m+1)} = \omega B_i^{-1}\{-A_iZ_{i+1}^{(m+1)} + -C_iZ_{i+1}^{(m)} + S_i - B_iZ_i^{(m)}\} + Z_i^{(m)},$$

$$1 \le i \le N,$$

where the vectors $Z_i^{(0)}$ are arbitrary. A more convenient computational form of (6.53) is

$$(6.54) \qquad B_i \hat{Z}_i^{(m+1)} = A_i Z_{i-1}^{(m+1)} + C_i Z_{i+1}^{(m)} + S_i, \qquad 1 \leq i \leq N,$$

and

$$(6.54') \qquad Z_i^{(m+1)} = \omega(\hat{Z}_i^{(m+1)} - Z_i^{(m)}) + Z_i^{(m)}, \qquad 1 \leq i \leq N.$$

Since the matrices $B_i$ are *tridiagonal* Stieltjes matrices, the matrix problems of (6.54) can be *directly* solved by the method of (6.46)–(6.46'). This particular block iterative method is often called the *successive line overrelaxation method* (SLOR), using the graphical idea that lines of mesh points are treated as units. In this sense, it is closely related to the Russian *"method of lines,"* due to Faddeeva (1949), where one approximates a partial differential equation by a coupled system of ordinary differential equations. See also Kantorovich and Krylov (1958), p. 321.

Our interest in passing from point to block iterative methods is motivated by our desire to obtain improved asymptotic rates of convergence. With our results of Sec. 3.6 on regular splitting of matrices, we can *definitely* say the block Jacobi iterative method is asymptotically iteratively faster than the point Jacobi method if at least one of the Stieltjes matrices $B_i$ of (6.48) is irreducible and of order $n \geq 2$. Now to complete the picture, we ask if there has been any increase in arithmetic effort in passing from point to line successive overrelaxation. Cuthill and Varga (1959) have shown that it is possible to normalize the equation in such a way that they both require *exactly* the same arithmetic computations per mesh point. Also, Blair (1960) has shown that this normalized method is stable with respect to the growth of rounding errors. In summary then, successive line overrelaxation is rigorously applicable, it is asymptotically faster than point overrelaxation and is numerically very stable with respect to rounding errors. Moreover, it can be performed in such a way so as to be no more costly arithmetically than point overrelaxation. There seems to be little reason[†] for *not* using such improved iterative methods for the numerical solution of the general problem of (6.50). For numerical results, see Appendix B.

Encouraged by the theoretical as well as practical results of direct inversion of tridiagonal matrices as an integral part of block iterative methods, effort has been expended in increasing the *size* of the blocks of **A** in

---

[†] Again, for the special cases already enumerated, point successive overrelaxation in two dimensions is itself quite attractive for binary machines, in that its arithmetic requirements in these special cases can be reduced to additions, shifts, and at most **one** multiplication per mesh point.

(6.48), which will be directly inverted. Of course, the theory for such extensions is fairly well established. Based on an idea by Heller (1960), we can consider other partitionings of the matrix $A$ which leave $A$ in block tridiagonal form. For example, consider the particular ordering of

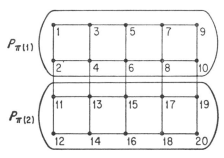

**Figure 26**

the mesh points in Figure 26, and partition the matrix $A$ in this case so that mesh points of *two* successive horizontal mesh lines are mesh points of successive blocks. Then, for a five-point formula,† the matrix $A$ is *block tridiagonal*, since the block directed graph of $A$ becomes

The results here are not special to second-order elliptic equations. In fact, Heller (1960) has shown that, by choosing the number of horizontal mesh lines per block sufficiently large, matrix problems arising from discrete approximations to elliptic partial differential equations of order $2n$, $n \geq 1$, can be partitioned to be *block tridiagonal*.

## EXERCISES

**1.** If we define the sequence of tridiagonal matrices

$$A_j = \begin{bmatrix} b_1 & c_1 & & & \bigcirc \\ a_2 & b_2 & c_2 & & \\ & \ddots & \ddots & \ddots & \\ & & & & c_{j-1} \\ \bigcirc & & & c_j & b_j \end{bmatrix}, \quad 1 \leq j \leq n,$$

---

† It is easy to see that if the matrix $A$ arises from a nine- or thirteen-point formula, this block tridiagonal feature is unchanged.

and assume that $\det A_j \neq 0$ for all $1 \leq j \leq n$, prove that the direct method of (6.46)–(6.46′) can be carried out. Moreover, characterize the quantities $b_i - a_i w_{i-1}, i > 1$ (Faddeeva (1959), p. 20).

2. Let $A$ be an $n \times n$ tridiagonal $M$-matrix. Using the result of the previous exercise, prove that the direct method of (6.46)–(6.46′) can always be used to solve $A\mathbf{z} = \mathbf{k}$.

3. Let $\qquad\qquad A =$

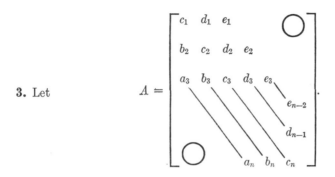

Derive an algorithm which generalizes (6.46)–(6.46′) for directly solving the matrix problem $A\mathbf{z} = \mathbf{k}$.

4. Consider the matrix problem $A(\epsilon)\mathbf{u} = \mathbf{k}$, described in Exercise 2 of Sec. 6.3. First, partition the matrix $A(\epsilon)$ by *horizontal* mesh lines, and let $B_H(\epsilon)$ be the associated block Jacobi matrix. Next, partition the matrix $A(\epsilon)$ by *vertical* mesh lines, and let $B_V(\epsilon)$ be the associated block Jacobi matrix. Prove that

$$\lim_{\epsilon \to 0} \rho(B_V(\epsilon)) = 1, \qquad \lim_{\epsilon \to 0} \rho(B_H(\epsilon)) \doteq 0.3.$$

Thus, the *orientation* of the mesh lines used in partitioning the matrix $A(\epsilon)$ is important for asymptotic rates of convergence.

5. Let $A$ be a Hermitian and positive definite $n \times n$ matrix that is partitioned as in (4.2). Prove that the block Jacobi matrix $B$ of (4.4) has real eigenvalues.

## 6.5. Asymptotic Convergence Rates for the Model Problem

Thus far we have derived finite difference approximations to certain partial differential equations, and we have compared at least theoretically various block iterative methods applied to such finite difference approximations. Now, we wish to obtain a more definite appraisal of the advantages of such block iterative methods relative to a fixed, simple, but nontrivial problem. To this end, consider the numerical solution of the Dirichlet problem,

$$(6.55) \qquad -u_{xx} - u_{yy} = 0, \qquad 0 < x, y < 1,$$

for the unit square $R$ with boundary $\Gamma$, subject to the boundary condition

$$(6.56) \qquad u(x, y) = \gamma(x, y), \qquad (x, y) \in \Gamma.$$

We shall call this the *model problem*. Note that this is the special case $P \equiv 1, \sigma = 0$ in (6.27), $\alpha \equiv 1, \beta \equiv 0$ in (6.28). Using a *uniform mesh*,

$$(6.57) \qquad x_i = ih, \quad y_j = jh, \quad 0 \le i, j \le N, \quad h = 1/N,$$

we see in this case that $R_h$ and $\Gamma_h$, as defined in Sec. 6.3, coincide with $R$ and $\Gamma$. Moreover, the finite difference approximation to (6.55), as deduced as the special case of (6.38), is

$$(6.58) \qquad 4u_{i,j} - (u_{i+1,j} + u_{i-1,j} + u_{i,j+1} + u_{i,j-1}) = h^2 \tau_{i,j},$$

$$1 \le i, j \le N - 1,$$

where we again use the notation $u_{i,j} \equiv u(x_i, y_j)$. The quantities $u_{0,j}$, $u_{N,j}$, $u_{i,0}$, $u_{i,N}$ are given values from (6.56). Thus, omitting the error term $\tau_{i,j}$ gives us the matrix problem

$$(6.59) \qquad\qquad\qquad A\mathbf{z} = \mathbf{k},$$

where $A$ is an $(N-1)^2 \times (N-1)^2$ real matrix satisfying the conclusion of Theorem 6.4. From (6.58), all the diagonal entries of the matrix $A$ are 4, so that we can obtain the *point Jacobi matrix* $B$ associated with $A$ simply from

$$(6.60) \qquad\qquad\qquad B = I - \tfrac{1}{4}A.$$

We can explicitly obtain the eigenvalues and eigenvectors of the point Jacobi matrix $B$ for the model problem here treated. Since $A$ is real, symmetric, and positive definite, then it is clear from (6.60) that $B$ is real, symmetric, and possesses real eigenvalues and real eigenvectors. If $\boldsymbol{\alpha}$ is an eigenvector of $B$ with components $\alpha_{i,j}$, then from (6.59) and (6.60)

$$(6.61) \qquad \tfrac{1}{4}\{\alpha_{i-1,j} + \alpha_{i+1,j} + \alpha_{i,j-1} + \alpha_{i,j+1}\} = \mu\alpha_{i,j}, \qquad 1 \le i, j \le N,$$

where $\alpha_{0,j} = \alpha_{N,j} = \alpha_{i,0} = \alpha_{i,N} = 0$. Let

$$\alpha_{i,j}^{(k,l)} \equiv \gamma_{k,l} \sin\left(\frac{k\pi i}{N}\right) \sin\left(\frac{l\pi j}{N}\right), \qquad 1 \le k, l \le N - 1, 1 \le i, j \le N - 1.$$

Substituting $\alpha_{i,j}^{(k;l)}$ in (6.61) shows that $\alpha^{(k,l)}$ is an eigenvector of $B$, with corresponding eigenvalue

$$(6.62) \qquad \mu_{k,l} = \frac{1}{2} \left\{ \cos\left(\frac{\pi k}{N}\right) + \cos\left(\frac{\pi l}{N}\right) \right\}.$$

The multiplicative constant $\gamma_{k,l}$ of $\alpha^{(k,l)}$ is now chosen so that the Euclidean norm $\| \alpha^{(k,l)} \|$ is unity. It is easy to see from (6.62) that

$$\mu_{k,l} \leq \mu_{1,1} = \cos\left(\frac{\pi}{N}\right);$$

but as $B$ is non-negative and irreducible, it has (from the Perron-Frobenius theory) a *unique* positive eigenvector with eigenvalue $\rho(B)$. Thus, as $\alpha^{(1,1)} > 0$, it follows that

$$(6.63) \qquad \rho(B) = \cos\left(\frac{\pi}{N}\right),$$

and

$$(6.64) \qquad R_\infty(B) = -\ln \cos\left(\frac{\pi}{N}\right)$$

is the asymptotic rate of convergence for the point Jacobi matrix $B$.

To determine the asymptotic rate of convergence of the point successive overrelaxation iterative method, we first make use of the results of Theorem 6.4, which states that there is an ordering (the natural ordering) for which $B$ is a consistently ordered cyclic matrix of index 2. As $B$ is symmetric and has real eigenvalues, we can rigorously state that, for this ordering, the optimum relaxation factor is given by (Theorem 4.4)

$$(6.65) \qquad \omega_b = \frac{2}{1 + \sqrt{1 - \rho^2(B)}} = \frac{2}{1 + \sin\left(\dfrac{\pi}{N}\right)}.$$

Now, the asymptotic rate of convergence of the point successive overrelaxation iterative method is given by

$$(6.66) \qquad R_\infty(\mathcal{L}_{\omega_b}^{(p\,t)}) = -\ln(\omega_b - 1).$$

Since

$$\ln(1+x) = x^2 - \frac{x^2}{2} + \frac{x^3}{3} + O(x^4), \qquad x \longrightarrow 0,$$

then, we compute easily that

$$(6.67) \qquad R_\infty(\mathcal{L}_{\omega_b}^{(p\,t)}) \sim \frac{2\pi}{N}, \quad R_\infty(B) \sim \frac{\pi^2}{2N^2}, \quad N \longrightarrow \infty,$$

and therefore

$$(6.68) \qquad \frac{R_\infty(\mathcal{L}_{\omega_b}^{(p\,t)})}{R_\infty(B)} \sim \frac{4N}{\pi}, \quad N \longrightarrow \infty.$$

Thus, for $N$ large, the point successive overrelaxation iterative method with optimum relaxation factor is an order of magnitude *faster* than the point Jacobi iterative method for this model problem.

The calculation we have just performed for point iterative methods is easily extended to the line successive overrelaxation iterative method. In this case, if the line Jacobi method has the eigenvector $\beta$ corresponding to the eigenvalue $\lambda$, then

$$(6.69) \qquad \lambda(4\beta_{i,j} - \beta_{i+1,j} - \beta_{i-1,j}) - (\beta_{i,j-1} + \beta_{i,j+1}) = 0;$$
$$0 \le i, j \le N.$$

Again, defining

$$\alpha_{i,j}^{(k,\,l)} \equiv \gamma_{k,l} \sin\left(\frac{k\pi i}{N}\right) \sin\left(\frac{l\pi j}{N}\right),$$

and substituting in (6.69) shows that

$$(6.70) \qquad \lambda_{k,l} = \frac{\cos\left(\dfrac{l\pi}{N}\right)}{2 - \cos\left(\dfrac{k\pi}{N}\right)} \le \frac{\cos\left(\dfrac{\pi}{N}\right)}{2 - \cos\left(\dfrac{\pi}{N}\right)}.$$

Again, we know that the line Jacobi matrix is non-negative, so that

$$\rho(B_{line}) = \frac{\cos\left(\dfrac{\pi}{N}\right)}{2 - \cos\left(\dfrac{\pi}{N}\right)}.$$

From this, we can easily calculate that

$$(6.71) \qquad R_\infty(\mathcal{L}_{\omega_b}^{(line)}) \sim \frac{2\sqrt{2}\pi}{N}, \quad N \longrightarrow \infty,$$

so that

$$(6.72) \qquad \frac{R_\infty(\mathcal{L}_{\omega_b}^{(line)})}{R_\infty(\mathcal{L}_{\omega_b}^{(p\,t)})} \sim \sqrt{2}, \quad N \longrightarrow \infty,$$

for this model problem. In other words, for large $N$, the line successive overrelaxation iterative method yields an *increase* of approximately 40 per cent in the asymptotic rate of convergence of the point successive over-relaxation method. The pertinent facts are that this gain *can be* achieved by means of normalized iterative methods, and that, generally speaking, more impressive gains can be made, especially in problems with nonuniform mesh spacings. Also, another very important conclusion we have reached here is that the gain or improvement (6.72) in the ratios of asymptotic rates of convergence is a fixed factor, independent of the mesh spacing $h = 1/N$, in contrast to the alternating direction methods of the next chapter.

The two-line iterative methods of Sec. 6.4 can also be analyzed for this model problem, and briefly, by means of the regular splitting theory of Sec. 3.6, we can similarly show again that

$$(6.73) \qquad \frac{R_\infty(\mathcal{L}_{\omega_b}^{(2line)})}{R_\infty(\mathcal{L}_{\omega_b}^{(line)})} \sim \sqrt{2}, \qquad N \longrightarrow \infty,$$

so that the use of multiline iterative methods may be indicated in practical studies.

More recently, Parter (1961b) has shown for problems which include the model problem as a special case that

$$(6.74) \qquad R_\infty(\mathcal{L}_{\omega_b}^{(k\ line)}) \sim 2\sqrt{2}\left(\frac{k}{2}\Lambda(\Delta y)^2\right)^{1/2},$$

where $\Lambda$ is the fundamental (or minimal) eigenvalue of the Helmholtz equation

$$(6.75) \qquad \begin{aligned} u_{xx} + u_{yy} + \Lambda u &= 0, \qquad (x, y) \in R, \\ u(x, y) &= 0, \qquad (x, y) \in \Gamma. \end{aligned}$$

For the model problem, $\Lambda = 2\pi^2$, and this result (6.74), generalizing (6.73), then shows how the ratio of the asymptotic convergence rates of different line successive overrelaxation iterative methods behaves as a function of $k$. Practically, little is gained, for example, by choosing $k = 10$ since arithmetic requirements now also increase linearly with $k$, as pointed out by Parter.

## BIBLIOGRAPHY AND DISCUSSION

**6.1.** The three different methods described in this section for deriving finite difference approximations to boundary-value problems have received varying
amounts of attention from authors. The Taylor's series method, perhaps the

oldest and best-known approach, is described at some length by Kantorovich and Krylov (1958), p. 191, and Fox (1957), p. 18. On the other hand, the integration method seems to have arisen more recently as a result of numerical investigations of heterogeneous reactor problems. See Stark (1956), Tikhonov and Samarskii (1956), Varga (1957b), Wachspress (1960), and Marchuk (1959), p. 104. The variational approach to the derivation of finite difference equations, which can be traced back at least to Courant, Friedrichs, and Lewy (1928), is summarized in Forsythe and Wasow (1960), p. 159. It is interesting to point out that the Ritz form of the variational method in (6.16) for deriving finite difference equations can be equivalently viewed as an application of Galerkin's method, since the functions $t_i(x)$ in (6.16) are linearly independent. For details, see Kantorovich and Krylov (1958), p. 262. The variational approach has also been explicitly used in one-dimensional numerical eigenvalue studies by Farrington, Gregory, and Taub (1957), as well as in the derivation of difference approximations to the biharmonic equation in two dimensions by Engeli, Ginsburg, Rutishauser, and Stiefel (1959).

The result of Theorem 6.2, which shows that the discrete approximations converge to the solution of the two-point boundary-value problem of (6.1)– (6.2), is a specialization of a more general result by Gerschgorin (1930). Most proofs of this and related results, however, are based on the *maximum principle*, whereas the proof given in the text has emphasized the role of matrix theory. Basic to questions of convergence of discrete approximations, of course, is the fundamental paper of Courant, Friedrichs, and Lewy (1928), where convergence is proved under the weaker assumption that $f(x)$ of (6.1) is continuous.

**6.2.** The identity of (6.22″), obtained from integration, is not the only one that is useful in deriving difference approximations to differential equations with discontinuous coefficients. As indicated in Exercise 9, Tikhonov and Samarskii (1956) have established more intricate identities that are useful in practical applications. Also, Hildebrand (1955) considers the application of variational methods to such problems with discontinuous coefficients.

In Theorem 6.3, it was shown that $M$-matrices occur naturally in three-point finite difference approximations of

$$- \frac{d}{dx}\left(p\,\frac{dy}{dx}\right) + q\,\frac{dy}{dx} + \sigma y = f,$$

and (Exercise 5) the corresponding tridiagonal matrix has positive real eigenvalues for all mesh spacings $h$ sufficiently small. However, it is also interesting that this matrix must have *distinct* real eigenvalues for sufficiently small mesh spacings. This is related to the topic of *oscillation matrices*, introduced by Gantmakher and Krein (1937). See also Gantmakher (1959), p. 126.

We have indicated in Exercise 7 that improvements in accuracy of finite difference approximations can be made by means of differentiating the differential equation provided, of course, that the coefficients are sufficiently smooth.

This is closely related to the iterative method of *difference corrections*, due to Fox (1957). The theory for this topic has been considerably extended in the recent paper of Bickley, Michaelson, and Osborne (1961). However, another useful notion is Richardson's (1910) *deferred approach to the limit*. See, for example, Henrici (1962), p. 376. It should be stated that experience with these latter methods have largely been gained with differential problems with continuous coefficients and uniform mesh spacings, and its recommended use for problems with discontinuous coefficients appears to be an open question.

**6.3.** The derivation of finite difference approximations of second-order elliptic partial differential equations and related topics has received a great deal of attention in the literature. For the omitted topic of the convergence of these approximations, see Forsythe and Wasow (1960), p. 283, for a clear exposition of results as well as an up-to-date collection of references.

The concept of a *mesh region* associated with each mesh point, formulated by MacNeal (1953), is useful in deriving finite difference approximations. Stiefel in his chapter of Engeli *et al.* (1959) refers to this as the *dual mesh*.

Although we have only briefly treated the derivation of finite difference approximations to boundary-value problems with physical interfaces, the accurate solution of such problems is quite important in practice. Sheldon (1958) considers the approximation of such problems with general curved internal interfaces, and the convergence of such discrete approximations has been studied by Downing (1960). Generally speaking, higher-order methods, such as described in Sec. 6.2, have not been practically applied in two- and higher-dimensional problems with interfaces, although there is some evidence by Mann, Bradshaw, and Cox (1957) that such applications are worthwhile. See also Nohel and Timlake (1959).

**6.4.** The direct method of (6.46)–(6.46′), although known as a special case of Gaussian elimination, was apparently advocated first for practical use for computers by J. von Neumann and L. H. Thomas, and later by Bruce, Peaceman, Rachford, and Rice (1953). The stability of this procedure with respect to rounding errors has been considered by Douglas (1959) and Wilkinson (1961). A closely related procedure was suggested by Cuthill and Varga (1959), mainly to increase the efficiency of the line successive overrelaxation iterative method. The rounding error analysis of Douglas has been extended by Blair (1960) for this procedure.

The extensions of the direct inversion of tridiagonal matrices to the direct inversion of block tridiagonal matrices has been considered more widely by Karlquist (1952), Cornock (1954), Potters (1955), Meyer and Hollingsworth (1957), Nohel and Timlake (1959), Schechter (1960), Egerváry (1960), and von Holdt (1962). Rounding errors are more serious in this application. Unfortunately, numerical experience with these methods is not extensive.

Line successive overrelaxation (SLOR), on the other hand, has been used much more extensively in the United States, and has been treated theoretically by Arms, Gates, and Zondek (1956), and Keller (1958) and (1960a). Two-line successive overrelaxation (S2LOR), first suggested by Heller (1960), has been investigated for efficient use by Parter (1959) and Varga (1960a). For second order equations, see Price and Varga (1962) for numerical results of the use and comparison of these different methods, as well as Appendix B. For applications of S2LOR to biharmonic problems, see Parter (1961c) and Callaghan, et al., (1960).

# CHAPTER 7

# ALTERNATING-DIRECTION IMPLICIT
# ITERATIVE METHODS

## 7.1. The Peaceman-Rachford Iterative Method

In the previous chapter we considered various block iterative methods, such as the line and 2-line successive overrelaxation iterative methods, which as practical methods involved the direct (or implicit) solution of particular lower-order matrix equations. Also, as a standard for comparison of the asymptotic rates of convergence of these methods, we considered the numerical solution of the Dirichlet problem for the unit square with uniform mesh spacings, calling this the *model problem*. This knowledge of block methods, as well as representative asymptotic rates of convergence for the model problem for the iterative methods considered thus far, serves as the basis for the introduction of the newer *alternating-direction implicit iterative methods*, due to Peaceman and Rachford (1955) and Douglas and Rachford (1956).

For simplicity, we consider first the partial differential equation

$$(7.1) \qquad -u_{xx}(x, y) - u_{yy}(x, y) + \sigma u(x, y) = S(x, y),$$

$$(x, y) \in R,$$

in the unit square $R: 0 < x, y < 1$ of the $x$-$y$ plane, where $\sigma \geq 0$. For boundary conditions, we assume that

$$(7.2) \qquad u(x, y) = \gamma(x, y), \qquad (x, y) \in \Gamma,$$

where $\gamma(x, y)$ is a prescribed function on the boundary $\Gamma$ of $R$. If a uniform mesh of length $h$ in each coordinate direction is imposed on $R$, then the partial differential equation (7.1) is approximated, using central differences, by

$$(7.3) \qquad \{-u(x_0 - h, y_0) + 2u(x_0, y_0) - u(x_0 + h, y_0)\}$$

$$+ \{-u(x_0, y_0 - h) + 2u(x_0, y_0) - u(x_0, y_0 + h)\}$$

$$+ \sigma\{h^2 u(x_0, y_0)\} = h^2 S(x_0, y_0).$$

The totality of these difference equations gives rise to $n$ linear equations in $n$ unknowns of the form

$$(7.4) \qquad A\mathbf{u} = \mathbf{k},$$

where it has been shown (Sec. 6.3) that the $n \times n$ matrix $A$ is an irreducible *Stieltjes matrix*, i.e., $A$ is a real symmetric and positive definite irreducible matrix with nonpositive off-diagonal entries. Moreover, it is clear from (7.3) that the matrix $A$ has at most five nonzero entries in any of its rows. The important observation we make is that the bracketed terms of (7.3) respectively define $n \times n$ matrices $H$, $V$, and $\Sigma$, in the sense that if $[H\mathbf{u}](x_0, y_0)$ denotes the component of the vector $H\mathbf{u}$ corresponding to the spatial mesh point $(x_0, y_0)$, then

$$[H\mathbf{u}](x_0, y_0) \equiv -u(x_0 - h, y_0) + 2u(x_0, y_0) - u(x_0 + h, y_0),$$

$$(7.5) \qquad [V\mathbf{u}](x_0, y_0) \equiv -u(x_0, y_0 - h) + 2u(x_0, y_0) - u(x_0, y_0 + h),$$

$$[\Sigma\,\mathbf{u}](x_0, y_0) \equiv \sigma h^2 u(x_0\ y_0),$$

when all the mesh points $(x_0 \pm h, y_0 \pm h)$ are mesh points of unknowns. Another way of saying this is that the three matrices $H$, $V$, and $\Sigma$ arise respectively as central difference approximations to the first three terms of the differential equation of (7.1), and give us the representation

$$(7.6) \qquad A = H + V + \Sigma.$$

The following properties of the matrices $H$, $V$, and $\Sigma$ are readily verified. They are all real symmetric $n \times n$ matrices. The matrix $\Sigma$ is a non-negative diagonal matrix, and is thus non-negative definite. The matrices $H$ and $V$ each have no more than three nonzero entries per row, and both $H$ and $V$ are diagonally dominant matrices with positive diagonal entries and nonpositive off-diagonal entries, and are thus Stieltjes matrices. Considering the directed graphs of the matrices $H$ and $V$, it becomes clear that $H$ and $V$ are each *completely reducible* (Sec. 2.2). In fact, from the very definition (7.5) of the matrices $H$ and $V$, we see that the entries of these matrices are respectively generated from discrete central difference approximations to the particular one-dimensional equation

$$-u_{ss} = 0$$

along different *horizontal* and *vertical* mesh lines of $R$. Thus, with the boundary conditions of (7.2) and Theorem 6.1, we can state that there exist $n \times n$ permutation matrices $P_1$ and $P_2$ such that $P_1 H P_1^T$ and $P_2 V P_2^T$ are the direct sum of *tridiagonal irreducible Stieltjes matrices*.

Now that we have determined properties of the matrices $H$, $V$, and $\Sigma$ defining a splitting (7.6) of the matrix $A$, it is easy to go back and generalize. First, it is clear that we can equally well consider the partial differential equation

$$(7.1') \qquad -(Pu_x)_x - (Pu_y)_y + \sigma(x, y)u(x, y) = f(x, y), \quad (x, y) \in R,$$

for the general bounded region $R$ in the $x$-$y$ plane with boundary conditions

$$(7.2') \qquad \alpha(x, y)u + \beta(x, y)\frac{\partial u}{\partial n} = \gamma(x, y), \quad (x, y) \in \Gamma,$$

on the boundary $\Gamma$ of $R$, where $\alpha$, $\beta$ satisfy (6.30), and $P$ and $\sigma$ are piecewise continuous in $\bar{R}$, the closure of $R$, and satisfy (6.29). Moreover, there is no need to be restricted to a uniform mesh in either coordinate direction. In essence, then, we can derive difference equations† for this more general class of boundary-value problems which results in the matrix problem $A\mathbf{u} = \mathbf{k}$, where $A = H + V + \Sigma$, and the $n \times n$ matrices $H$ and $V$ are, after suitable permutation of indices, direct sums of tridiagonal irreducible Stieltjes matrices, and $\Sigma$ is a non-negative diagonal $n \times n$ matrix. Thus, as the matrix $A$ is the sum of positive definite and non-negative definite Hermitian matrices, then $A$ is nonsingular, and the solution of the matrix equation $A\mathbf{u} = \mathbf{k}$ is unique.

To introduce one variant of these alternating-direction implicit methods, using (7.6) we can write the matrix equation $A\mathbf{u} = \mathbf{k}$ as a *pair* of matrix equations

$$(7.7) \qquad \begin{aligned} (H + \tfrac{1}{2}\Sigma + rI)\mathbf{u} &= (rI - V - \tfrac{1}{2}\Sigma)\mathbf{u} + \mathbf{k}, \\ (V + \tfrac{1}{2}\Sigma + rI)\mathbf{u} &= (rI - H - \tfrac{1}{2}\Sigma)\mathbf{u} + \mathbf{k}, \end{aligned}$$

for any *positive* scalar $r$. If

$$(7.8) \qquad H_1 \equiv H + \tfrac{1}{2}\Sigma, \qquad V_1 \equiv V + \tfrac{1}{2}\Sigma,$$

---

† See Sec. 6.3 for details of such a derivation.

then the *Peaceman-Rachford (1955) implicit alternating-direction iterative method* is the implicit process defined by

(7.9)
$$(H_1 + r_{m+1}I)\mathbf{u}^{(m+1/2)} = (r_{m+1}I - V_1)\mathbf{u}^{(m)} + \mathbf{k},$$

$$(V_1 + r_{m+1}I)\mathbf{u}^{(m+1)} = (r_{m+1}I - H_1)\mathbf{u}^{(m+1/2)} + \mathbf{k}, \qquad m \geq 0,$$

where $\mathbf{u}^{(0)}$ is an arbitrary initial vector approximation of the unique solution of (7.4), and the $r_m$'s are positive constants called *acceleration parameters*, which are to be chosen so as to make the convergence of this process rapid. Since the matrices $(H_1 + r_{m+1}I)$ and $(V_1 + r_{m+1}I)$ are, after suitable permutations, tridiagonal nonsingular matrices, the above implicit process can be directly carried out by the simple algorithm of Sec. 6.4 based on Gaussian elimination. In other words, this iterative method can be thought of as a line method with alternating directions. Indeed, the name *alternating-direction method* is derived from the observation that for the first equation of (7.9), we solve first along horizontal mesh lines, and then, for the second equation of (7.9), we solve along vertical mesh lines.

We remark that the vector $\mathbf{u}^{(m+1/2)}$ is treated as an auxiliary vector which is *not* retained from one complete iteration to the next. In contrast to the variants of the successive overrelaxation technique, it is also interesting to note from (7.9) that one does *not* always make use of components of the latest iterate of the vector $\mathbf{u}$ in this iterative procedure.

We combine the two equations of (7.9) into the form

(7.9′)
$$\mathbf{u}^{(m+1)} = T_{r_{m+1}}\mathbf{u}^{(m)} + \mathbf{g}_{r_{m+1}}(\mathbf{k}), \qquad m \geq 0,$$

where

(7.10)
$$T_r \equiv (V_1 + rI)^{-1}(rI - H_1)(H_1 + rI)^{-1}(rI - V_1),$$

and

(7.10′)
$$\mathbf{g}_r(\mathbf{k}) \equiv (V_1 + rI)^{-1}\{(rI - H_1)(H_1 + rI)^{-1} + I\}\mathbf{k}.$$

We call the matrix $T_r$ the *Peaceman-Rachford matrix*. If $\boldsymbol{\varepsilon}^{(m)} = \mathbf{u}^{(m)} - \mathbf{u}$ is the error vector associated with the vector iterate $\mathbf{u}^{(m)}$, then $\boldsymbol{\varepsilon}^{(m+1)} = T_{r_{m+1}}\boldsymbol{\varepsilon}^{(m)}$, and in general

(7.11)
$$\boldsymbol{\varepsilon}^{(m)} = \left(\prod_{j=1}^{m} T_{r_j}\right)\boldsymbol{\varepsilon}^{(0)}, \qquad m \geq 1,$$

where

$$\prod_{j=1}^{m} T_{r_j} \equiv T_{r_m} \cdot T_{r_{m-1}} \cdots T_1.$$

Like the Chebyshev semi-iterative method of Chapter 5, we see here that the acceleration parameters $r_m$ can be varied from one iteration to the next.

To indicate the convergence properties of $\left( \prod_{j=1}^{m} T_{r_j} \right)$, we first consider the simple case where all the constants $r_j$ are equal to the fixed constant $r > 0$. In this case, the matrix $\tilde{T}_r$, defined as

$$(7.12) \qquad \tilde{T}_r \equiv (V_1 + rI) \, T_r (V_1 + rI)^{-1},$$

is similar to $T_r$, and thus has the same eigenvalues as $T_r$. With (7.10), it follows that we can express $\tilde{T}_r$ as

$$(7.12') \qquad \tilde{T}_r = \{(rI - H_1)(H_1 + rI)^{-1}\} \cdot \{(rI - V_1)(V_1 + rI)^{-1}\}.$$

With our knowledge of matrix norms and spectral radii from Chapter 1, it is evident that

$$(7.13) \qquad \rho(T_r) = \rho(\tilde{T}_r) \leq \| \tilde{T}_r \| \leq \| (rI - H_1)(H_1 + rI)^{-1} \|$$
$$\cdot \| (rI - V_1)(V_1 + rI)^{-1} \|.$$

To bound the norms of the product matrices of (7.13), we recall that both $H_1$ and $V_1$ are $n \times n$ Stieltjes matrices, which implies that they are Hermitian and positive definite. Denoting the (positive) eigenvalues of $H_1$ by $\lambda_j$, $1 \leq j \leq n$, it is readily verified that the matrix $(rI - H_1)(rI + H_1)^{-1}$ is Hermitian, with eigenvalues

$$\left( \frac{r - \lambda_j}{r + \lambda_j} \right), \quad 1 \leq j \leq n.$$

Thus, from the Corollary of Theorem 1.3, we can write

$$\| (rI - H_1)(H_1 + rI)^{-1} \| = \max_{1 \leq j \leq n} \left| \frac{r - \lambda_j}{r + \lambda_j} \right| < 1.$$

The same argument applied to the corresponding matrix product with $V_1$ shows that $\| (rI - V_1)(V_1 + rI)^{-1} \| < 1$, and we therefore conclude from (7.13) that $\rho(T_r) < 1$. To obtain a more general result, note that it is sufficient in this development only to have $H_1$, say, be Hermitian and positive definite, and that $V_1$ be Hermitian and non-negative definite, which gives us

**Theorem 7.1.** *Let $H_1$ and $V_1$ be $n \times n$ Hermitian non-negative definite matrices, where at least one of the matrices $H_1$ and $V_1$ is positive definite. Then, for any $r > 0$, the Peaceman-Rachford matrix $T_r$ of (7.10) is convergent.*

It is clear that we could once again consider more general partial differential equations and boundary conditions, which, when approximated by finite differences, give rise to matrices $H_1 = H + \frac{1}{2}\Sigma$ and $V_1 = V + \frac{1}{2}\Sigma$, satisfying the hypotheses of Theorem 7.1. Such extensions will be considered in the exercises at the end of this section.

In many ways, this first convergence theorem for alternating-direction implicit methods follows the pattern of our previous discussion of iterative methods. For a fixed single parameter $r > 0$, it is analogous to the point successive overrelaxation iterative method where but one acceleration parameter is selected for rapid convergence. For this simple Peaceman-Rachford iterative procedure with a single acceleration parameter $r > 0$, we consider the model problem to see how it compares with the point successive overrelaxation iterative method. To make the comparison equitable, we minimize the spectral radius of each iteration matrix as a function of its single parameter.

For the model problem, the special case $\sigma = 0$ of (7.1), and (7.2), there are $N \equiv 1/h$ equal subdivisions in each coordinate direction, which gives $(N - 1)^2$ linear equations in $(N - 1)^2$ unknowns. If the vector $\alpha^{(k,l)}$ (see Sec. 6.5) is defined so that its component for the $i$th column and $j$th row of the mesh is

$$\alpha_{i,j}^{(k,l)} \equiv \gamma_{k,l} \sin\left(\frac{k\pi i}{N}\right) \cdot \sin\left(\frac{l\pi j}{N}\right),$$

$$1 \leq i, j \leq N - 1, \ 1 \leq k, l \leq N - 1,$$

then, using the definitions of the matrices $H$ and $V$ of (7.5), it follows that

$$H\alpha^{(k,l)} = 2\left\{1 - \cos\left(\frac{k\pi}{N}\right)\right\}\alpha^{(k,l)},$$

(7.14)

$$V\alpha^{(k,l)} = 2\left\{1 - \cos\left(\frac{l\pi}{N}\right)\right\}\alpha^{(k,l)},$$

for all $1 \leq k, l \leq N - 1$, or equivalently,

$$H\alpha^{(k,l)} = 4\sin^2\left(\frac{k\pi}{2N}\right)\alpha^{(k,l)},$$

(7.14')

$$V\alpha^{(k,l)} = 4\sin^2\left(\frac{l\pi}{2N}\right)\alpha^{(k,l)}.$$

It also follows from (7.14′) that each $\alpha^{(k,l)}$ is an eigenvector of the matrix $T_r$ of (7.10), and explicitly,

$$(7.15) \qquad T_r \alpha^{(k,l)} = \left( \frac{r - 4 \sin^2\left(\dfrac{l\pi}{2N}\right)}{r + 4 \sin^2\left(\dfrac{l\pi}{2N}\right)} \right) \left( \frac{r - 4 \sin^2\left(\dfrac{k\pi}{2N}\right)}{r + 4 \sin^2\left(\dfrac{k\pi}{2N}\right)} \right) \alpha^{(k,l)},$$

$$1 \leq k, l \leq N - 1.$$

Thus, as the vectors $\alpha^{(k,l)}$ form an orthonormal basis for this $(N - 1)^2$ dimensional vector space over the complex number field, we conclude that

$$(7.16) \qquad \rho(T_r) = \left\{ \max_{1 \leq l \leq N-1} \left| \frac{r - 4 \sin^2\left(\dfrac{l\pi}{2N}\right)}{r + 4 \sin^2\left(\dfrac{l\pi}{2N}\right)} \right| \right\}^2.$$

To minimize this as a function of $r$, first consider the simple function

$$(7.17) \qquad g_1(x; r) = \frac{r - x}{r + x}, \qquad r > 0,$$

where $0 < x_1 \leq x \leq x_2$. By direct computation, the derivative of $g_1(x; r)$ with respect to $x$ is negative for all $x \geq 0$, so that the maximum of $|g_1(x; r)|$ is taken on at one of the end points of the interval:

$$\max_{x_1 \leq x \leq x_2} |g_1(x; r)| = \max \left\{ \left| \frac{r - x_1}{r + x_1} \right|, \left| \frac{r - x_2}{r + x_2} \right| \right\}.$$

From this, it is easily verified that

$$\max_{x_1 \leq x \leq x_2} |g_1(x; r)| = \begin{cases} \dfrac{x_2 - r}{x_2 + r}, & 0 < r \leq \sqrt{x_1 x_2} \\[2ex] \dfrac{r - x_1}{r + x_1}, & r \geq \sqrt{x_1 x_2} \end{cases},$$

from which it follows that

$$(7.18) \qquad \min_{r > 0} \left\{ \max_{x_1 \leq x \leq x_2} |g_1(x; r)| \right\} = g_1(x_1; \sqrt{x_1 x_2})$$

$$= \frac{\sqrt{x_1 x_2} - x_1}{\sqrt{x_1 x_2} + x_1} = \frac{1 - (x_1/x_2)^{1/2}}{1 + (x_1/x_2)^{1/2}}.$$

To minimize the spectral radius $\rho(T_r)$ of (7.16) as a function of $r > 0$, we now simply set

$$x_1 = 4 \sin^2 \left(\frac{\pi}{2N}\right) \quad \text{and} \quad x_2 = 4 \sin^2 \left(\frac{\pi(N-1)}{2N}\right) = 4 \cos^2 \left(\frac{\pi}{2N}\right),$$

and thus, with $\hat{r} \equiv 4 \sin(\pi/2N) \cos(\pi/2N) = \sqrt{x_1 x_2}$, we have

$$(7.19) \qquad \min_{r>0} \rho(T_r) = \rho(T_{\hat{r}}) = \left(\frac{1 - \tan(\pi/2N)}{1 + \tan(\pi/2N)}\right)^2$$

$$= \left[\frac{1 + \cos(\pi/N) - \sin(\pi/N)}{1 + \cos(\pi/N) + \sin(\pi/N)}\right]^2.$$

But this expression is actually quite familiar to us. Indeed, as shown in Sec. 6.5, the spectral radius $\rho(B)$ of the point Jacobi matrix $B$ for the model problem is

$$\rho(B) = \cos\left(\frac{\pi}{N}\right),$$

and as this matrix $B$ is a convergent and cyclic matrix of index 2 with real eigenvalues, it follows (Theorem 4.4) that, for any consistent ordering of the mesh points of the model problem, the spectral radius $\rho(\mathcal{L}_\omega)$ for the corresponding point successive overrelaxation matrix $\mathcal{L}_\omega$ satisfies

$$(7.20) \qquad \min_\omega \rho(\mathcal{L}_\omega) = \rho(\mathcal{L}_{\omega_b}) = \omega_b - 1 = \left(\frac{\rho(B)}{1 + \sqrt{1 - \rho^2(B)}}\right)^2.$$

Now, with $\rho(B) = \cos(\pi/N)$, it can be verified that the two final terms of (7.20) and (7.19) are *equal*, and thus,

$$(7.21) \qquad \min_{r>0} \rho(T_r) = \rho(T_{\hat{r}}) = \min_\omega \rho(\mathcal{L}_\omega) = \rho(\mathcal{L}_{\omega_b}).$$

The conclusion we reach is that, as optimized one-parameter iterative methods, the Peaceman-Rachford iterative method and the point successive overrelaxation iterative method have *identical* asymptotic rates of convergence *for all $h > 0$* for the model problem. For the model problem, it should be pointed out that the actual application of one complete iteration of the Peaceman-Rachford iterative method requires a good deal more arithmetic work than does the point successive overrelaxation iterative method. This derived relationship (7.21) for the model problem should not, however, obscure the final evaluation of alternating-direction implicit methods. Indeed, the real power of the method is brought forth when one considers using a sequence of parameters $\{r_i\}$, which, unlike the Chebyshev semi-iterative method, materially affects the asymptotic convergence behavior of this method.

Finally, to complement this last result, we mention that the Peaceman-Rachford iterative method with a single optimized parameter $r$, like the point successive overrelaxation iterative method, enjoys a *monotonicity principle* with respect to increases of the fundamental region $R$. By this, we mean the following: Suppose that our fundamental region $R$, consisting of the union of squares of side $h$, is imbedded into a *square* region $\tilde{R}$, which is also the union

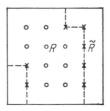

Figure 28

of such squares, as indicated in Figure 28. If the boundary-value problem for both mesh regions $R$ and $\tilde{R}$ is the Dirichlet problem, (7.1) and (7.2) with $\sigma = 0$, then it can be shown (Varga (1960b) that

$$(7.22) \qquad \min_r \rho(T_r(R)) \leq \min_r \rho(T_r(\tilde{R})) = \rho(T_{\tilde{r}}(\tilde{R})).$$

Analogously for the point Jacobi iterative method, it is clear that the associated point Jacobi matrices are non-negative and irreducible (Sec. 6.3), so that

$$\rho(B(R)) \leq \rho(B(\tilde{R})),$$

since in essence $B(R)$ is either a principal minor of $B(\tilde{R})$ or equal to $B(\tilde{R})$. Thus, in terms of the point successive overrelaxation iterative method, it follows from the monotonicity of $\omega_b$ as a function of $\rho(B)$ that

$$(7.23) \qquad \min_\omega \rho(\mathcal{L}_\omega(R)) = \omega_b(B(R)) - 1 \leq \omega_b(B(\tilde{R})) - 1,$$

and combining results, we conclude that

$$(7.24) \qquad \min_{r>0} \rho(T_r(R)) \leq \omega_b(B(\tilde{R})) - 1.$$

This is useful in determining bounds for the asymptotic rates of convergence of the Peaceman-Rachford iterative method. See Exercise 3.

## EXERCISES

**1.** For the $4 \times 4$ matrix $A$ of (1.6), derived from Laplace's equation over a uniform mesh, decompose $A$ in the manner of (7.6). Next, show for this case that the $4 \times 4$ matrices $H$ and $V$ are, after suitable permutations, direct sums of $2 \times 2$ irreducible Stieltjes matrices.

**2.** Derive a five-point approximation to Laplace's equation for the uniform mesh of the figure below where the circled mesh points are the unknowns of the problem. Decomposing the corresponding $6 \times 6$ matrix $A$ into the sum $H + V$ as in (7.6), show that the matrix $H$ is real, symmetric, and positive definite, but that the matrix $V$ is real, symmetric, and only non-negative definite. Whereas the Peaceman-Rachford matrix $T_r$ is convergent for any $r > 0$, determine analytically a value of $r$ which minimizes $\rho(T_r)$, and show that

$$\min_r \rho(T_r) \leq \frac{\sqrt{3} - 1}{\sqrt{3} + 1}.$$

(*Hint:* Try $\hat{r} = \sqrt{3}$.) Also, compute $HV - VH$.

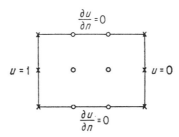

**3.** Derive a five-point difference approximation to the Dirichlet problem for the uniform mesh problem of the figure below where the boundary values are all unity. With $A = H + V$, show that

$$\min_r \rho(T_r) \leq \frac{2 - \sqrt{3}}{2 + \sqrt{3}}.$$

(*Hint:* Apply (7.24).) Also, compute $HV - VH$.

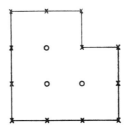

**4.** Let $H_1$ and $V_1$ be any $n \times n$ positive definite Hermitian matrices. Show that the Peaceman-Rachford matrix $T_r$ of (7.10) has *positive real eigenvalues* for all

a. $r > \max \{\rho(H_1), \rho(V_1)\}$,   and

b. $0 < r < \min \left\{ \dfrac{1}{\rho(H_1^{-1})}, \dfrac{1}{\rho(V_1^{-1})} \right\}.$

**5.** Let $H_1$ and $V_1$ be any $n \times n$ positive definite Hermitian matrices. Extend Theorem 7.1 by showing that $\rho\left(\prod_{i=1}^{m} T_{r_i}\right) < 1$ when $r_1 \geq r_2 \geq \cdots \geq r_m > 0$, and

$$\frac{r_1 - r_m}{2} \leq \frac{1}{\rho(V_1^{-1})}.$$

**6.** Let the real symmetric and positive definite $n \times n$ matrix $H$ be defined by

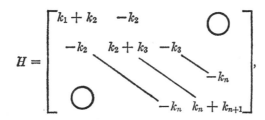

where $0 < \underline{K} \leq k_i \leq \bar{K}$ for $1 \leq i \leq n + 1$. If the eigenvalues of $H$ are $\lambda_j$, show that

$$0 < 2\underline{K}\left(1 - \cos\left(\frac{\pi}{n + 1}\right)\right) \leq \lambda_j \leq 2\bar{K}\left(1 + \cos\left(\frac{\pi}{n + 1}\right)\right).$$

**\*7.** Consider the numerical solution of the differential equation

$$-(K_1(x, y) u_x)_x - (K_2(x, y) u_y)_y = 0$$

in the unit square $R$ with Dirichlet boundary conditions, where

$$0 < \underline{K} \leq K_1(x, y), \quad K_2(x, y) \leq \bar{K}, \quad (x, y) \in \bar{R}.$$

Using a uniform mesh

$$\Delta x = \Delta y = h = \frac{1}{n + 1}$$

and the result of the previous exercise, show that the asymptotic rate of convergence of the Peaceman-Rachford matrix satisfies

$$\max_{r > 0} R_\infty(T_r) = O(h), \quad h \to 0.$$

## 7.2. The Commutative Case

The results of the application of the Peaceman-Rachford iterative method with a single acceleration parameter $r$ to the solution of the model problem suggest the possibility of even greater asymptotic rates of convergence through the use of a sequence of acceleration parameters $r_i$.

For this model problem, we observed that the matrices $H_1$ and $V_1$ of (7.8) possessed a common basis of orthogonal eigenvectors $\alpha^{(k,l)}$, where

$$\alpha_{i,j}^{(k,l)} = \gamma_{k,l} \sin\left(\frac{k\pi i}{N}\right) \sin\left(\frac{l\pi j}{N}\right),$$

and it is also readily verified that

$$(7.25) \qquad\qquad H_1 V_1 = V_1 H_1$$

for this model problem, which states that the matrices $H_1$ and $V_1$ *commute*.† The existence of such an orthonormal basis of eigenvectors $\alpha^{(k,l)}$ is actually equivalent to (7.25) by another classical theorem of *Frobenius*, which we now prove.

**Theorem 7.2.** *Let $H_1$ and $V_1$ be Hermitian $n \times n$ matrices. Then, there exists an orthonormal basis of eigenvectors $\{\alpha_i\}_{i=1}^n$ with $H_1\alpha_i = \sigma_i\alpha_i$ and $V_1\alpha_i = \tau_i\alpha_i$ for $1 \leq i \leq n$ if and only if $H_1V_1 = V_1H_1$.*

*Proof.* If such an orthonormal basis of eigenvectors $\{\alpha_i\}_{i=1}^n$ exists with $H_1\alpha_i = \sigma_i\alpha_i$ and $V_1\alpha_i = \tau_i\alpha_i$, then

$$H_1 V_1 \alpha_i = \sigma_i \tau_i \alpha_i = V_1 H_1 \alpha_i \qquad \text{for all } 1 \leq i \leq n.$$

Now, let $\mathbf{x}$ be *any* column vector of our $n$-dimensional vector space $V_n(C)$. As $\{\alpha_i\}_{i=1}^n$ is a basis for this vector space, there exist constants $c_i$ such that

$$\mathbf{x} = \sum_{i=1}^n c_i \alpha_i,$$

from which it follows that $H_1V_1\mathbf{x} = V_1H_1\mathbf{x}$. As this is true for all $\mathbf{x}$, we conclude that $H_1V_1 = V_1H_1$. Conversely, suppose that $H_1V_1 = V_1H_1$. As $H_1$ is Hermitian, let $U$ be an $n \times n$ unitary matrix which diagonalizes $H_1$, i.e., we assume

$$(7.26) \qquad UH_1U^* = \begin{bmatrix} \lambda_1 I_1 & & \bigcirc \\ & \lambda_2 I_2 & \\ & & \ddots \\ \bigcirc & & \lambda_r I_r \end{bmatrix} \equiv \hat{H},$$

† See Exercise 3 of Sec. 7.1 for another example where $H_1$ and $V_1$ do not commute.

where $\lambda_1 < \lambda_2 < \cdots < \lambda_r$ are the distinct eigenvalues of $H_1$, and $I_j$ is an identity matrix of order $n_j$ where $\sum_{j=1}^{r} n_j = n$. With $\hat{V} \equiv UV_1U^*$, partition the Hermitian matrix $\hat{V}$ relative to the partitioning of $\hat{H}$ of (7.26):

$$(7.27) \qquad \hat{V} = \begin{bmatrix} V_{1,1} & V_{1,2} & \cdots & V_{1,r} \\ V_{2,1} & V_{2,2} & \cdots & V_{2,r} \\ \cdot & \cdot & \cdot & \cdot \\ \cdot & \cdot & \cdot & \cdot \\ \cdot & \cdot & \cdot & \cdot \\ V_{r,1} & V_{r,2} & \cdots & V_{r,r} \end{bmatrix}.$$

It is clear that $H_1V_1 = V_1H_1$ implies that $\hat{H}\hat{V} = \hat{V}\hat{H}$. Carrying out this multiplication $\hat{H}\hat{V} = \hat{V}\hat{H}$ with the definitions of (7.26) and (7.27), we find that this in turn implies that $V_{j,k} = O$ for all $j \neq k$. Thus, the Hermitian matrix $\hat{V}$ is a block diagonal matrix, and each Hermitian submatrix $V_{j,j}$ possesses $n_j$ orthonormal eigenvectors which are simultaneously eigenvectors of the corresponding diagonal submatrix $\lambda_j I_j$ of $\hat{H}$. The totality of these eigenvectors then generate, in an obvious way, $n$ orthonormal vectors $\alpha_i$ with $H_1\alpha_i = \sigma_i\alpha_i$ and $V_1\alpha_i = \tau_i\alpha_i$, $1 \leq i \leq n$, completing the proof.

If $U$ is an $n \times n$ matrix whose columns are just the orthonormal vectors $\alpha_i$, $1 \leq i \leq n$, then $U$ is evidently unitary. Thus, we can restate Theorem 7.2 in the following equivalent form:

**Theorem 7.2'.** *Let $H_1$ and $V_1$ be Hermitian $n \times n$ matrices. Then, there exists an $n \times n$ unitary matrix $U$ for which $UH_1U^*$ and $UV_1U^*$ are both diagonal matrices if and only if $H_1V_1 = V_1H_1$.*

We henceforth assume in this section that $H_1V_1 = V_1H_1$, where $H_1$ and $V_1$ are $n \times n$ Hermitian and positive definite matrices. If $\{\alpha_i\}_{i=1}^{n}$ is an orthonormal basis of eigenvectors of $H_1$ and $V_1$, as guaranteed by Theorem 7.2, consider now the product $\prod_{j=1}^{m} T_{r_j}$ of $m$ Peaceman-Rachford matrices, corresponding to $m$ iterations of the Peaceman-Rachford method. Clearly,

$$(7.28) \qquad \left(\prod_{j=1}^{m} T_{r_j}\right)\alpha_i = \left\{\prod_{j=1}^{m}\left(\frac{r_j - \sigma_i}{r_j + \sigma_i}\right)\left(\frac{r_j - \tau_i}{r_j + \tau_i}\right)\right\}\alpha_i, \qquad 1 \leq i \leq n.$$

Using the $n \times n$ unitary matrix $U$ generated by the vectors $\boldsymbol{\alpha}_i$, we see that $U\left(\prod_{j=1}^{m} T_{r_j}\right)U^*$ is a real diagonal matrix, whose diagonal terms are obtained from (7.28). Thus, this product matrix is *Hermitian*, which gives us

**Theorem 7.3.** *Let $H_1$ and $V_1$ be $n \times n$ Hermitian and positive definite matrices with $H_1V_1 = V_1H_1$. Then if $\{r_j\}_{j=1}^{m}$ is any set of positive real numbers,*

$$(7.29) \qquad \left\|\prod_{j=1}^{m} T_{r_j}\right\| = \rho\left(\prod_{j=1}^{m} T_{r_j}\right) = \max_{1\leq i\leq n} \prod_{j=1}^{m} \left|\frac{r_j - \sigma_i}{r_j + \sigma_i}\right| \cdot \left|\frac{r_j - \tau_i}{r_j + \tau_i}\right| < 1.$$

*Moreover, if $r_{m+1} > 0$,*

$$(7.30) \qquad \left\|\prod_{j=1}^{m+1} T_{r_j}\right\| < \left\|\prod_{j=1}^{m} T_{r_j}\right\| < 1.$$

This monotone decreasing behavior of the norms of the products of Peaceman-Rachford matrices is similar to that encountered with Chebyshev semi-iterative methods. The similarity between these iterative methods is even more pronounced. First, we observe from (7.29) that if we knew *all* the eigenvalues of either the matrix $H_1$ or $V_1$ *a priori*, we could choose a sequence of positive real numbers $\{s_j\}_{j=1}^{n}$ so that

$$\left\|\prod_{j=1}^{n} T_{s_j}\right\| = 0,$$

and we would have a *direct method*, as opposed to an iterative method, just as in the Chebyshev semi-iterative methods.

Unfortunately, such a knowledge of all the eigenvalues of $H_1$ or $V_1$ is rarely encountered in practical problems, so we make the more practical assumption that we can estimate the spectral bounds $\alpha$ and $\beta$ for the eigenvalues $\sigma_j$ and $\tau_j$ of the positive definite matrices $H_1$ and $V_1$, i.e.,

$$(7.31) \qquad 0 < \alpha \leq \sigma_j, \quad \tau_j \leq \beta, \qquad 1 \leq j \leq n.$$

As

$$\max_{1\leq i\leq n} \prod_{j=1}^{m} \left|\frac{r_j - \sigma_i}{r_j + \sigma_i}\right| \cdot \left|\frac{r_j - \tau_i}{r_j + \tau_i}\right|$$

$$\leq \left\{\max_{1\leq i\leq n} \prod_{j=1}^{m} \left|\frac{r_j - \sigma_i}{r_j + \sigma_i}\right|\right\} \cdot \left\{\max_{1\leq i\leq n} \prod_{j=1}^{m} \left|\frac{r_j - \tau_i}{r_j + \tau_i}\right|\right\},$$

$$\leq \left\{\max_{\alpha\leq x\leq\beta} \prod_{j=1}^{m} \left|\frac{r_j - x}{r_j + x}\right|\right\}^2,$$

then

$$(7.32) \qquad \left\| \prod_{j=1}^{m} T_{r_j} \right\| \leq \left\{ \max_{\alpha \leq x \leq \beta} | g_m(x; r_j) | \right\}^2,$$

where

$$(7.33) \qquad g_m(x; r_j) \equiv \prod_{j=1}^{m} \left( \frac{r_j - x}{r_j + x} \right).$$

Note that this is simply the extension to $m$ parameters of the function considered in (7.17). In order to minimize the spectral norms of (7.32), we are led to the following min-max problem. If $S_m$ is the set of all functions $g_m(x; r_j)$, where $r_1, r_2, \cdots, r_m$ are positive or non-negative real numbers, let

$$(7.34) \qquad d_m[\alpha, \beta] \equiv \min_{g_m \epsilon S_m} \left\{ \max_{\alpha \leq x \leq \beta} | g_m(x; r_j) | \right\}.$$

We now state the solution of this min-max problem, based on the *Chebyshev principle* (see Achieser (1956), p. 51), which was first sketched by Wachspress (1957).†

**Theorem 7.4.** *There exists a unique function $g_m(x; \bar{r}_j) \in S_m$ for which*

$$(7.35) \qquad d_m[\alpha, \beta] = \max_{\alpha \leq x \leq \beta} | g_m(x; \bar{r}_j) |.$$

*The constants $\bar{r}_j$ are unique and distinct and satisfy $0 < \alpha < \bar{r}_j < \beta$, $1 \leq j \leq m$. The function $g_m(x; \bar{r}_j)$ is completely characterized by the property that $g_m(x; \bar{r}_j)$ takes on its maximum absolute value $d_m[\alpha, \beta]$, with alternating sign, in exactly $m + 1$ points $x_i$, where $\alpha = x_1 < x_2 < \cdots < x_{m+1} = \beta$.*

To illustrate this, recall the discussion in Sec. 7.1 of the minimization problem for $g_1(x; r)$ involving a single acceleration parameter $r$. In this case, $\bar{r} = \sqrt{\alpha\beta}$ minimized (7.34), and

$$g_1(\alpha, \bar{r}) = -g_1(\beta, \bar{r}) = d_1[\alpha, \beta].$$

Unlike the corresponding min-max problem in Sec. 5.1 for polynomials, a precise formula for the positive numbers $\bar{r}_1, \cdots, \bar{r}_m$ that give rise to the solution of the min-max problem of (7.34) is *not* known for *all* $m > 1$ as a general function of $\alpha$ and $\beta$. However, the following results of Wachspress (1962) give the precise formula in the special case that $m = 2^k$, $k \geq 0$.

† See also Wachspress (1962). The result of Theorem 7.4 can also be established from the recent deeper results of Rice (1960, 1961), concerning *unisolvent functions*.

**Corollary.** *Let* $\tilde{r}_1, \cdots, \tilde{r}_n$ *be the (unique) set of parameters which solve the min-max problem of* (7.34). *If* $\tilde{r}_j$ *is a parameter of this set, then so is* $\alpha\beta/\tilde{r}_j$. *Moreover,*

$$(7.36) \qquad | g_m(x; \tilde{r}_j) | = \left| g_m\left(\frac{\alpha\beta}{x}; \tilde{r}_j\right) \right|.$$

*Proof.* Let $y = \alpha\beta/x$. It then follows from the definition of $g_m(x; r_j)$ in (7.33) that

$$(7.37) \qquad g_m(x; \tilde{r}_j) = (-1)^m g_m\left(y; \frac{\alpha\beta}{\tilde{r}_j}\right).$$

But as $0 < \alpha \leq x \leq \beta$, then $y$ satisfies the same inequalities, and thus

$$\max_{\alpha \leq x \leq \beta} | g_m(x; \tilde{r}_j) | = \max_{\alpha \leq y \leq \beta} \left| g_m\left(y; \frac{\alpha\beta}{\tilde{r}_j}\right) \right|.$$

From the uniqueness of the parameters $\tilde{r}_j$ in Theorem 7.4, the result then follows.

For the case $m = 1$, note that it necessarily follows from this Corollary that the single parameter $\tilde{r}_1$ satisfies $\tilde{r}_1 = \alpha\beta/\tilde{r}_1$, which again gives $\tilde{r}_1 = \sqrt{\alpha\beta}$.

Suppose now that we can algebraically determine the $m$ optimum parameters which solve the min-max problem of (7.34) for any interval. This, as Wachspress (1962) has shown, leads to a precise determination of the $2m$ parameter problem. If $\tilde{r}_1 < \tilde{r}_2 < \cdots < \tilde{r}_{2m}$ are the parameters which solve the min-max problem for the interval $0 < \alpha \leq x \leq \beta$, then using the Corollary above, we can write

$$d_{2m}[\alpha, \beta] = \max_{\alpha \leq x \leq \beta} \left| \prod_{j=1}^{2m} \left(\frac{\tilde{r}_j - x}{\tilde{r}_j + x}\right) \right| = \max_{\alpha \leq x \leq \beta} \left| \prod_{j=1}^{m} \left(\frac{\tilde{r}_j - x}{\tilde{r}_j + x}\right)\left(\frac{\alpha\beta/\tilde{r}_j - x}{\alpha\beta/\tilde{r}_j + x}\right) \right|.$$

Note that the absolute value of the product is unchanged if $x$ is replaced by $\alpha\beta/x$. Thus, we can reduce the interval over which the maximum is taken to

$$d_{2m}[\alpha, \beta] = \max_{\alpha \leq x \leq \sqrt{\alpha\beta}} \left| \prod_{j=1}^{m} \left(\frac{\tilde{r}_j - x}{\tilde{r}_j + x}\right)\left(\frac{\alpha\beta/\tilde{r}_j - x}{\alpha\beta/\tilde{r}_j + x}\right) \right|.$$

Multiplying the factors and dividing through by $2x$ results in

$$d_{2m}[\alpha, \beta] = \max_{\alpha \leq x \leq \sqrt{\alpha\beta}} \left| \prod_{j=1}^{m} \left( \frac{\frac{\sqrt{\alpha\beta}}{2}\left(\frac{\sqrt{\alpha\beta}}{x} + \frac{x}{\sqrt{\alpha\beta}}\right) - \frac{1}{2}\left(\tilde{r}_j + \frac{\alpha\beta}{\tilde{r}_j}\right)}{\frac{\sqrt{\alpha\beta}}{2}\left(\frac{\sqrt{\alpha\beta}}{x} + \frac{x}{\sqrt{\alpha\beta}}\right) + \frac{1}{2}\left(\tilde{r}_j + \frac{\alpha\beta}{\tilde{r}_j}\right)} \right) \right|.$$

Letting

$$(7.38) \qquad y \equiv \frac{\sqrt{\alpha\beta}}{2}\left(\frac{\sqrt{\alpha\beta}}{x} + \frac{x}{\sqrt{\alpha\beta}}\right); \qquad \tilde{s}_j \equiv \frac{1}{2}\left(\tilde{r}_j + \frac{\alpha\beta}{\tilde{r}_j}\right),$$

we thus have

$$d_{2m}[\alpha, \beta] = \max_{\sqrt{\alpha\beta} \le y \le (\alpha+\beta)/2} \left| \prod_{j=1}^{m} \left(\frac{y - \tilde{s}_j}{y + \tilde{s}_j}\right) \right|$$

$$\ge d_m\left[\sqrt{\alpha\beta}, \frac{\alpha + \beta}{2}\right].$$

To show now that equality is valid, the argument above shows that $g_m(y, \tilde{s}_j)$ takes on its maximum absolute value $d_{2m}[\alpha, \beta]$ with alternating sign in $m + 1$ distinct points in $\sqrt{\alpha\beta} \le y \le (\alpha + \beta)/2$. Hence, by Theorem 7.4,

$$(7.39) \qquad d_{2m}[\alpha, \beta] = \max_{\sqrt{\alpha\beta} \le y \le (\alpha+\beta)/2} |g_m(y, \tilde{s}_j)| = d_m\left[\sqrt{\alpha\beta}, \frac{\alpha + \beta}{2}\right].$$

Since, by hypothesis, we can find the $m$ optimum parameters $\tilde{s}_j$ for the interval $\sqrt{\alpha\beta} \le x \le (\alpha + \beta)/2$, one then uses (7.38) to determine the corresponding $2m$ optimum parameters $\tilde{r}_j$ for the interval $\alpha \le x \le \beta$ by solving quadratic equations arising from (7.38).

Now, let $0 < \alpha < \beta$, and define

$$\alpha_0 = \alpha; \qquad \beta_0 = \beta;$$

$$(7.40)$$

$$\alpha_{i+1} = \sqrt{\alpha_i\beta_i}; \quad \beta_{i+1} = \frac{\alpha_i + \beta_i}{2}; \quad i \ge 0.$$

Repeated use of the relationship of (7.39) gives

$$(7.41) \qquad d_{2^k}[\alpha, \beta] = d_{2^{k-1}}[\alpha_1, \beta_1] = \cdots = d_{2^0=1}[\alpha_k, \beta_k].$$

Thus, we can reduce the problem of finding the $2^k$ optimum parameters for the interval $\alpha \le x \le \beta$ to the problem of finding a single optimum parameter for the interval $\alpha_k \le x \le \beta_k$, whose solution is obviously $\sqrt{\alpha_k\beta_k}$. Moreover, from Theorem 7.4, it follows that

$$d_1[\alpha_k, \beta_k] = g_1(\alpha_k; \sqrt{\alpha_k\beta_k}),$$

so that

$$(7.41') \qquad d_{2^k}[\alpha, \beta] = \frac{\sqrt{\beta_k} - \sqrt{\alpha_k}}{\sqrt{\beta_k} + \sqrt{\alpha_k}}, \qquad k \ge 0.$$

It is clear that this particularly elegant procedure of Wachspress for the case $m = 2^k$ is arithmetically very simple to use for practical computations.

In terms of spectral norms and average rates of convergence (Sec. 3.2), we now combine the result of (7.32) and (7.35) giving

$$
(7.42) \qquad \left\| \prod_{j=1}^{m} T_{\tilde{r}_j} \right\| \leq (d_m[\alpha, \beta])^2, \qquad m > 0,
$$

or equivalently, we have the *lower bounds* for the average rates of convergence

$$
(7.42') \qquad R\left( \prod_{j=1}^{m} T_{\tilde{r}_j} \right) \equiv \frac{-\ln \left\| \prod_{j=1}^{m} T_{\tilde{r}_j} \right\|}{m} \geq \frac{-2 \ln d_m[\alpha, \beta]}{m},
$$

$$
m > 0,
$$

and these lower bounds are readily calculated from (7.41') when $m = 2^k$.

In order to obtain lower bounds for the average rate of convergence of the Peaceman-Rachford iterative method in the commutative case for the *general* case of $m$ acceleration parameters, we now use a cruder analysis to estimate the optimum parameters $\tilde{r}_1, \cdots, \tilde{r}_m$. Divide the interval $0 < \alpha \leq x \leq \beta$ into $m$ subintervals $u_{k-1} \leq x \leq u_k$ where

$$
\alpha = u_0 < u_1 < u_2 < \cdots < u_m = \beta.
$$

In *each* subinterval $u_{k-1} \leq x \leq u_k$, we consider

$$
\min_{r_k > 0} \left\{ \max_{u_{k-1} \leq x \leq u_k} \left| \frac{r_k - x}{r_k + x} \right| \right\}, \qquad 1 \leq k \leq m,
$$

which, as we have seen, is solved by choosing $\hat{r}_k = \sqrt{u_{k-1} u_k}$. In order that

$$
\max_{u_{k-1} \leq x \leq u_k} \left| \frac{\hat{r}_k - x}{\hat{r}_k + x} \right| = \frac{1 - (u_{k-1}/u_k)^{1/2}}{1 + (u_{k-1}/u_k)^{1/2}} = \delta \qquad \text{for *all* } 1 \leq k \leq m,
$$

we set $\delta = (\gamma - 1)/(\gamma + 1)$ and inductively solve for $u_{k-1}$ in terms of $u_k$, which results in

$$
(7.43) \qquad u_k = \alpha \gamma^{2k}, \quad \hat{r}_k = \alpha \gamma^{2k-1}, \qquad 1 \leq k \leq m,
$$

where

$$
(7.43') \qquad \gamma \equiv \left( \frac{\beta}{\alpha} \right)^{1/2m}.
$$

We now let $g_m(x; \hat{r}_j)$ be our approximate solution of the min-max problem of (7.34). Simple arguments show that the maximum of the absolute value of the function $g_m(x; \hat{r}_j)$ is taken on at the end points of the interval $\alpha \leq x \leq \beta$, and

$$(7.44) \qquad \max_{\alpha \leq x \leq \beta} |\, g_m(x; \hat{r}_j) \,| \;=\; g_m(\alpha; \hat{r}_j) \;=\; (-1)^m g_m(\beta; \hat{r}_j)$$

$$= \prod_{k=1}^{m} \left( \frac{\gamma^{2k-1} - 1}{\gamma^{2k-1} + 1} \right).$$

From (7.44), the use of the particular acceleration parameters $\hat{r}_j$, coupled with (7.32), gives us the inequality

$$(7.45) \qquad \left\| \prod_{j=1}^{m} T_{\hat{r}_j} \right\| \leq \left( \prod_{k=1}^{m} \left( \frac{\gamma^{2k-1} - 1}{\gamma^{2k-1} + 1} \right) \right)^2$$

where $\gamma$ is defined in (7.43'). Thus, the average rate of convergence for $m$ iterations, as defined in Sec. 3.2, satisfies

$$(7.46) \qquad R\left( \prod_{j=1}^{m} T_{\hat{r}_j} \right) = \frac{-\ln \left\| \prod_{j=1}^{m} T_{\hat{r}_j} \right\|}{m} \geq \frac{2}{m} \sum_{k=1}^{m} \ln \left( \frac{\gamma^{2k-1} + 1}{\gamma^{2k-1} - 1} \right).$$

But as $\gamma^{2m} = (\beta/\alpha)$, then $2m \ln \gamma = \ln (\beta/\alpha)$, and substituting for $m$, we obtain

$$(7.47) \qquad R\left( \prod_{j=1}^{m} T_{\hat{r}_j} \right) \geq \left( \frac{4 \ln \gamma}{\ln (\beta/\alpha)} \right) \cdot \sum_{k=1}^{m} \ln \left( \frac{\gamma^{2k-1} + 1}{\gamma^{2k-1} - 1} \right).$$

We now observe that for $\gamma > 1$, the terms in the sum of (7.47) are *positive* and drop off rather rapidly with increasing $k$. Thus, by taking only the first term of this sum, we obtain the lower bound

$$(7.48) \qquad R\left( \prod_{j=1}^{m} T_{\hat{r}_j} \right) > \frac{4 \ln \gamma}{\ln (\beta/\alpha)} \ln \left( \frac{\gamma + 1}{\gamma - 1} \right) \equiv F_1(\gamma),$$

which is a function only of $\gamma$, where $\gamma$ itself is implicitly associated with $m$ through (7.43'). It can be shown (Young and Ehrlich (1960)) that for all $\gamma \geq 1$,

$$(7.49) \qquad F_1(\gamma) \leq F_1\left( \frac{1}{\sqrt{2} - 1} \right) = \frac{3.107}{\ln (\beta/\alpha)}.$$

Thus, with fixing $\hat{\gamma} = 1/(\sqrt{2} - 1) \doteq 2.41$, we have determined not only approximate acceleration parameters $\hat{r}_j$, but also the number of such parameters to be used cyclically. Again, with a knowledge of the eigenvalue bounds $\alpha$ and $\beta$, we determine a positive integer $m$ such that $(\hat{\gamma})^{2m} = (\beta/\alpha)$, and then our acceleration parameters $\hat{r}_k$ are generated from (7.43). Note that the ratio of successive acceleration parameters is approximately $(\hat{\gamma})^2 \doteq 5.8$.

To finally compare the average rate of convergence of (7.47) with that of the point successive overrelaxation method for the model problem, we now obtain, using $\gamma = \hat{\gamma}$,

$$(7.48') \qquad R\left(\prod_{j=1}^{m} T_{\hat{r}_j}\right) > F_1(\hat{\gamma}) = \frac{3.107}{\ln\,(\beta/\alpha)}.$$

For the model problem, the bounds for the eigenvalues of the matrices $H_1$ and $V_1$ are

$$\beta = 2\left(1 + \cos\frac{\pi}{N}\right) = 4\cos^2\left(\frac{\pi}{2N}\right)$$

and

$$\alpha = 2\left(1 - \cos\frac{\pi}{N}\right) = 4\sin^2\left(\frac{\pi}{2N}\right),$$

so that

$$(7.50) \qquad R\left(\prod_{j=1}^{m} T_{\hat{r}_j}\right) > \frac{3.107}{-2\ln\,[\tan\,(\pi/2N)]}.$$

As $\tan\,(\pi/2N) = \pi/2N + O(1/N^3)$ for large values of $N$,

$$(7.50') \qquad R\left(\prod_{j=1}^{m} T_{\hat{r}_j}\right) > \frac{3.107}{1.386 + 2\ln\,(N/\pi)}$$

for all $N$ sufficiently large. Recalling that the asymptotic rate of convergence of the point successive overrelaxation iterative method for the model problem is

$$R_\infty(\mathcal{L}_{\omega_b}) \sim \frac{2\pi}{N}, \qquad N \to \infty,$$

we see that there is an enormous difference in rates for very large $N$. To illustrate this, suppose that $N = 10^3$. Then, the ratio of these rates of convergence is

$$\frac{R\left(\prod\limits_{j=1}^{m} T_{\hat{r}_j}\right)}{R_\infty(\mathcal{L}_{\omega_b})} > 38.$$

Notice that in this analysis the number of acceleration parameters used cyclically changes with $N$.

Also of interest for the model problem is the fact that for *any* uniform mesh spacing $h = 1/N$, the Peaceman-Rachford iterative method with a *fixed* number $m$ of acceleration parameters $\hat{r}_j$ chosen according to (7.43) is always *iteratively faster* than the point successive overrelaxation method with optimum relaxation factor (see Exercise 8 of this section). Again, fixing the number of parameters $m$ for the model problem, then from (7.46) we have

$$R \left( \prod_{j=1}^{m} T_{\hat{r}_j} \right) > \frac{2}{m} \ln \left( \frac{\gamma + 1}{\gamma - 1} \right) = \frac{2}{m} \ln \left( 1 + \frac{2/\gamma}{1 - 1/\gamma} \right),$$

where $1/\gamma = (\tan (\pi/2N))^{1/m}$. For large values of $N$, we thus conclude (Young and Ehrlich (1960)) that

$$(7.51) \qquad R \left( \prod_{j=1}^{m} T_{\hat{r}_j} \right) > \frac{4}{m} \left( \frac{\pi}{2N} \right)^{1/m} = \frac{K}{N^{1/m}}$$

for all $N$ sufficiently large. Again, this means that when $N = 1/h$ is large, significantly faster convergence is to be had with the Peaceman-Rachford iterative method for any number $m > 1$ of acceleration parameters, as compared to the point successive overrelaxation iterative method for the model problem.

The analysis given above that establishes the remarkable rate of convergence in (7.50′) is crude and can certainly be improved. For example, letting

$$F_p(\gamma) \equiv \frac{4 \ln \gamma}{\ln (\beta/\alpha)} \sum_{k=1}^{p} \ln \left( \frac{\gamma^{2k-1} + 1}{\gamma^{2k-1} - 1} \right), \qquad p \geq 1,$$

one can similarly maximize $F_p(\gamma)$ as a function of $\gamma > 1$. In particular, with $p = 4$, $F_4(\gamma) \leq F_4(1.478) = 4.221$. This allows us to increase the constant 3.107 to 4.222, in (7.48′) provided that the approximate acceleration parameters $\hat{r}_j$ are determined from (7.43) where $\gamma_4 = 1.478$, as opposed to $\gamma_1 = 2.41$.

The results of the comparison between the Peaceman-Rachford iterative method and the successive overrelaxation iterative method for the model problem strongly suggest that these basic theoretical results would extend to the general case. Indeed, this has been our experience with the iterative methods considered thus far—namely, that theoretical results obtained for the model problem are indicative of theoretical results for the general case. The next section points out that the commutative property $H_1 V_1 = V_1 H_1$, which is valid for the model problem and basic to the theoretical development of this section, is actually quite exceptional.

## EXERCISES

**1.** Prove the following generalization of Theorem 7.2. Let $H_1$ and $V_1$ be two $n \times n$ matrices, each of which is similar to a diagonal matrix. Then $H_1 V_1 = V_1 H_1$ if and only if there exists a common basis of eigenvectors $\alpha_i$, $1 \leq i \leq n$, with $H_1 \alpha_i = \sigma_i \alpha_i$ and $V_1 \alpha_i = \tau_i \alpha_i$.

**2.** Using (7.38) and (7.39), show that the two optimum parameters $\tilde{r}_1$ and $\tilde{r}_2$ that solve the min-max problem $m = 2$ of (7.34) for the interval $0 < \alpha \leq x \leq \beta$, are given by

$$\tilde{r}_1 = \left[ \sqrt{\alpha\beta} \left( \frac{\alpha + \beta}{2} \right) \right]^{1/2} - \left[ \sqrt{\alpha\beta} \left( \frac{\alpha + \beta}{2} \right) - \alpha\beta \right]^{1/2},$$

$$\tilde{r}_2 = \left[ \sqrt{\alpha\beta} \left( \frac{\alpha + \beta}{2} \right) \right]^{1/2} + \left[ \sqrt{\alpha\beta} \left( \frac{\alpha + \beta}{2} \right) - \alpha\beta \right]^{1/2}.$$

**3.** For any $0 < \alpha < \beta$, using the definitions of (7.40), show that

$$\alpha_0 < \alpha_j < \alpha_{j+1} < \beta_{j+1} < \beta_j < \beta_0, \qquad j \geq 1,$$

and that $\lim_{j \to \infty} \alpha_j = \lim_{j \to \infty} \beta_j$.

**4.** Let $\alpha = 10^{-4}$ and $\beta = 1$. First, determine the $m$ optimum acceleration parameters $\tilde{r}_j$ that solve the min-max problem of (7.34) for each of the cases $m = 2^k$, $k = 0, 1, 2, 3, 4$. Next, using (7.41'), determine numerical lower bounds for the average rate of convergence

$$R\left( \prod_{j=1}^{m} T_{\tilde{r}_j} \right) \geq \frac{-2 \ln d_m[\alpha, \beta]}{m}, \qquad m = 2^k, \quad k = 0, 1, 2, 3, 4.$$

**5.** Prove that the infinite product

$$\prod_{k=1}^{\infty} \left( \frac{\gamma^{2k-1} + 1}{\gamma^{2k-1} - 1} \right)$$

is convergent for $\gamma > 1$.

**6.** Show for the model problem that equality is *valid* in (7.32) and, consequently, in (7.42') and (7.47).

**7.** Let the quantities $\tilde{r}_1$ and $\tilde{r}_2$, as given in Exercise 2, be the optimum parameters that solve the min-max problem of (7.34) for the case $m = 2$ for the interval $0 < \alpha \leq x \leq \beta$. Let $\hat{r}_1$ and $\hat{r}_2$ be their approximate values given by (7.43) for the case $m = 2$. Show for fixed $\beta$ and variable $\alpha > 0$ that

$$\lim_{\alpha \to 0} \frac{\tilde{r}_1}{\hat{r}_1} = \frac{1}{\sqrt{2}}, \qquad \lim_{\alpha \to 0} \frac{\tilde{r}_2}{\hat{r}_2} = \sqrt{2}.$$

8. For the point successive overrelaxation method, we know that

$$R_\infty(\mathcal{L}_{\omega b}) = \ln\left(\frac{1 + \sin \pi/N}{1 - \sin \pi/N}\right)$$

for the model problem. Using the results of Exercise 6 and (7.50), show that the ratio

$$R\left(\prod_{j=1}^{m} T_{\hat{r}_j}\right) \Big/ R_\infty(\mathcal{L}_{\omega b})$$

for each $m \geq 1$ is greater than unity for *all* $N \geq 2$.

9. For the model problem with $\alpha = 4 \sin^2 (\pi/2N)$, $\beta = 4 \cos^2 (\pi/2N)$, it was shown in (7.51) that

$$R\left(\prod_{j=1}^{2} T_{\hat{r}_j}\right) > \sqrt{2}\left(\frac{\pi}{N}\right)^{1/2}, \qquad N \to \infty.$$

Using the results of Exercise 6 and (7.41′), prove that the following sharper relationship is valid:

$$R\left(\prod_{j=1}^{2} T_{\hat{r}_j}\right) \sim 2\left(\frac{\pi}{N}\right)^{1/2}, \qquad N \to \infty.$$

Similarly, prove that

$$R\left(\prod_{j=1}^{4} T_{\hat{r}_j}\right) \sim 2\sqrt{2}\left(\frac{\pi}{N}\right)^{1/4}, \qquad N \to \infty,$$

which improves upon the inequality

$$R\left(\prod_{j=1}^{4} T_{\hat{r}_j}\right) > \frac{1}{2^{1/4}}\left(\frac{\pi}{N}\right)^{1/4}, \qquad N \to \infty.$$

10. For $0 < \alpha < \beta$, let the lower bounds for the average rates of convergence in (7.42′) be defined as

$$E_m(\alpha, \beta) \equiv \frac{-2 \ln d_m[\alpha, \beta]}{m}, \qquad m \geq 1.$$

Using Theorem 7.4, show that the subsequence $E_{2^k}(\alpha, \beta)$ is *strictly increasing* with increasing $k$. Thus, there is no finite value of $k \geq 1$ which maximizes the lower bounds (7.42′) for the average rates of convergence

$$R\left(\prod_{j=1}^{2^k} T_{\hat{r}_j}\right).$$

## 7.3. The Noncommutative Case

In order to rigorously apply the theoretical results of the previous section, we now ask under what conditions do the matrices $H_1$ and $V_1$ commute, i.e., $H_1V_1 = V_1H_1$. We investigate this problem relative to the elliptic partial differential equation

$$(7.52) \qquad -(K_1(x, y)u_x)_x - (K_2(x, y)u_y)_y$$

$$+ \sigma(x, y)u(x, y) = S(x, y), \qquad (x, y) \in R,$$

defined for a (connected) bounded region $R$, with boundary conditions

$$(7.53) \qquad u(x, y) = \gamma(x, y), \qquad (x, y) \in \Gamma,$$

where $\gamma(x, y)$ is a prescribed function on $\Gamma$, the external boundary of the region $R$. We assume that

$$K_1(x, y), \quad K_2(x, y), \sigma(x, y) \text{ are continuous in } \bar{R},$$

$$(7.54)$$

$$K_1(x, y) > 0, \quad K_2(x, y) > 0, \quad \sigma(x, y) \geq 0 \text{ in } \bar{R}.$$

For the discrete case, which leads to a definition of the matrices $H_1$ and $V_1$, a mesh of horizontal and vertical line segments is imposed on $\bar{R}$, and for convenience, we assume a uniform mesh of length $h$.† We denote the set of the interior mesh points of $R$ by $\Lambda_h$, corresponding to unknowns in the discrete case, and the discrete mesh points of $\Gamma$ by $\Gamma_h$. Considering only five-point discrete approximations to (7.52), we arrive typically at three-point difference expressions for the matrices $H$ and $V$. Although there are a variety of such approximations, we consider here the more usual central difference approximations:

$$(7.55) \qquad [H\mathbf{u}](x, y) = -K_1\left(x + \frac{h}{2}, y\right) u(x + h, y)$$

$$- K_1\left(x - \frac{h}{2}, y\right) u(x - h, y)$$

$$+ \left[K_1\left(x + \frac{h}{2}, y\right) + K_1\left(x - \frac{h}{2}, y\right)\right] u(x, y), \qquad (x, y) \in \Lambda_h,$$

† The case for nonuniform mesh spacings offers no additional difficulty. Similarly, mixed boundary conditions can also be treated. See Birkhoff and Varga (1959).

$$(7.55') \qquad [V\mathbf{u}](x, y) = -K_2\left(x, y + \frac{h}{2}\right) u(x, y + h)$$

$$- K_2\left(x, y - \frac{h}{2}\right) u(x, y - h)$$

$$+ \left[K_2\left(x, y + \frac{h}{2}\right) + K_2\left(x, y - \frac{h}{2}\right)\right] u(x, y), \qquad (x, y) \in \Delta_h,$$

$$(7.55'') \qquad [\Sigma \mathbf{u}](x, y) = h^2\sigma(x, y)\mathbf{u}(x, y), \qquad (x, y) \in \Delta_h,$$

which could also be deduced from the derivations of Sec. 6.3. If any of the mesh points $(x \pm h, y \pm h)$ are points of the boundary $\Gamma_h$, the corresponding terms in the definitions of the matrices above are set equal to zero. Since the region $R$ is connected, it is intuitively clear that the matrix $A \equiv H + V + \Sigma$ will be irreducible† for all $h$ sufficiently small, and we henceforth assume that $A$ is irreducible. Again, as in Sec. 7.1, we set $H_1 = H + \frac{1}{2}\Sigma$ and $V_1 = V + \frac{1}{2}\Sigma$.

**Lemma 7.1.** *If $H_1$ and $V_1$ commute, then*

$$K_1\left(x + \frac{h}{2}, y\right) = K_1\left(x + \frac{h}{2}, y + h\right)$$

*and*

$$K_2\left(x, y + \frac{h}{2}\right) = K_2\left(x + h, y + \frac{h}{2}\right)$$

*whenever $(x, y)$, $(x + h, y)$, $(x + h, y + h)$, and $(x, y + h)$ are mesh points of $\Delta_h$.*

*Proof.* Let the values of $K_1$ and $K_2$ at the four points of the statement of this lemma be denoted simply by the quantities $K_i^{(j)}$ of Figure 29. By

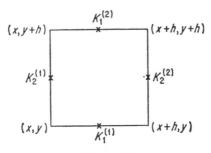

**Figure 29**

† One needs only to consider the directed graph of $A$ for this to become apparent.

direct computation,

$$[H_1V_1\mathbf{u}](x, y) = K_1^{(1)}K_2^{(2)}u(x + h, y + h) + \cdots,$$

and

$$[V_1H_1\mathbf{u}](x, y) = K_1^{(2)}K_2^{(1)}u(x + h, y + h) + \cdots.$$

Thus, as $H_1V_1 = V_1H_1$, we necessarily have that $K_1^{(1)}K_2^{(2)} = K_1^{(2)}K_2^{(1)}$. Similarly, $K_1^{(1)}K_2^{(1)} = K_1^{(2)}K_2^{(2)}$. Since these values $K_i^{(j)}$ are positive from (7.54), we can divide these two expressions, which gives

$$\frac{K_2^{(2)}}{K_2^{(1)}} = \frac{K_2^{(1)}}{K_2^{(2)}} \quad \text{or} \quad (K_2^{(1)})^2 = (K_2^{(2)})^2.$$

Using the positivity of $K_2(x, y)$, it follows that $K_2^{(2)} = K_2^{(1)}$, and similarly, $K_1^{(1)} = K_1^{(2)}$, which is the desired result.

**Lemma 7.2.** *If $H_1$ and $V_1$ commute and three mesh points of a square of side $h$ are mesh points of $\Lambda_h$, then so is the fourth.*

*Proof.* Using Figure 29, suppose that $(x, y + h)$, $(x + h, y + h)$, $(x + h, y)$ are mesh points of $\Lambda_h$ but that $(x, y)$ is a mesh point of $\Gamma_h$. Now,

$$[H_1V_1\mathbf{u}](x, y + h) = K_1^{(2)}K_2^{(2)}u(x + h, y) + \cdots.$$

On the other hand, $[V_1H_1\mathbf{u}](x, y + h)$ has no contribution to the mesh point $(x + h, y)$. Thus, as $K_1^{(2)}K_2^{(2)} > 0$, the matrices $H_1$ and $V_1$ do not commute, which is a contradiction.

It is interesting that the proof of this last lemma can be interpreted equivalently by *graph theory*. In essence, what is shown in the proof is that the directed graph for the matrix product $H_1V_1$ has a path of length unity from the node at $(x, y + h)$ to the node at $(x + h, y)$, whereas the directed graph for the matrix product $V_1H_1$ has no corresponding path.

**Corollary.** *If $H_1$ and $V_1$ commute, then the convex closure† of the mesh points of $\Lambda_h$ is a rectangle with sides parallel to the coordinate axes.*

*Proof.* By induction, it follows from Lemma 7.2 that the mesh points of $\Lambda_h$ form a rectangular array with sides parallel to the coordinate axes. Evidently, the convex closure of these points is a rectangle.

**Lemma 7.3.** *If $H_1$ and $V_1$ commute, then $\Sigma = cI$, where $c$ is a nonnegative scalar.*

---

† The convex closure of a finite point set is the smallest convex polygon containing the point set. It is identical with the convex hull of the finite point set. See Bonnesen and Fenchel (1934), p. 5.

*Proof.* Again referring to Figure 29, we have that

$$[V_1 H_1 \mathbf{u}](x, y + h)$$

$$= -K_2^{(1)} \left[ K_1 \left( x - \frac{h}{2}, y + h \right) + K_1^{(2)} + \frac{1}{2} h^2 \sigma(x, y + h) \right] u(x, y) + \cdots,$$

whereas

$$[H_1 V_1 \mathbf{u}](x, y + h)$$

$$= -K_2^{(1)} \left[ K_1 \left( x - \frac{h}{2}, y \right) + K_1^{(1)} + \frac{1}{2} h^2 \sigma(x, y) \right] u(x, y) + \cdots.$$

Thus, if $H_1$ and $V_1$ commute, then it follows from Lemma 7.1 that

$$\sigma(x, y) = \sigma(x, y + h).$$

Similarly,

$$\sigma(x, y) = \sigma(x + h, y).$$

From the Corollary above we can continue this argument for all mesh points of $\Lambda_h$, and thus the matrix $\Sigma$ is a non-negative scalar (because of (7.54)) multiple of the identity matrix, which completes the proof.

Combining the results of the lemmas and the Corollary, we have

**Theorem 7.5.** *The matrices* $(H + \frac{1}{2}\Sigma)$ *and* $(V + \frac{1}{2}\Sigma)$ *defined from* (7.55)–(7.55″) *commute if and only if the convex closure of the mesh points* $\Lambda_h$ *is a rectangle with sides parallel to the coordinate axes,* $\Sigma = cI$ *where* $c$ *is a non-negative scalar, and*

$$(7.56) \qquad K_1 \left( x + \frac{h}{2}, y \right) = f_1 \left( x + \frac{h}{2} \right), \qquad K_2 \left( x, y + \frac{h}{2} \right) = f_2 \left( y + \frac{h}{2} \right).$$

*Proof.* The necessity follows from the previous lemmas and the Corollary. The converse follows by direct computation.

It now becomes apparent that the basic assumption $H_1 V_1 = V_1 H_1$ made in Sec. 7.2 places rather severe limitations on the type of boundary-value problems which can be rigorously treated by the analysis of Sec. 7.2. Considering all mesh lengths $h > 0$ gives us, in analogy with Theorem 7.5,

**Theorem 7.6.** *Let* $K_1(x, y)$ *and* $K_2(x, y)$ *be any positive continuous functions defined on* $\bar{R}$. *Then, except for the case when* $K_1(x, y) = f_1(x)$, $K_2(x, y) = f_2(y)$, $\sigma(x, y)$ *is a constant, and* $\bar{R}$ *is a rectangle, for all sufficiently small mesh spacings* $h > 0$ *the matrices* $H_1$ *and* $V_1$ *defined from* (7.55)–(7.55″) *fail to commute.*

*Proof.* If the matrices $H_1$ and $V_1$ commute for all mesh spacings $h > 0$, it is clear from Lemma 7.1 that $K_1(x, y) = f_1(x)$, $K_2(x, y) = f_2(y)$ for all $(x, y) \in \bar{R}$. The remainder follows from Theorem 7.5.

It is appropriate to point out now that although the results of this section are essentially *negative* results for the Peaceman-Rachford iterative method, the positive result of Theorem 7.1 that proved the convergence of this method for a fixed acceleration parameter $r > 0$ was established *without* the assumption that $H_1 V_1 = V_1 H_1$. For further positive results that tie in the development of the Peaceman-Rachford iterative method with our previous concepts of cyclic and primitive matrices, we again focus our attention on the case where the acceleration parameters $r_m$ are all equal to a fixed positive real number $r$. From the definition of the Peaceman-Rachford method, we have from (7.9)

$$\mathbf{u}^{(m+1/2)} = (H_1 + rI)^{-1}(rI - V_1)\mathbf{u}^{(m)} + (H_1 + rI)^{-1}\mathbf{k},$$

(7.57)

$$\mathbf{u}^{(m+1)} = (V_1 + rI)^{-1}(rI - H_1)\mathbf{u}^{(m+1/2)} + (V_1 + rI)^{-1}\mathbf{k},$$

$$m \geq 0,$$

where we assume that $H_1$ and $V_1$ are $n \times n$ Hermitian and non-negative definite, and at least one of these matrices is positive definite. On the other hand, employing the device of Sec. 5.2, we can write the matrix equation $(H_1 + V_1)\mathbf{u} = \mathbf{k}$ as the *pair* of equations

$$(H_1 + rI)\mathbf{x} = (rI - V_1)\mathbf{y} + \mathbf{k},$$

(7.58)

$$(V_1 + rI)\mathbf{y} = (rI - H_1)\mathbf{x} + \mathbf{k},$$

and as neither $(H_1 + rI)$ nor $(V_1 + rI)$ is singular, we write this equivalently as

(7.59)
$$\begin{bmatrix} \mathbf{x} \\ \hline \mathbf{y} \end{bmatrix} = \begin{bmatrix} O & C_r \\ \hline D_r & O \end{bmatrix} \begin{bmatrix} \mathbf{x} \\ \hline \mathbf{y} \end{bmatrix} + \begin{bmatrix} (H_1 + rI)^{-1}\mathbf{k} \\ (V_1 + rI)^{-1}\mathbf{k} \end{bmatrix},$$

where

(7.59')    $C_r \equiv (H_1 + rI)^{-1}(rI - V_1), \quad D_r \equiv (V_1 + rI)^{-1}(rI - H_1).$

However, this can also be written as

(7.60)    $$\mathbf{z} = \hat{J}_r \mathbf{z} + \mathbf{h},$$

where

(7.61)    $$\mathbf{z} = \begin{bmatrix} \mathbf{x} \\ \mathbf{y} \end{bmatrix}, \quad \hat{J}_r = \begin{bmatrix} O & C_r \\ \hline D_r & O \end{bmatrix}.$$

It is now obvious that the $2n \times 2n$ matrix $\hat{J}_r$ is a *consistently ordered weakly cyclic matrix of index* 2 from its very definition. Moreover, it is clear that $\rho(\hat{J}_r^2) = \rho(T_r)$. As we already know (Theorem 7.1) that $T_r$ is a convergent matrix, then $\hat{J}_r$ is also a convergent matrix. Thus, there is a unique solution vector $\mathbf{z}$ of the matrix equation of (7.60), and evidently its components $\mathbf{x}$ and $\mathbf{y}$ are equal.

The block Gauss-Seidel iterative method applied to the matrix equation (7.60), with the particular partitioning of (7.61), is given by

(7.62)
$$\mathbf{x}^{(m+1)} = C_r \mathbf{y}^{(m)} + (H_1 + rI)^{-1}\mathbf{k}$$

$$\mathbf{y}^{(m+1)} = D_r \mathbf{x}^{(m+1)} + (V_1 + rI)^{-1}\mathbf{k}, \qquad m \geq 0,$$

which is necessarily convergent for any initial vector $\mathbf{y}^{(0)}$. But this application of the block Gauss-Seidel iterative method to (7.60) is *precisely* equivalent to the original definition of the Peaceman-Rachford iterative method; this is most easily seen by setting $\mathbf{x}^{(m+1)} \equiv \mathbf{y}^{(m+1/2)}$. In other words, for a fixed acceleration parameter $r$, the Peaceman-Rachford iterative method is equivalent to the application of the block Gauss-Seidel iterative method to the $2n \times 2n$ matrix equation of (7.60).

We can generalize this result to the case where $m$ positive acceleration parameters $r_1, r_2, \cdots, r_m$ are used cyclically in the Peaceman-Rachford iterative method. For this case, consider the $2mn \times 2mn$ matrix equation of

(7.63)
$$\mathbf{z} = \hat{J}_{r_1,\ldots,r_m}\mathbf{z} + \mathbf{h},$$

where

(7.64)
$$\hat{J}_{r_1,\ldots,r_m} \equiv \begin{bmatrix} 0 & 0 & 0 & \cdots & 0 & C_{r_1} \\ D_{r_1} & 0 & 0 & \cdots & 0 & 0 \\ 0 & C_{r_2} & 0 & \cdots & 0 & 0 \\ 0 & 0 & D_{r_2} & \ddots & 0 & 0 \\ \vdots & & & & & \\ 0 & 0 & 0 & \cdots & D_{r_m} & 0 \end{bmatrix},$$

and the matrices $C_r$ and $D_r$ are defined in (7.59′). Again, we see by inspection that $\hat{J}_{r_1,\ldots,r_m}$ is a *consistently ordered weakly cyclic matrix of index* $2m$. In an analogous way, we can show that

$$\rho\{(\hat{J}_{r_1,\ldots,r_m})^{2m}\} = \rho\left(\prod_{j=1}^{m} T_{r_j}\right),$$

and that the rigorous application of the block Gauss-Seidel iterative method to (7.63) is again the Peaceman-Rachford iterative method with $m$ acceleration parameters used cyclically, which gives us

**Theorem 7.7.** *The Peaceman-Rachford iterative method of (7.9) with $m$ positive acceleration parameters $r_j$, $1 \leq j \leq m$, used cyclically, is equivalent to the application of the Gauss-Seidel iterative method to the matrix equation (7.63), where the matrix $\hat{J}_{r_1,\ldots,r_m}$ of order $2mn$ is a consistently ordered weakly cyclic matrix of index $2m$.*

The theory of primitive matrices also has a nontrivial connection with the Peaceman-Rachford iterative method that is analogous to the observation that for $0 < \omega < 1$, the point successive overrelaxation matrix $\mathcal{L}_\omega$ is *primitive* if the corresponding point Jacobi matrix $B$ is non-negative and irreducible (see Exercise 4 of Sec. 3.6). If the $n \times n$ irreducible Stieltjes matrix $A$ of (7.6) is expressed as the sum of the two $n \times n$ matrices $H_1 = (h_{i,j})$ and $V_1 = (v_{i,j})$, each of which is, after suitable permutations, the direct sum of irreducible Stieltjes matrices as in Sec. 7.1, let the positive real number $\tau$ be defined by

$$(7.65) \qquad \tau \equiv \max_{1 \leq i \leq n} \{h_{i,i}; v_{i,i}\}.$$

**Theorem 7.8.** *Let $A = H_1 + V_1$ be an irreducible $n \times n$ matrix, where $H_1$ and $V_1$ are, after suitable permutations, the direct sum of Stieltjes matrices. For any acceleration parameter $r > \tau$, the Peaceman-Rachford matrix $T_r$ of (7.10) is a primitive matrix (i.e., the matrix $T_r$ is non-negative, irreducible, and noncyclic.)*

*Proof.* For any $r > 0$, the matrices $(H_1 + rI)^{-1}$ and $(V_1 + rI)^{-1}$ are non-negative matrices with positive diagonal entries. This follows from the Stieltjes character (Sec. 3.5) of $H_1$ and $V_1$. Next, using the fact that Stieltjes matrices have nonpositive off-diagonal entries, it follows that for $r > \tau$, the matrices $(rI - V_1)$ and $(rI - H_1)$ are non-negative matrices with positive diagonal entries. Thus, the product matrix $T_r$ is a non-negative matrix with positive diagonal entries. To prove that $T_r$ is primitive, from Lemma 2.8 and Theorem 2.5 of Sec. 2.2, it suffices to prove that the matrix $T_r = (t_{i,j})$ is irreducible. Clearly, the matrices $(rI - H_1) \equiv (h_{i,j}^{(1)})$, $(rI + H_1)^{-1} \equiv (h_{i,j}^{(2)})$, $(rI - V_1) \equiv (v_{i,j}^{(1)})$, $(rI + V_1)^{-1} = (v_{i,j}^{(2)})$ all have positive diagonal entries for $r > \tau$, and by definition

$$(7.66) \qquad t_{i,j} = \sum_{l_1,l_2,l_3} v_{i,l_1}^{(2)} h_{l_1,l_2}^{(1)} h_{l_2,l_3}^{(2)} v_{l_3,j}^{(1)}.$$

If $A = (a_{i,j})$, then $a_{i,j} \neq 0$ implies that either $v_{i,j}^{(1)} \neq 0$ or $h_{i,j}^{(1)} \neq 0$, and thus, from (7.66), $t_{i,j} \neq 0$. But as $A$ is irreducible by hypothesis, then so is $T_r$, which completes the proof.

An interesting application of this primitivity property is the following. If the mesh set $\Lambda_h$ for, say, the Dirichlet problem on a uniform mesh is a rectangle, then it can be shown (see Exercise 1 of this section) that $T_r > 0$ for $r > 2$. In other words, $T_r$ is primitive and the *index of primitivity*† $\gamma(T_r)$ is unity. This can be interpreted in a more geometrical way. Suppose in the case of this rectangle that some vector iterate $\mathbf{u}^{(m)}$ of the Peaceman-Rachford iterative method is such that its associated error vector $\boldsymbol{\varepsilon}^{(m)}$ has all its components zero, *save one* (which we assume is positive). The fact that $\gamma(T_r) = 1$ implies that the next complete iteration of the Peaceman-Rachford method gives rise to an error vector $\boldsymbol{\varepsilon}^{(m+1)}$, *all* of whose components are positive. In other words, one iteration distributes completely the error at a single point over the entire mesh. Although the point successive overrelaxation matrix $\mathcal{L}_\omega$ is also primitive for $0 < \omega < 1$ in the same case of a $p \times q$ rectangle, its index of primitivity is $p + q - 2$ (see Exercise 2 of this section), which means that a maximum of $p + q - 2$ iterations are required to equivalently distribute this error over the entire mesh. From this heuristic point of view, the successive overrrelaxation iterative method seems in comparison less attractive. In the next chapter, where iterative methods for solving elliptic difference equations are related to parabolic partial differential equations, we shall deduce (Theorem 8.6) the primitivity of certain iteration matrices as a consequence of the nature of their approximations to exponential matrices.

## EXERCISES

1. If the region $R$ is a $p \times q$ rectangle, prove for the Dirichlet problem on a uniform mesh that the Peaceman-Rachford matrix $T_r$ is primitive for $r > 2$, and $\gamma(T_r) = 1$.

2. For the same matrix problem as in the previous exercise, prove that the point successive overrelaxation matrix $\mathcal{L}_\omega$ is primitive for $0 < \omega < 1$, and $\gamma(\mathcal{L}_\omega) = p + q - 2$.

3. Consider the numerical solution of the Dirichlet problem for the case where the interior mesh set $\Lambda_h$ consists of the seven mesh points shown in the figure. Prove that $T_r$ is primitive for $r > 2$ and that $\gamma(T_r) = 2$. What is the index of primitivity for the point successive overrelaxation matrix?

† See Sec. 2.2.

## 7.4. Variants of the Peaceman-Rachford Iterative Method

The Peaceman-Rachford iterative method which we have investigated is merely one particular iterative method of the class of alternating-direction implicit iterative methods. To generate other related methods, we consider the first equation of (7.9) defining the auxiliary vector iterate $\mathbf{u}^{(m+1/2)}$:

$$(7.67) \qquad (H_1 + r_{m+1}I)\mathbf{u}^{(m+1/2)} = (r_{m+1}I - V_1)\mathbf{u}^{(m)} + \mathbf{k}, \qquad m \geq 0.$$

The second equation defining the Peaceman-Rachford iterative method is given by

$$(7.68) \qquad (V_1 + r_{m+1}I)\mathbf{u}^{(m+1)} = (r_{m+1}I - H_1)\mathbf{u}^{(m+1/2)} + \mathbf{k}, \qquad m \geq 0.$$

However, it is clear that we can write (7.68) in a form in which the matrix $H_1$ does not appear explicitly.† In fact, substituting for $H_1\mathbf{u}^{(m+1/2)}$ from (7.67) gives

$$(7.68') \qquad (V_1 + r_{m+1}I)\mathbf{u}^{(m+1)} = (V_1 - r_{m+1}I)\mathbf{u}^{(m)} + 2r_{m+1}\mathbf{u}^{(m+1/2)},$$

$$m \geq 0,$$

which could be heuristically derived from the simple identity

$$(7.69) \qquad (V_1 + r_{m+1}I)\mathbf{u} = (V_1 - r_{m+1}I)\mathbf{u} + 2r_{m+1}\mathbf{u}$$

by inserting iteration indices. The *Douglas-Rachford* (1956) *iterative method* is defined by (7.67) and

$$(7.70) \qquad (V_1 + r_{m+1}I)\mathbf{u}^{(m+1)} = V_1\mathbf{u}^{(m)} + r_{m+1}\mathbf{u}^{(m+1/2)}.$$

More generally, let us consider the iterative method defined from (7.67) and

$$(7.71) \qquad (V_1 + r_{m+1}I)\mathbf{u}^{(m+1)} = (V_1 - (1 - \omega)r_{m+1}I)\mathbf{u}^{(m)}$$

$$+ (2 - \omega)r_{m+1}\mathbf{u}^{(m+1/2)}, \qquad m \geq 0,$$

where $\omega$ is a fixed scalar. For $\omega = 0$ we have the Peaceman-Rachford iterative method, whereas for $\omega = 1$ we have the Douglas-Rachford iterative method. Combining (7.67) and (7.71), we can write this gen-

---

† Wachspress and Habetler (1960) make use of this observation to economize on arithmetic operations for the Peaceman-Rachford iterative method.

eralized alternating-direction implicit method as

(7.72)     $\mathbf{u}^{(m+1)} = W_{r_{m+1}}(\omega)\mathbf{u}^{(m+1)} + \mathbf{g}(r_{m+1}, \omega),$     $m \geq 0,$

where

(7.73)     $W_r(\omega) = (V_1 + rI)^{-1}(H_1 + rI)^{-1}$

$$\cdot \{H_1V_1 + (\omega - 1)r(H + V) + r^2I\},$$

and

(7.74)     $\mathbf{g}(r, \omega) = (2 - \omega)r(V_1 + rI)^{-1}(H_1 + rI)^{-1}\mathbf{k}.$

By direct verification, we can show that

(7.75)                $W_r(\omega) = \tfrac{1}{2}\{\omega I + (2 - \omega)T_r\},$

where $T_r$ is the Peaceman-Rachford matrix of (7.10); but it is easy to bound the spectral radius of the matrix $W_r(\omega)$ relative to that of the Peaceman-Rachford matrix $T_r$, since

(7.76)       $\rho(W_r(\omega)) \leq \tfrac{1}{2}\{\omega + (2 - \omega)\rho(T_r)\},$     $2 \geq \omega \geq 0.$

Thus, from Theorem 7.1, we immediately conclude

**Theorem 7.9.** *Let $H_1$ and $V_1$ be $n \times n$ Hermitian non-negative definite matrices, where at least one of the matrices $H_1$ and $V_1$ is positive definite. Then, for any $r > 0, 0 \leq \omega < 2$, the generalized alternating-direction implicit matrix $W_r(\omega)$ of (7.73) is convergent.*

**Corollary.** *With the hypotheses of Theorem 7.9, the Douglas-Rachford iterative method of (7.67) and (7.70) is convergent for any fixed $r > 0$ and any initial vector $\mathbf{u}^{(0)}$.*

In the commutative case $H_1V_1 = V_1H_1$, the relationship (7.75) between the eigenvalues of the Douglas-Rachford matrix $W_r(1)$ and the Peaceman-Rachford matrix $T_r$ allows one to show (Exercise 2) that

$$1 > \min_{r>0} \rho(W_r(1)) > \min_{r'>0} \rho(T_{r'}),$$

so that as optimized one-parameter iterative methods, the Peaceman-Rachford iterative method is always superior to the Douglas-Rachford iterative method. One of the claims stated for the Douglas-Rachford iterative method is that it can also be used to solve three-dimensional elliptic problems.†

† This was pointed out in the original paper of Douglas and Rachford.

A second important variant of these alternating-direction implicit methods has been given by Wachspress and Habetler (1960). Clearly, the identity matrix as used in (7.67) and (7.68) could be replaced by any positive real diagonal matrix $F$. This neither affects the algorithm for the direct solution of matrix equations such as

$$(H + r_{m+1}F)\mathbf{u}^{(m+1/2)} = \mathbf{h},$$

nor does it essentially increase the arithmetic work required in carrying out the Peaceman-Rachford iterative method. The reason for introducing this positive diagonal matrix $F$ is to "help" the associated matrices more nearly commute. Consider now the Wachspress-Habetler variant of the Peaceman-Rachford iterative method defined by

$$(7.77) \qquad (H_1 + r_{m+1}F)\mathbf{u}^{(m+1/2)} = (r_{m+1}F - V_1)\mathbf{u}^{(m)} + \mathbf{k}, \qquad m \geq 0,$$

and

$$(7.78) \qquad (V_1 + r_{m+1}F)\mathbf{u}^{(m+1)} = (r_{m+1}F - H_1)\mathbf{u}^{(m+1/2)} + \mathbf{k}, \qquad m \geq 0.$$

Defining $F^{1/2}\mathbf{u} \equiv \mathbf{v}$, this variant takes the more familiar form

$$(7.79) \qquad (\tilde{H}_1 + r_{m+1}I)\mathbf{v}^{(m+1/2)} = (r_{m+1}I - \tilde{V}_1)\mathbf{v}^{(m)} + F^{-1/2}\mathbf{k},$$
$$m \geq 0,$$

and

$$(7.80) \qquad (\tilde{V}_1 + r_{m+1}I)\mathbf{v}^{(m+1)} = (r_{m+1}I - \tilde{H}_1)\mathbf{v}^{(m+1/2)} + F^{-1/2}\mathbf{k},$$
$$m \geq 0,$$

where

$$(7.81) \qquad \tilde{H}_1 \equiv F^{-1/2}H_1F^{-1/2}; \qquad \tilde{V}_1 \equiv F^{-1/2}V_1F^{-1/2}.$$

In order to apply the theory for the commutative case of Sec. 7.2 to (7.79) and (7.80), we now ask if

$$(7.82) \qquad \tilde{H}_1\tilde{V}_1 = \tilde{V}_1\tilde{H}_1,$$

or equivalently, using (7.81), if

$$(7.83) \qquad H_1F^{-1}V_1 = V_1F^{-1}H_1.$$

In effect, as the diagonal entries of the matrix $F$ are parameters that are to be chosen, it is clear that there is a class of matrix problems for which $H_1$ and $V_1$ do *not* commute, but there exists a positive diagonal matrix $F$ for which (7.83) is satisfied. Indeed, Wachspress and Habetler (1960) have shown for the separable boundary-value problem of (7.1) and (7.2) that one can take *unequal* mesh spacings in each coordinate direction for the discrete case, and yet find a positive diagonal matrix $F$ for which

(7.83) is valid. On the other hand, the method of proof of Lemma 7.2 shows that such extensions of the commutative theory are still restricted to rectangular domains. Whether or not the discrete mesh region is rectangular, Wachspress and Habetler recommend the use of such a positive diagonal matrix $F$ because they consider the effect of this matrix in (7.81) as a *matrix conditioning* of the matrices $H_1$ and $V_1$.

In Sec. 8.4 we shall give another interpretation of the selection of this diagonal matrix $F$, which arises from matrix methods applied to the solution of parabolic partial differential equations. Young approaches this problem directly from the point of view of differential equations and links the selection of the matrix $F$ to appropriate multipliers of the differential equation.†

The possible use of a positive diagonal matrix $F$ in place of the identity matrix for the Peaceman-Rachford iterative method leads us to examine more carefully the role that the non-negative diagonal matrix $\Sigma$ plays in this iterative method. Recalling the definition of the matrices $H_1$ and $V_1$ in (7.8), we see that the matrix $\Sigma$ entered into the definition of the Wachspress-Habetler variant of the Peaceman-Rachford iterative method in a *symmetric* way. It seems appropriate to consider the following variant of this method, defined by

$$(7.84) \qquad (H + \Sigma + \tilde{r}_{m+1}F)\mathbf{u}^{(m+1/2)} = (\tilde{r}_{m+1}F - V)\mathbf{u}^{(m)} + \mathbf{k},$$

$$m \geq 0,$$

and

$$(7.84') \qquad (V + \Sigma + \tilde{r}_{m+1}F)\mathbf{u}^{(m+1)} = (\tilde{r}_{m+1}F - H)\mathbf{u}^{(m+1/2)} + \mathbf{k},$$

$$m \geq 0,$$

where $F$ is a positive diagonal matrix. Again, no logical or practical difficulties arise in this formulation because the matrix equations above can still be directly solved by the same basic algorithm. This method might be intuitively attractive since the matrices to be inverted are more "diagonally dominant" than their counterparts in (7.77) and (7.78), which would indicate greater stability against rounding errors in such direct inversions when $\Sigma$ is a positive diagonal matrix. With similar hypotheses, the method of proof of Theorem 7.1 can be extended (Wachspress-Habetler (1960)) to show that this method is also convergent for any fixed $r > 0$ (see Exercise 3).

In the case where $\Sigma \equiv cI$, where $c \geq 0$ and $F \equiv I$, it follows that the iterative method of (7.84)–(7.84') reduces *precisely* to the iterative

† To appear in Birkhoff, Varga, and Young (1962).

method of (7.77)–(7.78) upon identifying the corresponding acceleration parameters through the translation

$$r_{m+1} = \frac{c}{2} + \tilde{r}_{m+1}.$$

Note, then, that in the commutative case, $\Sigma$ is indeed a scalar multiple
of the identity matrix from Lemma 7.3, and thus this third variant above
offers no improvements in convergence rates over the previous variant in
the commutative case.

Thus far our investigations of the alternating-direction methods have
been restricted to the numerical solution of elliptic partial differential
equations with two space variables. As previously mentioned, the Douglas-
Rachford iterative method can be generalized for use in numerically solving
such problems with three space variables. It is also interesting to describe
the new method of Douglas (1962) that generalizes the Peaceman-Rachford iterative method to higher-dimensional problems and appears to be
more efficient than the corresponding Douglas-Rachford procedure.
Specifically, suppose that we wish to approximate the solution of the
Dirichlet problem

(7.85)     $u_{xx}(x, y, z) + u_{yy}(x, y, z) + u_{zz}(x, y, z) = 0,$

$$(x, y, z) \in R.$$

in a finite region $R$ with boundary $\Gamma$, where $u(x, y, z)$ is prescribed on $\Gamma$.
The matrix problem arising approximating (7.85) with a *seven-point
formula* takes the form

(7.86)     $(X + Y + Z)\mathbf{u} = \mathbf{k},$

where each of the matrices $X$, $Y$, and $Z$ is a Hermitian and positive definite
$n \times n$ matrix, and each is, after suitable permutation transformations, the
direct sum of *tridiagonal* matrices. Douglas (1962) proposes the iterative
method whose vector iterates $\mathbf{u}_m$ are defined by

$$(X + r_{m+1}I)\mathbf{u}^{(1)}_{m+1} = 2\mathbf{k} + (r_{m+1}I - X - 2Y - 2Z)\mathbf{u}_m,$$

$$(Y + r_{m+1}I)\mathbf{u}^{(2)}_{m+1} = 2\mathbf{k} + (r_{m+1}I - X - Y - 2Z)\mathbf{u}_m$$

(7.87)                                                          $$- X\mathbf{u}^{(1)}_{m+1},$$

$$(Z + r_{m+1}I)\mathbf{u}_{m+1} = 2\mathbf{k} + (r_{m+1}I - X - Y - Z)\mathbf{u}_m$$

$$- X\mathbf{u}^{(1)}_{m+1} - Y\mathbf{u}^{(2)}_{m+1},$$

where $r_{m+1} > 0$. Note that the determination of $\mathbf{u}_{m+1}$ consists of solving tridiagonal linear equations three times. It is then clear how to extend this procedure to $n$ dimensions. In the case where the matrices $X$, $Y$, and $Z$ all commute, Douglas (1962) has shown that the above procedure is convergent for any fixed $r > 0$.

Inevitably the person intent on solving physics and engineering problems asks which iterative methods are best for his problems—variants of the successive overrelaxation iterative method or variants of the alternating-direction implicit methods. Although there exists no clear-cut theoretical argument in the general case which rigorously gives one of the methods an advantage, many have experimented with these methods to shed light on their general usefulness and perhaps to indicate possible avenues of theoretical attacks. Although the Peaceman-Rachford iterative method has been widely used in industrial circles, the first published experimental work systematically examining the convergence of the Peaceman-Rachford iterative method was that of Young and Ehrlich (1960), which indicated how successful such ADI methods could be, even when it was apparent that the commutative theory on which the selection of parameters was based did *not* hold. Price and Varga (1962) extended the basic results of Young and Ehrlich in two dimensions to other numerical problems, especially those arising from problems in petroleum engineering and reactor physics, and compared the Peaceman-Rachford iterative method with more recent and powerful variants of the successive overrelaxation iterative method such as the 2-line cyclic Chebyshev semi-iterative method. These numerical investigation have produced problems for which *complex* eigenvalues of the matrix $T_r$ occur (Young-Ehrlich (1960)), as well as problems which *diverge* for two positive acceleration parameters used cyclically (Price-Varga (1962)). Generally speaking these numerical results do uniformly indicate the superiority of the alternating-direction implicit methods for *small* mesh spacings, as the observed convergence rates obtained are close to those predicted by the commutative theory of Sec. 7.2. More precisely, for each problem tested, the numerical results favor, in terms of *actual* machine time for comparable accuracy, the successive overrelaxation iterative variant of 2-line cyclic Chebyshev semi-iterative method for all mesh spacings $h \geq h^*$, and then favor a many-parameter (such as five or six) Peaceman-Rachford method for $h < h^*$. As one would expect, this crossover point $h^*$ varied from problem to problem. As an example, this crossover point occurred for the model problem at $h^* = \frac{1}{14}$, indicating that the Peaceman-Rachford method is efficient even for relatively coarse mesh spacings for this problem. For other problems, $h^*$ was much smaller.†

---

† Curiously enough, for the same model problem but with a square hole removed from the center of the unit square, the crossover point was $h^* = \frac{1}{70}$.

## EXERCISES

**1.** Verify $(7.75)$.

**2.** Let $H_1$ and $V_1$ be $n \times n$ positive definite Hermitian matrices, with $H_1 V_1 = V_1 H_1$. If $T_r$ is the Peaceman-Rachford matrix of $(7.10)$ and $W_r(1)$ is the Douglas-Rachford matrix of $(7.73)$, show using $(7.75)$ that

$$1 > \min_{r>0} \rho[W_r(1)] > \min_{r'>0} \rho(T_{r'}).$$

**3.** If $H$ and $V$ are $n \times n$ Hermitian and positive definite, and $F$ and $\Sigma$ are real positive diagonal $n \times n$ matrices, prove that, for any fixed acceleration parameter $\tilde{r} > 0$, the iterative method of $(7.84)$–$(7.84')$ is convergent (Wachspress-Habetler (1960)).

**4.** Let the matrices $X$, $Y$, and $Z$ in $(7.86)$ be Hermitian and positive definite. If these matrices all commute, i.e., $XY = YZ$, $XZ = ZX$, and $YZ = ZY$, show that the iterative method of $(7.87)$ is convergent for any fixed $r > 0$. Generalize this result to the case of $n \geq 3$ Hermitian and positive definite matrices (Douglas (1962)).

**5.** Let $H_1$ and $V_1$ be $n \times n$ Hermitian and positive definite matrices with $H_1 V_1 = V_1 H_1$. Show that for any fixed *complex* acceleration parameter $r = i\beta, \beta \neq 0$, the Douglas-Rachford iteration matrix $W_r(1)$ of $(7.75)$ is convergent.

**\*6.** The matrix problem arising from the nine-point approximation to the model problem (the Dirichlet problem for the unit square), using a uniform mesh $h = \Delta x = \Delta y = 1/(N+1)$, can be expressed as

$$(H + V + R)\mathbf{u} = \mathbf{k},$$

where the matrices $H$, $V$, and $R$ are defined by (see $(7.5)$):

$$[H\mathbf{u}](x_0, y_0) = -4u(x_0 - h, y_0) + 8u(x_0, y_0) - 4u(x_0 + h, y_0),$$

$$[V\mathbf{u}](x_0, y_0) = -4u(x_0, y_0 - h) + 8u(x_0, y_0) - 4u(x_0, y_0 + h),$$

$$[R\mathbf{u}](x_0, y_0) = -u(x_0 - h, y_0 - h) - u(x_0 - h, y_0 + h)$$
$$- u(x_0 + h, y_0 - h) - u(x_0 + h, y_0 + h) + 4u(x_0, y_0),$$

where all the mesh points $(x_0 \pm h, y_0 \pm h)$ are mesh points of unknowns. Show that the Douglas-Rachford-type iterative method defined by

$$(H + rI)\mathbf{u}^{(m+1/2)} = \mathbf{k} + (rI - V - R)\mathbf{u}^{(m)},$$

$$(V + rI)\mathbf{u}^{(m+1)} = V\mathbf{u}^{(m)} + r\mathbf{u}^{(m+1/2)},$$

is convergent for any $r > 0$. (*Hint*: Prove that the matrices $H$, $V$, and $R$ all commute.)

**7.** Let $H_1$ and $V_1$ be the Hermitian matrices of Exercise 2 of Sec. 7.1. Although $H_1V_1 \neq V_1H_1$, show that there exists a positive diagonal matrix $F$ such that (7.83) is valid. (*Hint*: Choose the diagonal entries of $F$ proportional to the areas of the mesh regions for the figure associated with that exercise.)

**8.** Let the $n \times n$ Hermitian matrices $H_1$ and $V_1$ be respectively positive definite and non-negative definite. Clearly, for $\alpha > 0$ sufficiently small, $\tilde{H}_1 \equiv H_1 - \alpha I$ and $\tilde{V}_1 \equiv V_1 + \alpha I$ are both Hermitian and positive definite. Show that the Peaceman-Rachford iterative method of (7.9) with acceleration parameters $r_{m+1}$ for the matrices $\tilde{H}_1$ and $\tilde{V}_1$ is equivalent to the variant defined by

$$(H_1 + s_{m+1}I)\mathbf{u}^{(m+1/2)} = (s_{m+1}I - V_1)\mathbf{u}^{(m)} + k,$$

$$(V_1 + \tilde{s}_{m+1}I)\mathbf{u}^{(m+1)} = (\tilde{s}_{m+1}I - H_1)\mathbf{u}^{(m+1/2)} + \mathbf{k}, \quad m \geq 0,$$

where $r_{m+1} - \alpha = s_{m+1}$, $r_{m+1} + \alpha = \tilde{s}_{m+1}$.

## BIBLIOGRAPHY AND DISCUSSION

**7.1.** The first alternating-direction implicit iterative method was given by Peaceman and Rachford (1955). Shortly thereafter, the closely related iterative method (7.67)–(7.70) was given by Douglas and Rachford (1956). It is interesting to note that these iterative methods for solving elliptic difference equations were initially derived as by-products of numerical methods for approximating the solution of parabolic equations (see Chapter 8).

Peaceman and Rachford (1955), like Frankel (1950), analyzed their new iterative method only for the solution of the model problem in two dimensions, i.e., the Dirichlet problem for the unit square, but they nevertheless established for this *model problem* the remarkable asymptotic convergence rate of $O(1/|\ln h|)$ as $h \to 0$ for a particular selection of acceleration parameters $r_i$. See also Douglas (1957) and Young (1956b).

For the partial differential equation (7.1′)–(7.2′) for a general region with nonuniform mesh spacings, the proof that the Peaceman-Rachford (and Douglas-Rachford) iterative method is convergent for any fixed acceleration parameter $r > 0$ is due to Sheldon and Wachspress (see Wachspress (1957)). See also Birkhoff and Varga (1959), Pearcy (1962), and Lees (1962). As optimized one-parameter problems, the comparison (7.21) of the Peaceman-Rachford iterative method with the successive overrelaxation iterative method for the model problem, as well as the monotonicity principle of Sec. 7.1, was given by Varga (1960b). Extensions of this monotonicity principle to discrete approximations of more general differential equations have been obtained by Birkhoff, Varga, and Young (1962), where it is proved under general conditions that the asymptotic rate of convergence $R_\infty(T_r)$ for a single optimized parameter $r$ is $O(h)$ as $h \to 0$ (see Exercises 7 and 8).

**7.2.** The first step in extending the original analysis of these alternating-direction methods beyond the model problem was taken by Wachspress (1957), who considered the case of commutative matrices (7.25). Theorem 7.3 in terms of commutative matrices is given by Wachspress (1957), and Birkhoff and Varga (1959). Wachspress (1957) also studied the min-max problem of (7.34), whose rigorous solution follows from the recent work of Rice (1960, 1961) on unisolvent families of functions. Basic to such theoretical results is the so-called *Chebyshev principle* discussed in Achieser (1956), Motzkin (1949), and Tornheim (1950). The particularly elegant and simple algorithm for selecting the best acceleration parameters $r_i$, $1 \le i \le m$, in the min-max sense for the case $m = 2^p$ is due to Wachspress (1962), although an independent proof was simultaneously given (unpublished) by Gastinel.

Bounds of the form (7.48′) and (7.51) for the average rate of convergence of the Peaceman-Rachford iterative method in the commutative case have been given by Douglas (1957), Young (1956b), Young and Ehrlich (1960), and Wachspress and Habetler (1960). We have not exhausted all possible methods for choosing acceleration parameters. See also Wachspress and Habetler (1960) for a related method based again on geometric means, as well as de Boor and Rice (1962).

**7.3.** The results of this section are taken from the paper by Birkhoff and Varga (1959). In essence, these results show that the property of commutation of the matrices $H$ and $V$ is quite special, and holds only for rectangular regions. This does not mean, however, that these iterative methods are not useful for problems involving general domains. Further results, extending the material of this section, are given by Young in Birkhoff, Varga, and Young (1962). That the $p$-cyclic theory of matrices has applications to alternating-direction implicit methods (Theorem 7.7) was given by Varga (1959a).

**7.4.** We have considered the Douglas-Rachford (1956) alternating-direction implicit iterative method as a variant of the Peaceman-Rachford iterative method. Other variants, such as the use of a positive diagonal matrix $F$ and various splittings of the matrix $\Sigma$, are due to Wachspress and Habetler (1960). See also Habetler (1959). The generalization of the Peaceman-Rachford iterative method to higher dimensions is due to Douglas (1962).

For numerical results comparing the Peaceman-Rachford iterative method and variants of the successive overrelaxation iterative method, see Young and Ehrlich (1960), and Price and Varga (1962). Young and Ehrlich give an example in which *complex eigenvalues* occur. Price and Varga give an example in which *divergence* is obtained for two acceleration parameters used cyclically. In general, these numerical results indicate that the Peaceman-Rachford iterative method is highly successful for sufficiently fine mesh problems.

Alternating-direction implicit methods have also been applied to the numerical solution of biharmonic problems. See Conte and Dames (1958, 1960) and Heller (1960). There, Heller considers more generally the application of these methods to $(2p)$th order elliptic difference equations. Attempts to combine alternating-direction implicit methods with successive overrelaxation iterative method, forming a two-parameter iterative method, have been considered by Evans (1960) and Varga (1960b). For numerical results of such combined methods, see Evans (1960).

Our discussion of alternating-direction implicit methods has centered on two-dimensional elliptic differential equations. Whereas Douglas and Rachford (1956) proposed a variant to be useful in three-dimensional problems, Douglas 1962) has discovered what might be called a three-dimensional variant of the Peaceman-Rachford method that would appear to be more efficient in practice than the original method proposed in Douglas-Rachford (1956). The analogue of Theorem 7.1, however, fails in higher dimensions, i.e., three dimensional mesh regions can be exhibited for which the Douglas-Rachford method (7.67)–(7.70) and the method of (7.87) *diverge* even for certain fixed $r > 0$.

For applications of alternating-direction implicit iterative methods to mildly nonlinear elliptic differential equations, see Douglas (1961a).

# CHAPTER 8

# MATRIX METHODS FOR PARABOLIC
# PARTIAL DIFFERENTIAL EQUATIONS

## 8.1. Semi-Discrete Approximation

Many of the problems of physics and engineering that require numerical approximations are special cases of the following second-order linear parabolic partial differential equation:

$$(8.1) \qquad \phi(\mathbf{x})u_t(\mathbf{x}; t) \;=\; \sum_{i=1}^{n} (K_i(\mathbf{x})u_{x_i})_{x_i}$$

$$+ \sum_{i=1}^{n} G_i(\mathbf{x})u_{x_i} \;-\; \sigma(\mathbf{x})u(\mathbf{x}; t) \;+\; S(\mathbf{x}; t), \quad \mathbf{x} \in R, t > 0,$$

where $R$ is a given finite (connected) region in Euclidean $n$-dimensional space, with (external) boundary conditions

$$(8.2) \qquad \alpha(\mathbf{x})u(\mathbf{x}; t) \;+\; \beta(\mathbf{x}) \frac{\partial u(\mathbf{x}; t)}{\partial n} \;=\; \gamma(\mathbf{x}), \qquad \mathbf{x} \in \Gamma, t > 0,$$

where $\Gamma$ is the external boundary of $R$. Characteristic of such problems is the additional *initial* condition

$$(8.3) \qquad u(\mathbf{x}; 0) \;=\; g(\mathbf{x}); \qquad \mathbf{x} \in R.$$

We assume for simplicity that the given functions $\phi$, $K_i$, $G_i$, $\sigma$, $S$, and $g$

are continuous† in $\bar{R}$ and satisfy the following conditions:

1. $\phi$,   $K_i$ are strictly positive in $\bar{R}$,   $1 \leq i \leq n$,

(8.4)

2. $\sigma$ is non-negative in $\bar{R}$.

For the external boundary condition of (8.2), we assume that the functions $\alpha$, $\beta$, and $\gamma$ are piecewise continuous on $\Gamma$, and, as in Chapter 6, satisfy

$$(8.5) \qquad \alpha(\mathbf{x}) \geq 0, \quad \beta(\mathbf{x}) \geq 0, \quad \alpha(\mathbf{x}) + \beta(\mathbf{x}) > 0, \quad \mathbf{x} \in \Gamma.$$

These assumptions cover important physics and engineering problems. For example, in reactor physics the time-dependent density of neutrons $u(\mathbf{x}; t)$ of a particular average energy in a reactor satisfies, in a diffusion approximation,

$$\frac{1}{v} u_t(\mathbf{x}; t) = \sum_{i=1}^{n} (K(\mathbf{x})u_{x_i})_{x_i} - \sigma(\mathbf{x})u(\mathbf{x}; t) + S(\mathbf{x}; t),$$

which physically represents a conservation of neutrons. Here, $v$ is the average velocity of these neutrons, $K(\mathbf{x})$ is the *diffusion coefficient*, and $\sigma(\mathbf{x})$ is the *total removal cross section*.‡ In petroleum engineering, the flow of a compressible fluid in a homogeneous porous medium is given by

$$\phi u_t(\mathbf{x}; t) = \nabla^2 u(\mathbf{x}; t) = \sum_{i=1}^{n} u_{x_i x_i}(\mathbf{x}; t),$$

where $\phi$ depends on the density, compressibility, and viscosity of the fluid, as well as the permeability and porosity of the medium.§

More widely known, however, is the *heat* or *diffusion equation*. To be specific, the temperature $u(x, t)$ in an infinite thin rod, satisfies

$$(8.1') \qquad u_t(x, t) = Ku_{xx}(x, t), \qquad -\infty < x < +\infty, t > 0,$$

where the constant $K > 0$ is the *diffusivity*¶ of the rod. For an initial condition, we are given the initial temperature distribution

$$(8.3') \qquad u(x, 0) = g(x), \qquad -\infty < x < \infty,$$

---

† It will be clear from the discussion that follows that problems with internal interfaces, where the functions $\phi$, $K_i$, $G_i$, $\sigma$, $S$, and $g$ are only *piecewise* continuous in $\bar{R}$, can similarly be treated.

‡ See Glasstone and Edlund (1952), p. 291.

§ See M. Muskat (1937), p. 627. In actual petroleum problems, $\phi$ can be flow-dependent, which makes this partial differential equation nonlinear.

¶ See, for example, Carslaw and Jaeger (1959), p. 9.

where $g(x)$ is continuous and uniformly bounded on $-\infty < x < \infty$. It is well known that the solution of this problem is *unique* and is explicitly given by†

$$(8.6) \qquad u(x, t) = \frac{1}{\sqrt{4\pi Kt}} \int_{-\infty}^{+\infty} \exp\left(\frac{-(x - x')^2}{4Kt}\right) g(x') \, dx',$$

$$-\infty < x < +\infty, t > 0.$$

We can also express this solution in the form

$$(8.6') \qquad u(x, t + \Delta t) = \frac{1}{\sqrt{4\pi K \, \Delta t}} \int_{-\infty}^{+\infty} \exp\left(\frac{-(x - x')^2}{4K \, \Delta t}\right) u(x', t) \, dx',$$

where $t \geq 0$, $\Delta t > 0$. The point of this last representation is that $u(x, t + \Delta t)$ is linked (through positive weights) to $u(x', t)$ for *all* $x'$. Hence, we might expect in this case that matrix equations arising from finite difference approximations of $(8.1')$ would preserve some form of this particular coupling, i.e.,

$$u_m^{n+1} \equiv u(m \, \Delta x, (n + 1) \, \Delta t)$$

would be coupled through *non-negative* coefficients to each $u_j^n$. Clearly, this suggests that the Perron-Frobenius theory of non-negative matrices would again be useful in finite difference approximations of (8.1). Indeed, one of the main objectives of this chapter is to show that this property of non-negative couplings, as noted for the infinite thin rod, is a general characteristic of discrete approximations to (8.1)–(8.3) and gives a further association of the Perron-Frobenius theory of non-negative matrices to the numerical solution of partial differential equations.

To obtain finite difference approximations of the general initial value problem (8.1)–(8.3), we discretize first only the spatial variables, leaving the time variable *continuous*. In this *semi-discrete* form,‡ matrix properties of the resulting system of ordinary differential equations are studied. Then, when the time variable is finally discretized, the concept of *stability* is introduced. In this way, stability, or lack thereof, is seen to be a property of matrix approximations only. Again, our primary concern in this chapter is with the analysis of matrix methods rather than with questions of convergence of such finite difference approximations to the solution of (8.1)–(8.3).

Returning to the parabolic partial differential equation of (8.1) with boundary conditions given in (8.2), we now assume, for simplicity,

---

† See Carslaw and Jaeger (1959), p. 35. For a discussion of existence and uniqueness for the more general problem of (8.1)–(8.3), see, for example, Bernstein (1950) and Hadamard (1952).

‡ These approximations are also called *semi-explicit*. See Lees (1961).

that the number of spatial variables is $n = 2$, although it will be clear that the results to be obtained extend to the general case. Letting $R_h$ denote a general two-dimensional (nonuniform) Cartesian mesh region which approximates $\bar{R}$, we derive spatial equations, as in Sec. 6.3, by integrating the differential equation (8.1) at each mesh point $(x_i, y_i)$ over its corresponding two-dimensional mesh rectangle $r_{i,j}$. For the integration of the left side of (8.1), we make the approximation†

$$(8.7) \qquad \int\int_{r_{i,j}} \phi(x, y)\, \frac{\partial u}{\partial t}\, dx\, dy \;\doteq\; \frac{du(x_i, y_i; t)}{dt} \int\int_{r_{i,j}} \phi(x, y)\, dx\, dy.$$

The five-point approximation to the right side of (8.1), by means of integration, is carried out as in Sec. 6.3, and we thus obtain the *ordinary* matrix differential equation

$$(8.8) \qquad C\, \frac{d\mathbf{u}(t)}{dt} = -A\mathbf{u}(t) + \mathbf{s}(t) + \tilde{\tau}(t), \qquad t > 0,$$

where $C$ and $A$ are $n \times n$ real matrices with time-independent entries,‡ and

$$(8.9) \qquad\qquad\qquad \mathbf{u}(0) = \mathbf{g}.$$

Note that the integration method of Sec. 6.3 directly utilizes the boundary conditions of (8.2), which are incorporated in the vector $\mathbf{s}(t)$ and the matrix $A$, and that the vector $\mathbf{g}$ of (8.9) is some integral average on $R_h$ of the function $g(x, y)$ defined on $\bar{R}$.

For properties of the matrices $C$ and $A$, we have from (8.7) that $C$ is by construction a diagonal matrix whose diagonal entries are integrals of the function $\phi$. But from (8.4), $C$ is then a positive diagonal matrix. For the matrix $A$, we have, by virtue of the five-point approximation, that there are at most five nonzero entries in any row of $A$. Also, using the hypotheses of (8.4) and (8.5), we have for all *sufficiently fine* Cartesian mesh regions $R_h$ that $A$ is irreducibly diagonally dominant§ with nonpositive off-diagonal entries, and is thus an irreducible $M$-matrix. Multiplying on the left by $C^{-1}$ in (8.8), we have

$$(8.8') \qquad \frac{d\mathbf{u}(t)}{dt} = -C^{-1}A\mathbf{u}(t) + C^{-1}\mathbf{s}(t) + \tau(t), \qquad t > 0,$$

---

† Since $\phi$ is a known function in $\bar{R}$, we can *in principle* exactly evaluate the integral of $\phi$ over each rectangular mesh region $\bar{R}_i$.

‡ See Exercise 7 for cases where the matrices $A$ and $C$ have time-independent entries even when the functions $\alpha$, $\beta$, and $\gamma$ of (8.2) are time-dependent.

§ This requires the additional hypothesis that $\sigma(\mathbf{x}) \equiv 0$ in $\bar{R}$ and $\alpha(\mathbf{x}) \equiv 0$ on $\Gamma$ cannot simultaneously occur.

where $\tau(t) \equiv C^{-1}\tilde{\tau}(t)$. With the above properties for the matrices $C$ and $A$, it follows that $C^{-1}A$ is also an *irreducible M-matrix* for sufficiently fine mesh regions $R_h$.

To give a particular estimate for $\tau(t)$, consider the following special case of (8.1):

$$u_t(x, y; t) = u_{xx}(x, y; t) + u_{yy}(x, y; t), \qquad (x, y) \in R, t > 0,$$

in a rectangle $R$. If the Cartesian mesh $R_h$ which approximates $\bar{R}$ is uniform, i.e., $\Delta x = \Delta y = h$ and $\beta \equiv 0$ in (8.2), it can be verified by means of Taylor's series expansions that for all $0 \leq t \leq T$,

$$(8.10) \qquad \tau_i(t) = O(h^2)$$

if the fourth derivatives $u_{xxxx}$ and $u_{yyyy}$ are continuous and bounded† in $\bar{R}$ for $0 \leq t \leq T$.

To solve the ordinary matrix differential equation of (8.8′), we make use of the exponential of a matrix:

$$\exp{(M)} = I + M + \frac{M^2}{2!} + \cdots,$$

which is convergent‡ for *any* $n \times n$ matrix $M$. With this definition, it follows that§

$$(8.11) \qquad \mathbf{u}(t) = \exp{(-tC^{-1}A)}\mathbf{u}(0) + \exp{(-tC^{-1}A)}$$

$$\cdot \int_0^t \exp{(\lambda C^{-1}A)}(C^{-1}\mathbf{s}(\lambda) + \tau(\lambda))\, d\lambda, \qquad t \geq 0,$$

with $\mathbf{u}(0) \equiv \mathbf{g}$, is the solution of (8.8)–(8.9).

For convenience of exposition of the remaining sections, we now assume that the source term $S(x, y; t)$ of (8.1) is *time-independent*. Because of this, the vector $\mathbf{s}$ of (8.8) is also time-independent, and the solution of (8.8) and (8.9) takes the simpler form

$$(8.11′) \qquad \mathbf{u}(t) = A^{-1}\mathbf{s} + \exp{(-tC^{-1}A)}\{\mathbf{u}(0) - A^{-1}\mathbf{s}\}$$

$$+ \exp{(-tC^{-1}A)} \int_0^t \exp{(\lambda C^{-1}A)}\tau(\lambda)\, d\lambda, \qquad t \geq 0.$$

---

† For example, see Douglas (1961b) for more general results concerning discretization errors.

‡ See Exercise 1 of Sec. 3.5.

§ By $\int_\alpha^\beta \mathbf{g}(x)\, dx$, we mean a vector with components $\int_\alpha^\beta g_i(x)\, dx$.

Neglecting the term involving the vector $\tau(t)$, the previous equation gives rise to the vector approximation

$$(8.12) \qquad \mathbf{v}(t) = A^{-1}\mathbf{s} + \exp(-tC^{-1}A)\{\mathbf{v}(0) - A^{-1}\mathbf{s}\},$$

where

$$(8.13) \qquad\qquad\qquad \mathbf{v}(0) = \mathbf{g}.$$

Note that this is just the solution of

$$(8.8'') \qquad\qquad C\frac{d\mathbf{v}(t)}{dt} = -A\mathbf{v}(t) + \mathbf{s}.$$

Using (8.11′) and (8.12), it follows that

$$(8.14) \qquad \mathbf{u}(t) - \mathbf{v}(t) = \int_0^t \exp\left[-(t-\lambda)C^{-1}A\right]\tau(\lambda)\,d\lambda, \qquad t \geq 0,$$

which can be used to estimate $\| \mathbf{u}(t) - \mathbf{v}(t) \|$. (See Exercise 6.)

By analogy with the physical problem, the expression of (8.12) suggests that the terms of the right side respectively correspond to the steady-state solution of (8.8″) and a transient term. The matrix properties of $\exp(-tC^{-1}A)$, which correctly establish this, will be developed in the next section.

## EXERCISES

**1.** If $C$ and $D$ are both $n \times n$ matrices, show that

$$\exp(tC + tD) = \exp(tC)\cdot\exp(tD), \qquad t > 0,$$

if and only if $CD = DC$. As a consequence, if $A$ is the matrix of (1.6) and $B$ is the matrix of (1.7), conclude that

$$\exp(-tA) = \exp(-t)\cdot\exp(tB), \qquad \text{for all } t.$$

**2.** Let $A$ be any $n \times n$ complex matrix. Prove that
a. $\exp(A)\cdot\exp(-A) = I$. Thus, $\exp(A)$ is always nonsingular.
b. $\| \exp(A) \| \leq \exp(\| A \|)$.

**3.** If $[A, B] \equiv AB - BA$, show for small values of $t$ that
$\exp\{t(C + D)\} - \exp(tC)\cdot\exp(tD)$

$$= \frac{t^2}{2}[D, C] + \frac{t^3}{6}\{[CD, C] + [D^2, C] + [D, C^2] + [D, CD]\} + O(t^4).$$

**4.** Verify that (8.11) is the solution of (8.8) and (8.9).

**5.** If the matrix $C^{-1}A$ is Hermitian and positive definite, using (8.12) show that

$$\| \mathbf{v}(t) \| \leq \| A^{-1}\mathbf{s} \| + \| \mathbf{v}(0) - A^{-1}\mathbf{s} \| \qquad \textit{for all } t \geq 0.$$

For the vector norms $|| \mathbf{x} ||_1$ and $|| \mathbf{x} ||_\infty$ of Exercise 1, Sec. 1.3, show that $|| \mathbf{v}(t) ||_1$ and $|| \mathbf{v}(t) ||_\infty$ are also *uniformly bounded* for all $t \geq 0$.

**6.** If the matrix $C^{-1}A$ is Hermitian and positive definite, and $|| \boldsymbol{\tau}(t) || \leq M(\Delta x)^2$ for all $0 \leq t \leq T$, show from (8.14) that

$$|| \mathbf{u}(t) - \mathbf{v}(t) || \leq TM (\Delta x)^2, \qquad 0 \leq t \leq T.$$

**7.** Suppose that the functions $\alpha$, $\beta$, and $\gamma$ of the boundary condition (8.2) satisfy (8.5), but are *time-dependent*. Show that the matrices $A$ and $C$, derived in the manner of this section, still have time-independent entries, if

a. whenever $\alpha(\mathbf{x}; t) > 0$, then $\dfrac{\beta(\mathbf{x}; t)}{\alpha(\mathbf{x}; t)}$ is time-independent;

b. whenever $\beta(\mathbf{x}; t) > 0$, then $\dfrac{\alpha(\mathbf{x}; t)}{\beta(\mathbf{x}; t)}$ is time-independent.

**8.** To show that choices other than diagonal matrices are possible for the matrix $C$ of (8.8), consider the problem

$$\frac{\partial u(x, t)}{\partial t} = \frac{\partial^2 u(x, t)}{\partial x^2}, \qquad 0 < x < 1, t > 0,$$

where $u(0; t) = \alpha_1$, $u(1; t) = \alpha_2$ and $u(x; 0) = g(x)$, $0 < x < 1$, with $\alpha_1$ and $\alpha_2$ scalars. Letting $x_i = ih$ where $h = 1/(N + 1)$ for $0 \leq i \leq N + 1$ and $u(x_i; t) \equiv u_i(t)$, show that

$$\frac{d}{dt} \left[ \frac{1}{12} \{10u_i(t) + u_{i-1}(t) + u_{i+1}(t)\} \right]$$

$$= \frac{u_{i-1}(t) - 2u_i(t) + u_{i+1}(t)}{h^2} + \tilde{\tau}_i(t),$$

$1 \leq i \leq N$, where $\tilde{\tau}_i(t) = O(h^4)$ as $h \to 0$ for $0 \leq t \leq T$ if $\partial^6 u(x, t)/\partial x^6$ is continuous and bounded in $0 \leq x \leq 1, 0 \leq t \leq T$. In this case, the matrix $C$ is tridiagonal.

**9.** Let $B \equiv \left[ \begin{array}{c|c} O & F \\ \hline F^* & 0 \end{array} \right]$. Show that

$$\exp (tB) = \left[ \begin{array}{c|c} \cosh \left(\sqrt{tFF^*}\right) & \left(\dfrac{\sinh \left(\sqrt{tFF^*}\right)}{\sqrt{FF^*}}\right) F \\ \hline \left(\dfrac{\sinh \left(\sqrt{tF^*F}\right)}{\sqrt{F^*F}}\right) F^* & \cosh \left(\sqrt{tF^*F}\right) \end{array} \right].$$

(Note that $\cosh \sqrt{z}$ and $(\sinh \sqrt{z})/\sqrt{z}$ are entire functions of $z$, so that from Exercise 1, Sec. 3.5, the matrices $\cosh \sqrt{A}$ and $(\sinh \sqrt{A})/\sqrt{A}$ are defined for any square complex matrix $A$.)

## 8.2. Essentially Positive Matrices

The exponential matrix $\exp(-tC^{-1}A)$, which resulted from the semi-discrete approximations of the previous section, has interesting matrix properties. We begin with

DEFINITION 8.1. A real $n \times n$ matrix $Q = (q_{i,j})$ is *essentially positive* if $q_{i,j} \geq 0$ for all $i \neq j$, $1 \leq i, j \leq n$, and $Q$ is irreducible.

With this definition, it is clear that $Q$ is an essentially positive matrix if and only if $(Q + sI)$ is a non-negative, irreducible, and primitive matrix for some real $s$.

**Theorem 8.1.** *A matrix $Q$ is essentially positive if and only if $\exp(tQ) > O$ for all $t > 0$.*

*Proof.* If the matrix $\exp(tQ) = I + tQ + t^2Q^2/2! + \cdots$ has positive entries for all $t > 0$, then $Q$ is evidently irreducible. Otherwise, $Q$ and all its powers are reducible, which implies that $\exp(tQ)$ would have some zero entries. If there exists a $q_{i,j} < 0$ for some $i \neq j$, then clearly $\exp(tQ) = I + tQ + \cdots$ has a negative entry for all sufficiently small $t > 0$. Thus, $\exp(tQ) > O$ implies that $Q$ is essentially positive. On the other hand, assume that $Q$ is essentially positive. Then $Q + sI$ is non-negative, irreducible, and primitive for all sufficiently large $s > 0$. Since the powers of $Q + sI$ are non-negative and all sufficiently high powers of $Q + sI$ are positive by Frobenius' Theorem 2.5, then $\exp(Q + sI) > O$. But it is easily verified† that

$$\exp(tQ) = \exp(-st)\exp(t(Q + sI)) > O,$$

which completes the proof.

As an immediate application of this result, note that as the matrix $C^{-1}A$ of (8.8′) is by construction an irreducible $M$-matrix for all sufficiently fine mesh regions $R_h$, then $Q \equiv -C^{-1}A$ is necessarily essentially positive. Thus, the semi-discrete approximation of (8.12),

(8.15)    $\mathbf{v}(t) = \exp(-tC^{-1}A)\mathbf{v}(0) + \{I - \exp(-tC^{-1}A)\}A^{-1}\mathbf{s},$

for sufficiently fine mesh regions $R_h$, couples (through positive coefficients) *each* component of $\mathbf{v}(t)$ to every component of $\mathbf{v}(0)$ for all $t > 0$, and is the matrix analogue of the behavior noted for the infinite thin rod.

† See Exercise 1 of Sec. 8.1.

Closely related to the Perron-Frobenius Theorem 2.1 is

**Theorem 8.2.** *Let $Q$ be an essentially positive matrix. Then $Q$ has a real eigenvalue $\zeta(Q)$ such that*

1. *To $\zeta(Q)$ there corresponds an engenvector $\mathbf{x} > \mathbf{0}$.*
2. *If $\alpha$ is any other eigenvalue of $Q$, then $Re \; \alpha < \zeta(Q)$.*
3. *$\zeta(Q)$ increases when any element of $Q$ increases.*

*Proof.* If $Q$ is essentially positive, then the matrix $T \equiv Q + sI$ is non-negative, irreducible, and primitive for some real $s > 0$. Thus, from the Perron-Frobenius Theorem 2.1, there exists a vector $\mathbf{x} > \mathbf{0}$ such that $T\mathbf{x} = \rho(T)\mathbf{x}$. But this implies that

$$Q\mathbf{x} = (\rho(T) - s)\mathbf{x} \equiv \zeta(Q)\mathbf{x}.$$

The other conclusions of this theorem follow similarly from the Perron-Frobenius theorem.

Physically, using the infinite thin rod of Sec. 8.1 as an example, we know that the temperature of the rod is bounded for all $t \geq 0$ and that there is a steady-state temperature of the rod, i.e., $\lim_{t\to\infty} u(x; t)$ exists for all $-\infty < x < +\infty$. Analogously, we ask if the semi-discrete approximations of Sec. 8.1 possess these basic properties. In order to answer this, we first determine the general asymptotic behavior of $\exp(tQ)$ for essentially positive matrices $Q$ in terms of norms. The following lemma is readily established in the manner of Lemma 3.1 of Sec. 3.2.

**Lemma 8.1.**   *Let $J$ be the upper bi-diagonal $p \times p$ complex matrix of the form*

(8.16)
$$J = \begin{bmatrix} \lambda & 1 & & & O \\ & \lambda & 1 & & \\ & & \ddots & \ddots & \\ & & & & 1 \\ O & & & & \lambda \end{bmatrix}.$$

*Then,*

(8.17)   $$\| \exp(tJ) \| \sim \frac{t^{p-1}}{(p-1)!} \exp(t \, Re \, \lambda), \qquad t \to +\infty.$$

Let $Q$ be an $n \times n$ essentially positive matrix. If $S$ is the nonsingular matrix such that $Q = SJS^{-1}$, where $J$ is the Jordan normal form of $Q$, then we apply Lemma 8.1 to the diagonal blocks of the matrix $J$. Since $Q$ is essentially positive, the eigenvalue of $Q$ with the largest real part, $\zeta(Q)$, is simple from Theorem 8.2, which in turn implies that the corresponding diagonal submatrix of $J$ is $1 \times 1$. Thus, we have, in analogy with Theorem 3.1 of Sec. 3.2,

**Theorem 8.3.** *Let $Q$ be an $n \times n$ essentially positive matrix. If $\zeta(Q)$ is the eigenvalue of Theorem 8.2, then*

$$(8.18) \qquad \| \exp (tQ) \| \sim K \exp (t\zeta(Q)), \qquad t \to +\infty,$$

*where $K$ is a positive constant independent of $t$.*

As $\zeta(Q)$ then dictates the asymptotic behavior of $\| \exp (tQ) \|$ for large $t$ when $Q$ is essentially positive, we accordingly make†

DEFINITION 8.2. If $Q$ is essentially positive, then $Q$ is *supercritical*, *critical*, or *subcritical* if $\zeta(Q)$ is respectively positive, zero, or negative.

Consider now the nonhomogeneous ordinary matrix differential equation

$$(8.19) \qquad \frac{d\mathbf{v}(t)}{dt} = Q\mathbf{v}(t) + \mathbf{r},$$

where $\mathbf{v}(0)$ is the given initial vector condition, and $\mathbf{r}$ is a time-independent vector. If $Q$ is nonsingular, then the unique solution of (8.19) satisfying the initial vector condition is

$$(8.20) \qquad \mathbf{v}(t) = -Q^{-1}\mathbf{r} + \exp (tQ) \cdot \{\mathbf{v}(0) + Q^{-1}\mathbf{r}\}, \qquad t \geq 0.$$

**Theorem 8.4.** *Let the $n \times n$ matrix $Q$ of (8.19) be essentially positive and nonsingular. If $Q$ is supercritical, then for certain initial vectors $\mathbf{v}(0)$, the solution $\mathbf{v}(t)$ of (8.19) satisfies*

$$(8.21) \qquad \lim_{t \to +\infty} \| \mathbf{v}(t) \| = +\infty.$$

*If $Q$ is subcritical, then the solution vector $\mathbf{v}(t)$ of (8.19) is uniformly bounded in norm for all $t \geq 0$, and satisfies*

$$(8.22) \qquad \lim_{t \to +\infty} \mathbf{v}(t) = -Q^{-1}\mathbf{r}.$$

*Proof.* Since $Q$ has no zero eigenvalues, then $Q$ is either supercritical or subcritical. The result then follows from Theorem 8.3 and (8.20). Note that the vector $-Q^{-1}\mathbf{r}$, when $Q$ is nonsingular, is just the solution of (8.19) with the derivative term set to zero.

† These terms are the outgrowth of investigations concerning mathematical models for nuclear reactor theory. See, for example, Birkhoff and Varga (1958).

With the previous theorem, we arrive at the final result in this section, which connects the theory in this section to the ordinary matrix differential equation of (8.8″).

**Corollary.** *If $C$ is an $n \times n$ positive diagonal matrix, and $A$ is an irreducible $n \times n$ M-matrix, then the unique vector solution of*

$$(8.23) \qquad C \frac{d\mathbf{v}(t)}{dt} = -A\mathbf{v}(t) + \mathbf{s}, \qquad t \geq 0,$$

*subject to the initial vector condition $\mathbf{v}(0)$, is uniformly bounded in norm for all $t \geq 0$, and satisfies*

$$(8.24) \qquad \lim_{t \to +\infty} \mathbf{v}(t) = A^{-1}\mathbf{s}.$$

*Proof.* Let $Q \equiv -C^{-1}A$. Then $Q$ is a nonsingular essentially positive matrix. If $Q\mathbf{x} = \varsigma(Q)\mathbf{x}$ where $\mathbf{x} > 0$, it follows that

$$A^{-1}C\mathbf{x} = \left(\frac{-1}{\varsigma(Q)}\right)\mathbf{x}.$$

But by the hypotheses, $A^{-1}C$ is a positive matrix, so that $\varsigma(Q)$ is necessarily a negative real number, which proves that $Q$ is subcritical. The remainder then follows from Theorem 8.4.

We have thus shown that for sufficiently fine mesh regions $R_h$, the semi-discrete approximations (8.23) as introduced in Sec. 8.1 possess solutions which are *uniformly bounded* in norm for all $t \geq 0$, and possess finite steady-state solutions, as suggested by the example of the infinite thin rod. Furthermore, the semi-discrete approximations for the partial differential equation of (8.1) possess the positive couplings noted for this physical example.

## EXERCISES

**1.** A real $n \times n$ matrix $Q = (q_{i,j})$ is *essentially non-negative* if $q_{i,j} \geq 0$ for all $i \neq j$, $1 \leq i, j \leq n$. Prove that $Q$ is essentially non-negative if and only if $\exp(tQ) \geq O$ for all $t \geq 0$.

**2.** Prove that $Q$ is essentially positive if and only if $-Q + sI$ is an irreducible M-matrix for some real $s$.

**3.** Complete the proof of Theorem 8.2.

**4.** Prove Lemma 8.1.

**5.** Let $Q$ be an arbitrary $n \times n$ complex matrix with eigenvalues $\lambda_1, \lambda_2, \cdots, \lambda_n$, where $\operatorname{Re} \lambda_1 \leq \operatorname{Re} \lambda_2 \leq \cdots \leq \operatorname{Re} \lambda_n$. Extending Lemma 8.1, find a necessary and sufficient condition that $\| \exp(tQ) \|$ be bounded for *all* $t \geq 0$.

**6.** Let $Q$ be an essentially positive matrix, and let $\mathbf{y}$ be the positive eigenvector of $\exp(tQ^T)$, where $\|\mathbf{y}\| = 1$. If $\mathbf{v}(t)$ is the solution of

$$\frac{d\mathbf{v}(t)}{dt} = Q\mathbf{v}(t),$$

where $\mathbf{v}(0) = \mathbf{g}$ is the initial vector condition, show that

$$\mathbf{y}^T\mathbf{v}(t) = \exp(t\zeta(Q))\mathbf{y}^T\mathbf{g}$$

for all $t \geq 0$.

**\*7.** Let $Q$ be an essentially positive $n \times n$ matrix with eigenvalues $\lambda_1, \lambda_2, \cdots, \lambda_n$, where $\operatorname{Re}\lambda_1 \leq \operatorname{Re}\lambda_2 \leq \cdots < \operatorname{Re}\lambda_n = \zeta(Q)$, and let the initial vector $\mathbf{g}$ of Exercise 6 have positive components. If $\mathbf{y}$ is the positive eigenvector of $\exp(tQ^T)$, show that the solution of

$$\frac{d\mathbf{v}(t)}{dt} = Q\mathbf{v}(t)$$

can be expressed as

$$\mathbf{v}(t) = \tilde{K}\exp(t\zeta(Q))\mathbf{x} + O(\exp(t\mu)), \qquad t \to +\infty,$$

where $\mathbf{x}$ is the positive vector of Theorem 8.2,

$$\zeta(Q) > \mu > \max_{1 \leq j < n}|\operatorname{Re}\lambda_j| \quad \text{and} \quad \tilde{K} \equiv (\mathbf{y}^T\mathbf{g}/\mathbf{y}^T\mathbf{x}) > 0$$

(Birkhoff and Varga (1958)).

**8.** For Theorem 8.3, show that the result of (8.18) is valid for the matrix norms $\|A\|_1$ and $\|A\|_\infty$ defined in Exercise 2 of Sec. 1.3. Characterize the constant $K$ in each case.

**9.** Let $Q$ be any $n \times n$ complex matrix such that $\|\exp(tQ)\|$ is uniformly bounded for all $t \geq 0$. Show that there exists a positive scalar $K$ such that

$$\|(Q - zI)^{-1}\| \leq \frac{K}{\operatorname{Re} z}$$

for any complex $z$ with $\operatorname{Re} z > 0$ (Kreiss (1959a)).

**10.** Let $Q$ be an essentially positive $n \times n$ matrix. If $\zeta(Q)$ is the eigenvalue of $Q$ characterized in Theorem 8.2, show that $\zeta(Q)$ has a min-max representation analogous to that of Theorem 2.2:

$$\max_{\mathbf{x}\in P^*}\left\{\min_{1\leq i\leq n}\frac{\displaystyle\sum_{j=1}^{n}a_{i,j}x_j}{x_i}\right\} = \zeta(Q) = \min_{\mathbf{x}\in P^*}\left\{\max_{1\leq i\leq n}\frac{\displaystyle\sum_{j=1}^{n}a_{i,j}x_j}{x_i}\right\},$$

where $P^*$ is the hyperoctant of vectors $\mathbf{x} > 0$ (Beckenbach and Bellman (1961), p. 84).

**8.3.** Matrix Approximations for exp $(-tS)$

The exponentials of matrices introduced in this chapter to solve the semi-discrete approximations of Sec. 8.1 had the attractive features of uniform boundedness in norm, as well as the positive couplings from one time step to the next. Unfortunately, the direct determination of these exponentials of matrices relative to even large computers seems impractical in two- or three-dimensional problems† because of the enormous storage problem created by the positivity of these matrices. To illustrate this, suppose that a two-dimensional problem (8.1) is approximated on a mesh consisting of 50 subdivisions in each coordinate direction. To completely specify the matrix $A$ of (8.8) would require (using symmetry) only 7500 nonzero coefficients. To specify the matrix exp $(-tC^{-1}A)$, on the other hand, would require approximately $3 \cdot 10^6$ nonzero coefficients.

We now ask about matrix approximations of exp $(-tS)$, where $S \equiv C^{-1}A$, from the construction of Sec. 8.1, is necessarily a sparse matrix. First, we shall consider some well-known numerical methods for solving parabolic partial differential equations. As we shall see, we can consider these methods as either matrix approximations for the matrix exp $(-\Delta t\, S)$ in

$$(8.25) \quad \mathbf{v}(t_0 + \Delta t) = \exp(-\Delta t\, S)\mathbf{v}(t_0) + \{I - \exp(-\Delta t\, S)\}A^{-1}\mathbf{s},$$

or discrete approximations in time of the ordinary matrix differential equation

$$(8.26) \qquad C\frac{d\mathbf{v}(t)}{dt} = -A\mathbf{v}(t) + \mathbf{s}.$$

*The Forward Difference Method.*

The matrix approximation

$$(8.27) \qquad \exp(-\Delta t\, S) \doteq I - \Delta t\, S, \qquad S \equiv C^{-1}A,$$

obtained by taking the first two terms of the expansion for exp $(-\Delta t\, S)$, when substituted in (8.25), results in

$$(8.27') \qquad \mathbf{w}(t_0 + \Delta t) = (I - \Delta t\, S)\mathbf{w}(t_0) + \Delta t\, C^{-1}\mathbf{s}.$$

By rearranging, this can be written as

$$(8.27'') \qquad C\left\{\frac{\mathbf{w}(t_0 + \Delta t) - \mathbf{w}(t_0)}{\Delta t}\right\} = -A\mathbf{w}(t_0) + \mathbf{s}.$$

† For one-space variable problems, it may not be so impractical.

This last equation then appears as a discrete approximation of (8.26). Starting with the given initial vector $\mathbf{w}(0)$, one can explicitly step ahead, finding successively $\mathbf{w}(\Delta t)$, $\mathbf{w}(2\,\Delta t)$, etc. This well-known method is appropriately called the *forward difference* or *explicit method*. Note that the time increments need *not* be constant. From (8.27″), we observe that only the positive diagonal matrix $C$ must be inverted to carry out this method.

### The Backward Difference Implicit Method.

Consider now the matrix approximation

(8.28) $$\exp\,(-\Delta t\,S) \doteq (I + \Delta t\,S)^{-1}, \qquad \Delta t \geq 0.$$

Since $S = C^{-1}A$ is an irreducible $M$-matrix, all eigenvalues of $S$ have their real parts positive.† Thus, $I + \Delta t\,S$ is nonsingular for all $\Delta t \geq 0$. For $\Delta t$ small, we can write

$$(I + \Delta t\,S)^{-1} = I - \Delta t\,S + (\Delta t\,S)^2 - \cdots,$$

which shows that the matrix approximation of (8.28) also agrees through linear terms with $\exp\,(-\Delta t\,S)$. Substituting in (8.25) gives

(8.28′) $$(I + \Delta t\,S)\mathbf{w}(t_0 + \Delta t) = \mathbf{w}(t_0) + \Delta t\,C^{-1}\mathbf{s},$$

which can be written equivalently as

(8.28″) $$C\left\{\frac{\mathbf{w}(t_0 + \Delta t) - \mathbf{w}(t_0)}{\Delta t}\right\} = -A\mathbf{w}(t_0 + \Delta t) + \mathbf{s}.$$

The name *backward difference method* for this procedure stems from the observation that one can explicitly use (8.28′) to step *backward* in time to calculate $\mathbf{w}(t_0)$ from $\mathbf{w}(t_0 + \Delta t)$. Note that this would involve only matrix multiplications and vector additions. Used, however, as a numerical procedure to calculate $\mathbf{w}(t_0 + \Delta t)$ from $\mathbf{w}(t_0)$, (8.28′) shows this method to be implicit, since it requires the solution of a matrix problem.

### The Crank-Nicolson Implicit Method.

With $S$ an irreducible $M$-matrix, we can also form the approximation

(8.29) $$\exp\,(-\Delta t\,S) \doteq \left(I + \frac{\Delta t}{2}\,S\right)^{-1}\left(I - \frac{\Delta t}{2}\,S\right), \qquad \Delta t \geq 0,$$

† See Exercise 4 of Sec. 3.5.

which for small $\Delta t$ gives an expansion

$$\left(I + \frac{\Delta t}{2} S\right)^{-1} \left(I - \frac{\Delta t}{2} S\right) = I - \Delta t \, S + \frac{(\Delta t \, S)^2}{2} - \frac{(\Delta t \, S)^3}{4} + \cdots.$$

This now agrees through *quadratic* terms with the expansion for $\exp(-\Delta t \, S)$. Substituting in (8.25), we have

$$(8.29') \qquad \left(I + \frac{\Delta t}{2} S\right) \mathbf{w}(t_0 + \Delta t) = \left(I - \frac{\Delta t}{2} S\right) \mathbf{w}(t_0) + \Delta t \, C^{-1} \mathbf{s},$$

which can be written as

$$(8.29'') \qquad C \left\{ \frac{\mathbf{w}(t_0 + \Delta t) - \mathbf{w}(t_0)}{\Delta t} \right\} = -\frac{A}{2} \{\mathbf{w}(t_0 + \Delta t) + \mathbf{w}(t_0)\} + \mathbf{s}.$$

It is evident that the above approximation in (8.29''), unlike (8.28'') and (8.27''), is a *central difference approximation* to (8.26) for $t = t_0 + \Delta t/2$. From (8.29'), we see that this method for generating $\mathbf{w}(t_0 + \Delta t)$ from $\mathbf{w}(t_0)$ is also implicit. Crank and Nicholson (1947) first considered this approximation in the numerical solution of the one-dimensional heat equation (8.1'). In this case, the matrix $S = C^{-1}A$ is a nonsingular, real, tridiagonal matrix, and the matrix equation (8.29') can be efficiently solved by Gaussian elimination, as described in Sec. 6.4. Note that the same is true of the numerical solution of (8.28') for the backward difference method.

The three methods just described are among the best known and most widely used numerical methods for approximating the solution of the parabolic partial differential equation of (8.1).

If the matrix $S = C^{-1}A$ is an irreducible $M$-matrix, then $-S$ is a subcritical essentially positive matrix. From Theorem 8.3 the matrix norms $\| \exp(-\Delta t \, S) \|$ are *uniformly bounded* for *all* $\Delta t \geq 0$, and this implies (the Corollary of Sec. 8.2) that the vectors $\mathbf{v}(t)$ of (8.23) are uniformly bounded in norm for all $t \geq 0$. The different approximations for $\exp(-\Delta t \, S)$ that we have just considered do *not* all share this boundedness property. For example, the forward difference approximation $(I - \Delta t \, S)$ has its spectral norm bounded below by

$$\| (I - \Delta t \, S) \| \geq \rho(I - \Delta t \, S) = \max_{1 \leq i \leq n} | 1 - \Delta t \, \lambda_i |,$$

where the $\lambda_i$'s are eigenvalues of $S$ which satisfy $0 < \alpha \leq \operatorname{Re} \lambda_i \leq \beta$. It is evident that $\rho(I - \Delta t \, S)$ is *not* uniformly bounded for all $\Delta t \geq 0$.

We now concentrate on those approximations for $\exp(-\Delta t \, S)$ which have their spectral radii less than unity for certain choices of $\Delta t \geq 0$.

The reason that these approximations are of interest lies simply in the following: Let $T(\Delta t)$ be any matrix approximation of exp $(-\Delta t\, S)$. Then, we would approximate the vector $\mathbf{v}(t_0 + \Delta t)$ of (8.25) by

$$(8.30) \qquad \mathbf{w}(t_0 + \Delta t) = T(\Delta t) \cdot \{\mathbf{w}(t_0) - A^{-1}\mathbf{s}\} + A^{-1}\mathbf{s}, \qquad \Delta t > 0.$$

Starting with $t_0 = 0$, we arrive by induction at

$$(8.30') \qquad \mathbf{w}(m\, \Delta t) = [T\,(\Delta t)]^m\{\mathbf{w}(0) - A^{-1}\mathbf{s}\} + A^{-1}\mathbf{s}, \qquad m \geq 0.$$

If $\rho(T(\Delta t)) \geq 1$, we see that the sequence of vectors $\mathbf{w}(m\,\Delta t)$ will *not* in general be bounded in norm for all $\mathbf{w}(0)$. Actually, what we are considering here is the topic of *stability*. It has been recognized for some time that although in practice one wishes to take longer time steps ($\Delta t$ large) in order to arrive at an approximation $\mathbf{w}(t)$ for the solution $u(\mathbf{x};\, t)$ of (8.1) with as *few* arithmetic computations as possible, one must restrict the size of $\Delta t = t/m$ in order to insure that $\mathbf{w}(m\, \Delta t)$ is a reasonable approximation to $u(\mathbf{x};\, t)$. This brings us to

DEFINITION 8.3. The matrix $T(t)$ is *stable* for $0 \leq t \leq t_0$ if $\rho(T(t)) \leq 1$ for this interval. It is *unconditionally stable* if $\rho(T(t)) < 1$ for all $t > 0$.

Note that the definition of stability is *independent* of the closeness of approximation of $T(t)$ to exp $(-tS)$.

For stability intervals for the approximations considered, we have

**Theorem 8.5.** *Let $S$ be an $n \times n$ matrix whose eigenvalues $\lambda_i$ satisfy $0 < \alpha \leq Re\ \lambda_i \leq \beta, 1 \leq i \leq n$. Then, the forward difference matrix approximation $I - \Delta t\ S$ is stable for*

$$(8.31) \qquad\qquad 0 \leq \Delta t \leq \min_{1 \leq i \leq n} \left\{ \frac{2\ Re\ \lambda_i}{|\lambda_i|^2} \right\}.$$

*On the other hand, the matrix approximations*

$$(I + \Delta t\ S)^{-1} \quad and \quad \left( I + \frac{\Delta t}{2} S \right)^{-1} \cdot \left( I - \frac{\Delta t}{2} S \right)$$

*for the backward difference and Crank-Nicolson matrix approximations are unconditionally stable.*

*Proof.* The result of (8.31) follows by direct computation. For the backward difference matrix approximation $(I + \Delta t\ S)^{-1}$, the eigenvalues of this matrix are $(1 + \Delta t\ \lambda_i)^{-1}$. Since the real part of $(1 + \Delta t\ \lambda_i)$ is greater than unity for *all* $\Delta t \geq 0$, then this matrix approximation is unconditionally stable. The proof for the Crank-Nicolson matrix approximation follows similarly. (See Exercise 7.)

One might ask how the approximations of $\exp(-\Delta t\, S)$ which we have discussed, as well as other approximations, are generated. This is most simply described in terms of Padé rational approximations† from classical analysis. To begin, let $f(z)$ be any analytic function in the neighborhood of the origin:

$$(8.32) \qquad f(z) = a_0 + a_1 z + a_2 z^2 + \cdots.$$

We consider approximations to $f(z)$ defined by

$$(8.33) \qquad f(z) \doteq \frac{n_{p,q}(z)}{d_{p,q}(z)},$$

where $n_{p,q}(z)$ and $d_{p,q}(z)$ are respectively polynomials of degree $q$ and $p$ in $z$, and we assume that $d_{p,q}(0) \neq 0$. We now select for *each* pair of non-negative integers $p$ and $q$ those polynomials $n_{p,q}(z)$ and $d_{p,q}(z)$ such that the Taylor's series expansion of $n_{p,q}(z)/d_{p,q}(z)$ about the origin agrees with as many leading terms of $f(z)$ of (8.32) as possible. Since the ratio $n_{p,q}(z)/d_{p,q}(z)$ contains $p + q + 1$ *essential* unknown coefficients, it is evident that the expression

$$d_{p,q}(z)\, f(z) - n_{p,q}(z) = O(|z|^{p+q+1}), \qquad |z| \to 0,$$

gives rise to $p + q + 1$ linear equations in these essential unknowns, whose solution determines these unknown coefficients. In this way, one can generate the double-entry *Padé table* for $f(z)$. It can be verified‡ that the first few entries of the Padé table for $f(z) = \exp(-z)$ are given by:

| | | $q = 0$ | $q = 1$ | $q = 2$ |
|---|---|---|---|---|
| | $p = 0$ | $1$ | $1 - z$ | $1 - z + \dfrac{z^2}{2}$ |
| $\exp(-z):$ | $p = 1$ | $\dfrac{1}{1 + z}$ | $\dfrac{2 - z}{2 + z}$ | $\dfrac{6 - 4z + z^2}{6 + 2z}$ |
| | $p = 2$ | $\dfrac{1}{1 + z + \dfrac{z^2}{2}}$ | $\dfrac{6 - 2z}{6 + 4z + z^2}$ | $\dfrac{12 - 6z + z^2}{12 + 6z + z^2}$ |

† Due to Padé (1892). See, for example, Wall (1948), Chap. XX, for an up-to-date treatment.

‡ See Exercise 3.

These rational approximations for $\exp(-z)$ generate matrix approximations of $\exp(-\Delta t\, S)$ in an obvious way. *Formally*, we merely replace the variable $z$ by the matrix $\Delta t\, S$, and we let

$$(8.34) \qquad \exp(-\Delta t\, S) \doteq [d_{p,q}(\Delta t\, S)]^{-1}[n_{p,q}(\Delta t\, S)] \equiv E_{p,q}(\Delta t\, S)$$

be the $p$, $q$ Padé matrix approximation of $\exp(-\Delta t\, S)$. From the entries of the Padé table for $\exp(-z)$, we see that the matrix approximations for $\exp(-\Delta t\, S)$ corresponding to the forward difference, backward difference, and Crank-Nicolson methods are *exactly* the Padé matrix approximations $E_{0,1}(\Delta t\, S)$, $E_{1,0}(\Delta t\, S)$, and $E_{1,1}(\Delta t\, S)$, respectively.

Our purpose in introducing these Padé approximations is threefold. First, we see that the matrix approximations associated with the forward difference, backward difference, and Crank-Nicolson methods are special cases of Padé matrix approximations of $\exp(-\Delta t\, S)$. Second, these Padé approximations generate many other useful matrix approximations of $\exp(-\Delta t\, S)$. Although it is true that such higher-order Padé approximations have not as yet been used in practical applications, it is entirely possible that they may be of use on larger computers where higher-order implicit methods could be feasible. Third, the direct association of these matrix approximations with the classical analysis topic of rational approximation of analytic functions gives a powerful tool for the analysis of stability of these approximations. For example, it can be shown† that if the eigenvalues of $S$ are positive real numbers, then the Padé matrix approximation $E_{p,q}(\Delta t\, S)$ in (8.34) is *unconditionally stable* if and only if $p \geq q$.

Another important concept, due to Lax and Richtmyer (1956), on approximations of the matrix $\exp(-\Delta t\, S)$ is given by

DEFINITION 8.4. The matrix $T(t)$ is a *consistent* approximation of $\exp(-tS)$ if $T(t)$ has a matrix power series development about $t = 0$ that agrees through at least linear terms with the expansion of $\exp(-tS)$.

Note that all Padé matrix approximations for $p + q > 0$ are by definition consistent approximations of $\exp(-tS)$. (See also Exercise 4.)

Finally, we end this section with a result further linking the Perron-Frobenius theory of non-negative matrices to the numerical solution of the parabolic problem (8.1).

**Theorem 8.6.** *Let* $T(t)$ *be any consistent matrix approximation of* $\exp(-tS)$ *where* $S$ *is an irreducible M-matrix. If* $T(t) \geq 0$ *for* $0 \leq t \leq t_0$, *where* $t_0 > 0$, *then there exists a* $t_1 > 0$ *such that* $T(t)$ *is primitive (i.e., non-negative, irreducible, and noncyclic) for* $0 < t < t_1$.

† See Varga (1961).

*Proof.* Since $T(t)$ is a consistent approximation, then

$$T(t) = I - tS + O(t^2), \quad \text{as } t \to 0.$$

Since $T(t) \geq O$ for positive $t$ sufficiently small, then as $S$ is an irreducible $M$-matrix, $T(t)$ is non-negative, irreducible, with positive diagonal entries for positive $t$ sufficiently small. But from Lemma 2.8, $T(t)$ is evidently primitive for some interval in $t$, completing the proof.

The physical, as well as mathematical, significance of this result is that, from (8.30′),

$$(8.35) \qquad \mathbf{w}(m \, \Delta t) = [T \, (\Delta t)]^m \{\mathbf{w}(0) - A^{-1}\mathbf{s}\} + A^{-1}\mathbf{s}, \qquad m \geq 0.$$

Thus, for all $\Delta t > 0$ sufficiently small and $m$ sufficiently large, each component $w_i(m \, \Delta t)$ is coupled through positive entries to *every* component $w_j(0)$, $1 \leq j \leq n$, so that the *primitivity* of the matrix approximation $T \, (\Delta t)$ of $\exp \, (-\Delta t \, S)$ is the natural matrix analogue of the positive coupling noted for the heat equation.

## EXERCISES

**1.** Consider the heat equation of (8.1′) for the finite interval $0 < x < 1$, with boundary conditions $u(0, t) \equiv \mu_1$, $u(1, t) \equiv \mu_2$, where $\mu_1$ and $\mu_2$ are positive constants. Using a uniform mesh $\Delta x = 1/(n + 1)$ and setting $x_i = i \, \Delta x$, $0 \leq i \leq n + 1$, show first that the $n \times n$ matrix $S = C^{-1}A$ is given explicitly by

$$S = \frac{K}{(\Delta x)^2} \begin{bmatrix} 2 & -1 & & & \\ -1 & 2 & -1 & & \\ & -1 & \ddots & \ddots & \\ & & \ddots & \ddots & -1 \\ & & & -1 & 2 \end{bmatrix}.$$

Show next that the forward difference method approximation $(I - \Delta t \, S)$ of $\exp \, (-\Delta t \, S)$ is stable for

$$0 \leq \Delta t \leq \frac{(\Delta x)^2}{2K}.$$

(*Hint*: Use Corollary 1 of Theorem 1.5 to estimate $\rho(I - \Delta t \, S)$.) (Courant, Friedrichs, Lewy (1928).)

**2.** Consider the heat equation

$$\frac{\partial u}{\partial t} = K \sum_{i=1}^{n} \frac{\partial^2 u}{\partial x_i^2}$$

for the unit $n$-dimensional hypercube, $0 < x_i < 1$, $1 \leq i \leq n$, where $u(\mathbf{x}, t)$

is specified for $\mathbf{x} \in \Gamma$, the boundary of this hypercube. Using a uniform mesh $\Delta x_i = 1/(N+1)$ in each coordinate direction, show that the forward difference method approximation $(I - \Delta t\, S)$ of $\exp(-\Delta t\, S)$ is stable for

$$0 \le \Delta t \le \frac{(\Delta x)^2}{2nK}.$$

(*Hint*: Generalize the result of the previous exercise.)

*3. Show that the $p, q$ entry of the Padé table for $\exp(-z)$ is composed of

$$n_{p,q}(z) = \sum_{k=0}^{q} \frac{(p+q-k)!\,q!}{(p+q)!\,k!(q-k)!} (-z)^k$$

and

$$d_{p,q}(z) = \sum_{k=0}^{p} \frac{(p+q-k)!\,p!}{(p+q)!\,k!(p-k)!} z^k$$

(Hummel and Seebeck (1949)).

**4.** Show that $\exp(-z) - n_{p,q}(z)/d_{p,q}(z) \sim c_{p,q} z^{p+q+1}$ as $|z| \to 0$, and determine $c_{p,q}$.

*5. Show that $\exp(-z) - n_{p,q}(z)/d_{p,q}(z)$ is of one sign for all $z \ge 0$ (Hummel-Seebeck (1949)).

*6. It is known (Wall (1948), p. 348) that $\exp(-z)$ has the continued fraction expansion

$$\exp(-z) = \cfrac{1}{1 + z \cfrac{}{1 - z \cfrac{}{2 + z \cfrac{}{3 - z \cfrac{}{2 + z \cfrac{}{5 - z \cfrac{}{2 + \cdots}}}}}}}$$

Prove that the approximants,

$$1, \quad \frac{1}{1+z}, \quad \cfrac{1}{1 + \cfrac{z}{1 - z/2}}, \quad \cdots,$$

are particular entries of the Padé table for $\exp(-z)$.

7. Let $S$ be any $n \times n$ complex matrix with eigenvalues $\lambda_i$ satisfying $\operatorname{Re} \lambda_i > 0$. Show that the matrix $(I + (\Delta t/2) S)^{-1}(I - (\Delta t/2) S)$ is an unconditionally stable and consistent approximation of $\exp(-\Delta t\, S)$, for $\Delta t > 0$.

8. Let $S$ be an irreducible $M$-matrix. Prove that the Padé matrix approximations $E_{0,1}(\Delta t\, S)$, $E_{1,0}(\Delta t\, S)$, and $E_{1,1}(\Delta t\, S)$ of $\exp(-\Delta t\, S)$ of (8.34) are *non-negative* matrices for certain intervals in $\Delta t \geq 0$. What are these intervals? In particular, show for the matrix $S$ of Exercise 1 that the interval for which $I - \Delta t\, S$ is non-negative is *exactly* the same as the stability interval for $I - \Delta t\, S$.

9. Let $T(\Delta t)$ be a *strictly* stable approximation (i.e., $\rho(T(\Delta t)) < 1$) of $\exp(-\Delta t\, S)$. If $\mathbf{w}(m\, \Delta t)$ is given by (8.35), show that there exists a positive constant $M$ such that

$$\| \mathbf{w}(m\, \Delta t) \| \leq \| A^{-1}\mathbf{s} \| + M \| \mathbf{w}(0) - A^{-1}\mathbf{s} \|$$

for *all* $m \geq 0$. Show that this inequality is also true for the vector norms $\| \mathbf{x} \|_1$ and $\| \mathbf{x} \|_\infty$ of Exercise 1, Sec. 1.3. If $S$ is the matrix of Exercise 1 of this section, show that the constant $M$ can be chosen to be unity for $0 \leq \Delta t \leq (\Delta x)^2/2K$, for *all* these vector norms.

## 8.4. Relationship with Iterative Methods for Solving Elliptic Difference Equations

The form of equation (8.30),

$$(8.36) \qquad \mathbf{w}((n+1)\Delta t) = T(\Delta t)\mathbf{w}(n\, \Delta t) + \{I - T(\Delta t)\}A^{-1}\mathbf{s},$$

$$\Delta t > 0,$$

where $\mathbf{w}(0) = \mathbf{v}(0) = \mathbf{g}$, suggests that matrix approximations of $\exp(-\Delta t\, S)$ *induce* abstract iterative methods for solving the matrix problem

$$(8.37) \qquad\qquad\qquad A\mathbf{w} = \mathbf{s}.$$

In fact, if we let $\mathbf{w}(n\, \Delta t) \equiv \mathbf{w}^{(n)}$, it is clear that (8.36) can be written equivalently as

$$(8.36') \qquad \mathbf{w}^{(n+1)} = T(\Delta t)\mathbf{w}^{(n)} + (I - T(\Delta t))A^{-1}\mathbf{s},$$

which we see defines an iterative method with $T(\Delta t)$ as the corresponding iteration matrix. If $\rho(T(\Delta t)) < 1$, this iterative procedure converges for any initial vector $\mathbf{w}^{(0)}$, and it necessarily follows in this case from (8.36') that

$$(8.38) \qquad\qquad\qquad \lim_{n\to\infty} \mathbf{w}^{(n)} = A^{-1}\mathbf{s}.$$

The relationship of the iterative method of (8.36) or (8.36′) with the solution of the matrix differential equation

$$(8.39) \qquad \frac{d\mathbf{v}(t)}{dt} = -C^{-1}A\mathbf{v}(t) + C^{-1}\mathbf{s}, \qquad t \geq 0,$$

where $\mathbf{v}(0) = \mathbf{g}$ becomes clearer if we recall that, when $C^{-1}A$ is an irreducible $M$-matrix, the solution vector $\mathbf{v}(t)$ of (8.39) tends to the steady-state solution, $A^{-1}\mathbf{s}$, of (8.39). Thus, we have established that any stable approximation $T(\Delta t)$ of $\exp(-\Delta t \, C^{-1}A)$ with $\rho(T(\Delta t)) < 1$ gives rise to a convergent iterative method (8.36′), whose solution is the steady-state solution of (8.39).

What is of interest to us now is essentially the converse of the above observation. For all the iterative methods previously described in this book, we shall now show that we can regard these iterative methods specifically in terms of discrete approximations to parabolic differential equations, so that the actual process of iteration can be viewed as simply marking progress in time to the steady-state solution of an equation of the form (8.39). That such a relationship exists is intuitively very plausible, and in reality an old idea. In order to bring out this relationship, we are first reminded that in an iterative procedure, such as in (8.36′), the initial vector $\mathbf{w}^{(0)}$ is some vector approximation of the solution of $A\mathbf{x} = \mathbf{s}$. Although this initial approximation corresponds to the initial condition for a parabolic partial differential equation, it is, however, *not* prescribed in contrast to the matrix differential equation of (8.8)–(8.9). Next, the solution of (8.37) is *independent* of the matrix $C$ of (8.39). This gives us the additional freedom of *choosing*, to our advantage, any nonsingular matrix $C$ which might lead to more rapidly convergent iterative methods.

### The Point Jacobi Iterative Method.

We first express the $n \times n$ matrix $A$ in the form

$$(8.40) \qquad A = D - E - F,$$

where $D$ is a positive diagonal matrix and $E$ and $F$ are respectively strictly lower and upper triangular matrices. Setting $C \equiv D$, consider the consistent Padé matrix approximation $E_{0,1}(\Delta t \, S)$ for $\exp(-\Delta t \, S)$ where $S = C^{-1}A$:

$$(8.41) \qquad \exp(-\Delta t \, S) \doteq I - \Delta t \, S = I - \Delta t \, D^{-1}(D - E - F)$$

$$\equiv T_1(\Delta t).$$

With $L \equiv D^{-1}E$ and $U \equiv D^{-1}F$, the matrix $T_1 (\Delta t)$ of (8.41) gives rise to the iterative procedure

$$(8.41') \qquad \mathbf{w}((n+1) \Delta t) = \{(1 - \Delta t)I + \Delta t(L + U)\}\mathbf{w}(n \Delta t)$$

$$+ \Delta t \, D^{-1}\mathbf{s}.$$

Clearly, the choice $\Delta t = 1$ gives us the familiar *point Jacobi iterative method*. To extend this to the block Jacobi method, we partition the matrix $A$ as in (3.58), and now let the matrix $C$ be defined as the block diagonal matrix $D$ of (3.59). Again, the choice of $\Delta t = 1$ in the approximation of (8.41) gives the associated block Jacobi iterative method.

To carry the analogy further, we now use a sequence of time steps $\Delta t_i$, $1 \leq i \leq m$, in (8.41'). With $B \equiv L + U$, this defines an iterative procedure whose error vectors $\boldsymbol{\varepsilon}^{(j)}$ satisfy

$$(8.41'') \qquad \boldsymbol{\varepsilon}^{(j)} = \prod_{i=1}^{j} \{(1 - \Delta t_i)I + \Delta t_i \, B\}\boldsymbol{\varepsilon}^{(0)}, \qquad 1 \leq j \leq m.$$

It is possible to select the $\Delta t_i$'s such that

$$\prod_{i=1}^{m} \{(1 - \Delta t_i)I + \Delta t_i \, B\}$$

is the polynomial $\tilde{p}_m(B)$ of Sec. 5.1. In other words, we can generate the Chebyshev semi-iterative method with respect to the Jacobi method simply by taking proper nonuniform time increments $\Delta t_i$.

*The Point Successive Overrelaxation Iterative Method.*

With $A = D - E - F$ and $C \equiv D$, consider the matrix approximation

$$(8.42) \qquad \exp(-\Delta t \, S) \doteqdot (I - \Delta t \, L)^{-1}\{\Delta t \, U + (1 - \Delta t)I\} \equiv T_2 (\Delta t).$$

For $\Delta t > 0$ sufficiently small, the expansion of $T_2 (\Delta t)$ agrees through linear terms with the expansion of $\exp(-\Delta t \, S)$. Thus, $T_2 (\Delta t)$ is a *consistent* approximation of $\exp(-\Delta t \, S)$. The matrix $T_2 (\Delta t)$ gives rise to the iterative method

$$(8.42') \qquad \mathbf{w}((n+1) \Delta t) = (I - \Delta t \, L)^{-1}\{\Delta t \, U + (1 - \Delta t)I\}\mathbf{w}(n \Delta t)$$

$$+ \Delta t(I - \Delta t \, L)^{-1}D^{-1}\mathbf{s}.$$

It is clear why this is called the *point successive overrelaxation iterative method*, since the matrix $T_2 (\Delta t)$ is precisely the point successive overrelaxation matrix $\mathcal{L}_{\Delta t}$ of (3.15). Note that the relaxation factor $\omega$ corresponds *exactly* to the time increment $\Delta t$. Again, the application of block techniques to successive overrelaxation iterative methods can be similarly realized in terms of consistent approximations for an exponential matrix.

Again, let $A = D - E - F$ and $C \equiv D$. Now, we suppose that $S = C^{-1}A = I - B$ is such that $B$ is weakly cyclic of index 2 and has the form

$$B = \left[\begin{array}{c|c} O & F \\ \hline F^* & O \end{array}\right].$$

If we take the consistent Padé matrix approximation $E_{0,2}\,(\Delta t\, S)$, then

$$(8.43) \qquad \exp\,(-\Delta t\, S) \doteq I - \Delta t\, S + \frac{(\Delta t)^2}{2}\, S^2 \equiv \hat{T}_2\,(\Delta t),$$

where

$$(8.43') \quad \hat{T}_2(t) = \left[\begin{array}{c|c} \left(1 - t + \dfrac{t^2}{2}\right) I + \dfrac{t^2}{2}\, FF^* & tF - t^2 F \\[3ex] \hline \\ tF^* - t^2 F^* & \left(1 - t + \dfrac{t^2}{2}\right) I + \dfrac{t^2}{2}\, F^*F \end{array}\right],$$

and we immediately see that the choice $t = 1$ is such that $\hat{T}_2(1)$ is *completely reducible*. This complete reducibility is the basis for the cyclic reduction iterative methods of Sec. 5.4.

## The Peaceman-Rachford Iterative Method.

The iterative methods described above have all been linked with Padé matrix approximations $p = 0$ and $q \geq 1$, which are *explicit*, and not in general unconditionally stable. For example, if $A$ is Hermitian and positive definite, then from Theorem 3.6 the matrix approximation $T_2\,(\Delta t)$ of (8.42) is stable only for the finite interval $0 \leq \Delta t \leq 2$. The Peaceman-Rachford variant of the alternating-direction implicit methods, however, is related to the implicit (and unconditionally stable) Padé matrix approximation $E_{1,1}\,(\Delta t\, S)$, which we have called the *Crank-Nicolson method*. Recalling from Sec. 7.1 that

$$(8.44) \qquad\qquad A = H + V + \Sigma = H_1 + V_1,$$

where

$$(8.44') \qquad\qquad H_1 = H + \tfrac{1}{2}\Sigma, \qquad V_1 = V + \tfrac{1}{2}\Sigma,$$

we now consider matrix approximations for $\exp(-\Delta t\, C^{-1}A)$ where $C \equiv I$. From (8.44), we have

$$(8.45) \qquad \exp(-\Delta t\, A) = \exp(-\Delta t[H_1 + V_1])$$

$$\doteq \exp(-\Delta t\, H_1)\cdot\exp(-\Delta t\, V_1)$$

where equality is *valid* if $H_1$ and $V_1$ commute. The nature of the matrices $H_1$ and $V_1$ was such that each was, after suitable permutations, the direct sum of tridiagonal matrices. The Crank-Nicolson Padé matrix approximation $E_{1,1}(\Delta t\, H_1)$ of the factor $\exp(-\Delta t\, H_1)$ is

$$(8.46) \qquad \exp(-\Delta t\, H_1) \doteq \left(I + \frac{\Delta t}{2} H_1\right)^{-1}\left(I - \frac{\Delta t}{2} H_1\right).$$

Thus, making the same approximation for $\exp(-\Delta t\, V_1)$, we arrive at

$$(8.47) \qquad \exp(-\Delta t\, A) \doteq \left(I + \frac{\Delta t}{2} H_1\right)^{-1}\left(I - \frac{\Delta t}{2} H_1\right)\left(I + \frac{\Delta t}{2} V_1\right)^{-1}$$

$$\cdot\left(I - \frac{\Delta t}{2} V_1\right).$$

Permuting the factors, which of course is valid when $H_1$ and $V_1$ commute, we then have

$$(8.48) \qquad \exp(-\Delta t\, A) \doteq \left(I + \frac{\Delta t}{2} V_1\right)^{-1}\left(I - \frac{\Delta t}{2} H_1\right)\left(I + \frac{\Delta t}{2} H_1\right)^{-1}$$

$$\cdot\left(I - \frac{\Delta t}{2} V_1\right) \equiv T_3(\Delta t).$$

Since $T_3(\Delta t)$ can also be written for $\Delta t > 0$ as

$$(8.48') \qquad T_3(\Delta t) = \left(V_1 + \frac{2}{\Delta t} I\right)^{-1}\left(\frac{2}{\Delta t} I - H_1\right)\left(H_1 + \frac{2}{\Delta t} I\right)^{-1}$$

$$\cdot\left(\frac{2}{\Delta t} I - V_1\right),$$

we see then that $T_3(\Delta t)$ is the Peaceman-Rachford matrix of (7.10). Note that the acceleration parameter $r$ for the Peaceman-Rachford iterative method is given by

$$(8.49) \qquad\qquad\qquad r = \frac{2}{\Delta t}.$$

Several items are of interest here. First, it can be verified that the matrix $T_3 (\Delta t)$ is a consistent approximation of $\exp (-\Delta t\ A)$. Moreover, if $\Delta t$ is small, then the expression of $T_3 (\Delta t)$ agrees through *quadratic terms* with the expansion of $\exp (-\Delta t\ A)$. Second, we know (Theorem 8.5) that the Crank-Nicolson Padé matrix approximation $E_{1,1} (\Delta t\ A)$ is unconditionally stable when $A$ has its eigenvalues $\lambda_i$ satisfying $\mathrm{Re}\ \lambda_i > 0$. It is not surprising, then, that the interval in $r$ for which the Peaceman-Rachford matrix for a single parameter is convergent is $r > 0$. (See Theorem 7.1.)

Using the idea that different choices of the diagonal matrix $C$ in (8.39) can be useful in determining properties of iterative methods, it is now interesting to describe the Wachspress-Habetler variant (see Sec. 7.4) of the Peaceman-Rachford iterative method in terms of this approach. Let $F$ be the positive diagonal matrix corresponding to $\phi \equiv 1$ in (8.1). In other words, the positive diagonal entries of $F$ are from Sec. 8.1 *exactly* the areas of the mesh rectangles $r_{i,j}$ of Sec. 8.1. Thus, from

$$(8.50) \qquad F \frac{d\mathbf{v}(t)}{dt} = -(H_1 + V_1)\mathbf{v}(t) + \mathbf{s} = -A\mathbf{v}(t) + \mathbf{s},$$

where $\mathbf{v}(0)$ will be our initial approximation to $A^{-1}\mathbf{s}$, we consider matrix approximations to $\exp (-\Delta t\ F^{-1}A)$ of the form

$$(8.51) \qquad \exp (-\Delta t\ F^{-1}A) \doteq \left(F + \frac{\Delta t}{2} V_1\right)^{-1} \left(F - \frac{\Delta t}{2} H_1\right)$$

$$\cdot \left(F + \frac{\Delta t}{2} H_1\right)^{-1} \left(F - \frac{\Delta t}{2} V_1\right) \equiv T_4(\Delta t),$$

which can be shown to be a consistent approximation to $\exp (-\Delta t\ F^{-1}A)$. The matrix $T_4 (\Delta t)$ turns out, of course, to be the Wachspress-Habetler variant $(7.77)-(7.78)$ of the Peaceman-Rachford iterative method, corresponding to the acceleration parameter $r = 2/\Delta t$. As previously mentioned, this choice of the diagonal matrix $C = F$, which appears to be a very natural choice in solving (8.1) with $\phi \equiv 1$, allows one to rigorously apply the commutative theory (Sec. 7.2) for alternating-direction implicit methods to the numerical solution of the Dirichlet problem in a rectangle with nonuniform mesh spacings.

## The Symmetric Successive Overrelaxation Iterative Method.

We now assume that the matrix $A = D - E - E^*$ is Hermitian and positive definite, that the associated matrix $D$ is Hermitian and positive definite, and that $D - \omega E$ is nonsingular for $0 \leq \omega \leq 2$. Setting $C \equiv D$, consider the matrix approximation

$$(8.52) \qquad \exp{(-tC^{-1}A)} \doteq \left(D - \frac{t}{2} E^*\right)^{-1} \left(\frac{t}{2} E + \left(1 - \frac{t}{2}\right)D\right)$$

$$\cdot \left(D - \frac{t}{2} E\right)^{-1} \left(\frac{t}{2} E^* + \left(1 - \frac{t}{2}\right) D\right) \equiv T_5(t).$$

Although $T_5(t)$ can be verified to be a consistent approximation of $\exp{(-tC^{-1}A)}$, what interests us here is that $T_5(t)$ can be expressed as

$$(8.53) \qquad T_5(t) = \left(I - \frac{t}{2} U\right)^{-1} \left(\frac{t}{2} L + \left(1 - \frac{t}{2}\right) I\right)\left(I - \frac{t}{2} L\right)^{-1}$$

$$\cdot \left(\frac{t}{2} U + \left(1 - \frac{t}{2}\right) I\right),$$

where $L = D^{-1}E$ and $U = D^{-1}E^*$. Since the inner two bracketed terms commute and can be interchanged, then $T_5(t)$ is similar to

$$\tilde{T}_5(t) \equiv \left(I - \frac{t}{2} U\right) T_5(t) \left(I - \frac{t}{2} U\right)^{-1},$$

which can be written as

$$(8.54) \qquad \tilde{T}_5(t) = \left(I - \frac{t}{2} L\right)^{-1} \left(\frac{t}{2} L + \left(1 - \frac{t}{2}\right) I\right)$$

$$\cdot \left(\frac{t}{2} U + \left(1 - \frac{t}{2}\right) I\right)\left(I - \frac{t}{2} U\right)^{-1},$$

which can be expressed equivalently as

$$(8.54') \qquad \tilde{T}_5(t) = \left(D - \frac{t}{2} E\right)^{-1} \left(\frac{t}{2} E + \left(1 - \frac{t}{2}\right) D\right)$$

$$\cdot \left(\frac{t}{2} E^* + \left(1 - \frac{t}{2}\right) D\right)\left(D - \frac{t}{2} E^*\right)^{-1}$$

$$= Q^*Q,$$

where

$$(8.55) \qquad Q \equiv \left(\frac{t}{2} E^* + \left(1 - \frac{t}{2}\right) D\right)\left(D - \frac{t}{2} E^*\right)^{-1}.$$

Thus, $\tilde{T}_5(t)$ is a non-negative definite Hermitian matrix, which proves that the eigenvalues of $T_5(t)$ are non-negative real numbers. Moreover, in a similar manner it can be shown that the eigenvalues of $T_5(t)$ are less than unity for $0 < t/2 < 2$. (See Exercise 6.)

For the iterative method resulting from the matrix approximation $T_5(t)$ of $\exp(-tC^{-1}A)$, this can be written as a two-step method:

$$\mathbf{w}^{(n+1/2)} = \left(I - \frac{t}{2} L\right)^{-1} \left\{\frac{t}{2} U + \left(1 - \frac{t}{2}\right) I\right\} \mathbf{w}^{(n)}$$

$$+ \frac{t}{2} \left(I - \frac{t}{2} L\right)^{-1} D^{-1} \mathbf{s},$$

(8.56)

$$\mathbf{w}^{(n+1)} = \left(I - \frac{t}{2} U\right)^{-1} \left\{\frac{t}{2} L + \left(1 - \frac{t}{2}\right) I\right\} \mathbf{w}^{(n+1/2)}$$

$$+ \frac{t}{2} \left(I - \frac{t}{2} U\right)^{-1} D^{-1} \mathbf{s},$$

which is called the *symmetric successive overrelaxation iterative method* and has been considered by Aitken (1950), Sheldon (1955), and Habetler and Wachspress (1961). This iterative method corresponds to sweeping the mesh in one direction, and then reversing the direction for another sweep of the mesh.† Sheldon (1955) observed that since the resulting iteration matrix $T_5(t)$ has non-negative real eigenvalues less than unity for $0 < t/2 < 2$, one can rigorously apply the Chebyshev semi-iterative method of Sec. 5.1 to (8.56) to accelerate convergence further. In effect, then, one first selects the optimum value of $t$ which gives the smallest spectral radius for the matrix $T_5(t)$ and then generates a three-term Chebyshev recurrence relation for the vectors $\mathbf{w}^{(n)}$.

As a final result in this chapter, we recall that in our discussions of the successive overrelaxation iterative method and the Peaceman-Rachford variant of the alternating-direction implicit iterative methods, each of the associated iteration matrices was shown to be *primitive* for certain choices of acceleration parameters.‡ However, these isolated and seemingly independent results are merely corollaries to Theorem 8.6, as the associated matrix approximations of (8.42) and (8.48) are non-negative consistent approximations to an irreducible $M$-matrix in each case.

† For this reason, this method is sometimes called the *to-fro* or *forward-backward* iterative method.

‡ See Exercise 4 of Sec. 3.6 and Theorem 7.8 of Sec. 7.3.

In summary, the main purpose of this section is to establish a correspondence between the iterative methods previously described and matrix methods for approximating the solution of parabolic partial differential equations. Although it is true that all these iterative methods could be adapted for use in approximating the solution of parabolic partial differential equations, the variants of the alternating-direction implicit method are in fact most widely used in practical problems, mainly because of their inherent unconditional stability.

## EXERCISES

**1.** Show that one can select time increments $\Delta t_i$, $1 \leq i \leq m$, such that the matrix of (8.41''),

$$\prod_{i=1}^{m} \{(1 - \Delta t_i)I + \Delta t_i B\},$$

is precisely the polynomial $\tilde{p}_m(B)$ of (5.19).

**2.** Let $A = D - E - F$ be an $n \times n$ matrix where $D$ is a positive diagonal matrix and $E$ and $F$ are respectively strictly lower and upper triangular matrices. For the splittings

$$M_1 = D, \qquad\qquad N_1 = E + F,$$

$$M_2 = D - E, \qquad\qquad N_2 = F,$$

$$M_3 = \frac{1}{\omega}(D - \omega E), \qquad N_3 = \frac{(1 - \omega)}{\omega}D + F, \qquad \omega \neq 0,$$

let $C_i \equiv M_i$, $i = 1, 2, 3$, and consider respectively the Padé matrix approximation $E_{0,1}(\Delta t \, C_i^{-1}A)$ for $\exp(-\Delta t \, C_i^{-1}A)$. Show for $\Delta t = 1$ that these Padé approximations exactly correspond to the point Jacobi, point Gauss-Seidel, and point successive overrelaxation iterative methods, respectively.

**3.** Verify that the matrices $T_i(\Delta t)$, $2 \leq i \leq 5$, of (8.42), (8.48), (8.51), and (8.52) are each consistent approximations of $\exp(-\Delta t \, S)$. Next, for small $\Delta t$ show that the expansion of $T_3(\Delta t)$ in (8.48) agrees through *quadratic* terms of $\exp(-\Delta t \, A)$ without assuming that $H_1$ and $V_1$ commute.

**4.** Let $A \equiv I - B$, where

$$B = \left[\begin{array}{c|c} O & F \\ \hline F^* & O \end{array}\right]$$

is convergent, and set $C \equiv I$. Show that the optimum value of $t/2 = \omega$ which minimizes the spectral radius of $T_5(t)$ of (8.52) is $\omega = 1$ (Kahan (1958)).

*5. Consider two iterative methods for solving the matrix problem $Ax = k$, where $A \equiv I - B$ and $B$ is the convergent matrix of the previous Exercise 4. The first iterative method is defined by applying the Chebyshev semi-iterative method of Sec. 5.1 to the optimized basic iteration matrix $T_5(t)$ of (8.52) with $t/2 = \omega = 1$ and $C \equiv I$. The second iterative method is defined by applying the Chebyshev semi-iterative method to the basic cyclic reduction iterative method of Sec. 5.4. Show that the second method is iteratively faster for $m$ iterations than the first for *every* $m > 1$.

*6. Let $A = D - E - E^*$ and $D$ be $n \times n$ Hermitian and positive definite matrices. Prove that the eigenvalues $\lambda_i$ of $T_5(t)$ of (8.52) satisfy $0 < \lambda_i < 1$ for all $1 \leq i \leq n$ when $0 < t/2 < 2$. (*Hint*: Apply Stein's result of Exercise 6, Sec. 1.3) (Habetler and Wachspress (1961)).

## BIBLIOGRAPHY AND DISCUSSION

**8.1.** It is interesting that various forms of semi-discrete approximations have been considered independently by several authors. The first such approach was taken by Hartree and Womersley (1937), who, in numerically approximating the solution of the one-dimensional heat equation (8.1'), discretized the *time* variable but left the space variable continuous. The main reason for this approach was to solve the resulting system of ordinary differential equations by means of a differential analyzer. This general semi-discrete approach, of course, is closely related to the Russian "method of lines," where an elliptic partial differential equation in two space variables is approximated by a system of ordinary differential equations. See Faddeeva (1949) and Kantorovich and Krylov (1958), p. 321. Franklin (1959b), Lees (1961), and Varga (1961) all consider approximations to parabolic partial differential equations in which the spatial variables are discretized but the time variable is kept continuous.

The explicit representation of the solution (8.11) of the particular ordinary matrix differential equation of (8.8'), using exponentials of a square matrix, is a special case of results to be found in Bellman (1953), p. 12, and is further connected to the abstract theory of semigroups of operators, as the matrix $A$ is the *infinitesimal generator* for the semigroup of matrix operators $\exp(tA)$. For the theory of semigroups, see Phillips (1961) and Hille and Phillips (1957).

Bellman (1953) also gives explicit representations for the solution of the more general differential equation

$$\frac{d\mathbf{v}(t)}{dt} = -A(t)\mathbf{v}(t) + \mathbf{s}(t),$$

where the matrix $A(t)$ has time-dependent entries. Because of this, a corresponding discussion of semi-discrete approximations to the partial differential equation (8.1) in which the coefficients $\phi(\mathbf{x}; t)$, $K_i(\mathbf{x}; t)$, etc., are functions of time can also be given.

Results similar to those of Exercises 5 and 6, in which bounds for the norm of the approximate solution $\mathbf{v}(t)$ as well as bounds for the norm of $(\mathbf{u}(t) - \mathbf{v}(t))$ are obtained, are sometimes derived from the *maximum principle* for such parabolic problems. See Douglas (1961b), Keller (1960b), and Lees (1960).

**8.2.** The terms *essentially positive, subcritical, critical,* and *supercritical* were introduced by Birkhoff and Varga (1958) in the study of the numerical solution of the time-dependent multigroup diffusion equations of reactor physics, and the results of this section are largely drawn from this reference. For generalizations to more general operators, see Birkhoff (1961) and Habetler and Martino (1961), and references given therein.

*Essentially positive* matrices, as defined in this section, are also called *input-output* matrices and *Leontieff matrices* in problems arising from economic considerations. For a detailed discussion of such matrices and their economic applications, see Karlin (1959b), and references cited there. See also Beckenbach and Bellman (1961, pp. 83 and 94), where the result of Theorem 8.2, for example, is given in a slightly weaker form.

From Theorem 8.3 we deduce that the norms $\| \exp(tQ) \|$ of a subcritical essentially positive matrix are uniformly bounded for all $t \geq 0$. More generally, one can deduce from Lemma 8.1 necessary and sufficient conditions that $\| \exp(tQ) \|$ be bounded for all $t \geq 0$ for a *general* $n \times n$ complex matrix. For such extensions, see Bellman (1953), p. 25, and Kreiss (1959a), as well as Exercises 5 and 9. It should also be stated that the result of Exercise 9 is clearly related to the Hille-Yosida theorem of semigroup theory. See Phillips (1961).

**8.3.** The analyses of the forward difference, backward difference, and Crank-Nicholson methods for numerically approximating the solution of parabolic partial differential equations have received a great deal of attention. For excellent bibliographical coverage, see Todd (1956), Richtmyer (1957), Young (1961), and Douglas (1961b).

The initial works of Courant, Friedrichs, and Lewy (1928), von Neumann (see O'Brien, Hyman, and Kaplan (1951)), Laasonen (1949), Crank and Nicolson (1947), and many subsequent articles give analyses of these finite difference methods for uniform mesh spacings, and generally one finds in the literature that Fourier series methods are a commonly used tool. Although the goal in this section, following Varga (1961), has been to show the relevance of these basic numerical methods to particular Padé matrix approximations of $\exp(-tA)$, the stability analysis of these Padé approximations in Theorem 8.5 for $M$-matrices is applicable under fairly wide circumstances, since nonhomogeneous, nonrectangular problems with internal interfaces and nonuniform spatial meshes do not require special treatment. See also Householder (1958) for a related use of norms in studying stability criteria.

Not all known methods for approximating solutions of parabolic partial differential equations can be obtained from Padé matrix approximations of $\exp(-tA)$. For example, multilevel approximations such as the unconditionally stable method of DuFort and Frankel (1953) and the unconditionally *unstable* method of Richardson (1910) require special treatment. See, for example, Douglas (1961b), Todd (1956), and Richtmyer (1957).

Padé rational approximations are only a particular type of rational approximation of $\exp(-tA)$. Franklin (1959b) considers consistent explicit polynomial approximations of high order in order to maximize the stability interval in $\Delta t$. Varga (1961) considers Chebyshev rational approximations of $\exp(-tA)$ that are aimed at obtaining useful vector approximations in just *one* time step.

It is interesting to note that several definitions of stability exist in the literature. For example, O'Brien, Hyman, and Kaplan (1951) relate stability to growth of rounding errors, whereas Lax and Richtmyer (1956) define an operator $C(\Delta t)$ to be stable if for some $\tau > 0$, the set of norms $\| C^m(\Delta t) \|$ is uniformly bounded for $0 < \Delta t \leq \tau$, $0 \leq m\, \Delta t \leq T$. However, if the size of $m\, \Delta t$ is unrestricted, a supposition of interest in problems admitting a finite steady-state solution, this definition of Lax and Richtmyer reduces to that given in Definition 8.3. See also Richtmyer (1957) and Esch (1960).

Because we have omitted the entire topic of *convergence* of the difference approximations to the solution of the differential equation, the basic results of Lax and Richtmyer (1956) and Douglas (1956), which in essence show the equivalence of the concepts of stability and convergence, are not discussed here. These analyses consider the combined errors due to spatial as well as time discretizations. In this regard, the article of John (1952) is basic. See also Lees (1959), Strang (1960), and Kreiss (1959b).

The matrix result of Theorem 8.6, linking the Perron-Frobenius theory of non-negative matrices to the solutions of discrete approximations to certain parabolic partial differential equations, is apparently new.

It must also be mentioned that our treatment in Sec. 8.3 does *not* cover the parabolic partial differential equations for which the coefficients of the differential equation are functions of time, or functions of the unknown solution, nor does it cover coupled systems of parabolic partial differential equations. For the numerical solution of such problems, we recommend the survey article by Douglas (1961b).

**8.4.** Many authors, including Frankel (1950) and Garabedian (1956), have either indicated or made use of the heuristic analogy between iterative methods for solving elliptic differential equations and numerical methods for solving parabolic partial differential equations. For the alternating-direction implicit iterative methods of Peaceman and Rachford (1955) and Douglas and Rachford (1956), this association is quite clear. The conclusion in this section that all such iterative methods can be obtained as consistent approximations of exponential matrices seems to be new, but other approaches have also produced interesting results. Notably, Garabedian (1956) has linked the successive overrelaxation iterative method with *hyperbolic* partial differential equations, which will be described in Chapter 9.

The application of the results of Theorem 8.6, further tying in the Perron-Frobenius theory of non-negative primitive matrices, also appears to be new.

The *symmetric* successive overrelaxation iterative method is compared with the successive overrelaxation iterative method by Habetler and Wachspress (1961), where it is shown that the symmetric successive overrelaxation iterative method is superior in problems with limited application.

# CHAPTER 9

# ESTIMATION OF ACCELERATION PARAMETERS

**9.1.** Application of the Theory of Non-negative Matrices

Our investigations into solving systems of linear equations thus far have centered on the theoretical aspects of variants of the successive overrelaxation iterative method and the alternating-direction implicit methods. In all cases, the theoretical selection of associated optimum acceleration parameters was based on knowledge of certain key eigenvalues. Specifically, only the spectral radius $\rho(B)$ of the block Jacobi matrix $B$ is needed (Theorem 4.4) to determine the optimum relaxation factor

$$(9.1) \qquad \omega_b = \frac{2}{1 + \sqrt{1 - \rho^2(B)}}$$

when the successive overrelaxation iterative method is applied in cases where $B$ is a consistently ordered weakly cyclic matrix (of index 2) with real eigenvalues. Similarly, when the block Jacobi matrix has the special form

$$B = \left[ \begin{array}{c|c} O & F \\ \hline F^* & O \end{array} \right],$$

the cyclic Chebyshev semi-iterative method of Sec. 5.3 makes use of acceleration parameters $\omega_m$ defined by

$$(9.2) \qquad \omega_m = \frac{2C_{m-1}(1/\rho(B))}{\rho(B)\,C_m(1/\rho(B))}, \qquad m > 1,$$

where $C_m(x)$ is the Chebyshev polynomial of degree $m$. Again, the determination of optimum parameters depends only on the knowledge of $\rho(B)$. The situation regarding variants of the alternating-direction implicit methods is not essentially different. In the commutative case $H_1V_1 = V_1H_1$ of Sec. 7.2, only the knowledge of the *bounds* $\alpha$ and $\beta$,

$$(9.3) \qquad\qquad 0 < \alpha \leq \sigma_j, \quad \tau_j \leq \beta,$$

of the eigenvalues $\sigma_j$ and $\tau_j$ of $H_1$ and $V_1$ was essential to the theoretical problem of selecting optimum acceleration parameters $\tilde{r}_j$. However, to dismiss the practical problem of estimating key parameters because in actual computations one needs but one or two eigenvalue estimates would be overly optimistic. Those who have had even limited experience in solving such matrix problems by iterative methods have learned (often painfully) that small changes in the estimates of these key eigenvalues can sometimes drastically affect rates of convergence.† In this section, we shall show that further use can be made of the Perron-Frobenius theory of non-negative matrices in estimating these key eigenvalues.

We assume now that the $n \times n$ block Jacobi matrix $B$ for the successive overrelaxation variants is non-negative, irreducible, and further that $B$ is a consistently ordered weakly cyclic matrix of index 2. These assumptions are satisfied (Theorems 6.3 and 6.4) for a large class of discrete approximations to boundary-value problems, as we have seen. What is sought are estimates of the spectral radius $\rho(B)$ of $B$, and the discussion in Sec. 4.3 tells us that we seek preferably estimates *not less* than $\rho(B)$. If we write $B = L + U$, where $L$ and $U$ are respectively strictly lower and upper triangular matrices, the associated block Gauss-Seidel matrix $\mathcal{L}_1 \equiv (I - L)^{-1}U$ is also a non-negative matrix, and as $B$ is a consistently ordered weakly cyclic matrix of index 2, we then have (from the Corollary of Theorem 4.3)

$$(9.4) \qquad\qquad \rho(\mathcal{L}_1) = \rho^2(B).$$

From this expression, we then can reduce our problem to one of estimating the spectral radius of $\mathcal{L}_1$. Let $\mathbf{x}^{(0)}$ be any vector with positive components, and define

$$(9.5) \qquad\qquad \mathbf{x}^{(m+1)} = \mathcal{L}_1\mathbf{x}^{(m)}, \qquad m \geq 0.$$

It can be shown (Exercise 7 of Sec. 3.3) that all the vectors $\mathbf{x}^{(m)}$ have positive components. Because of this, we can define the positive numbers

$$(9.6) \qquad \underline{\lambda}_m \equiv \min_{1 \leq i \leq n}\left(\frac{x_i^{(m+1)}}{x_i^{(m)}}\right), \qquad \bar{\lambda}_m \equiv \max_{1 \leq i \leq n}\left(\frac{x_i^{(m+1)}}{x_i^{(m)}}\right), \qquad m \geq 0,$$

† See Exercise 1. For more complete numerical results, see Young (1955) and Stark (1956).

and from an extension (Exercises 2 and 3 of Sec. 2.3) of the min-max characterization (Theorem 2.2) of the spectral radius of a non-negative irreducible matrix, we can deduce that

$$(9.7) \qquad \underline{\lambda}_m \leq \underline{\lambda}_{m+1} \leq \rho(\mathcal{L}_1) \leq \bar{\lambda}_{m+1} \leq \bar{\lambda}_m, \qquad m \geq 0.$$

Moreover, familiar arguments show† that

$$(9.8) \qquad \lim_{m \to \infty} \underline{\lambda}_m = \lim_{m \to \infty} \bar{\lambda}_m = \rho(\mathcal{L}_1).$$

Summarizing, we have

**Theorem 9.1.** *Let $B$ be an $n \times n$ block Jacobi matrix which is non-negative, irreducible, and a consistently ordered weakly cyclic matrix of index 2. If $\mathbf{x}^{(0)}$ is any positive vector, then the sequences of positive numbers $\{\underline{\lambda}_m\}_{m=0}^{\infty}$ and $\{\bar{\lambda}_m\}_{m=0}^{\infty}$ defined in (9.6) are respectively nondecreasing and nonincreasing, and have the common limit $\rho^2(B)$.*

**Corollary.** *In addition, let the matrix $B$ be convergent with real eigenvalues. If $\bar{\lambda}_m \leq 1$ for $m \geq m_0$, and*

$$(9.9) \qquad \underline{\omega}_m \equiv \frac{2}{1 + \sqrt{1 - \underline{\lambda}_m}}, \qquad \bar{\omega}_m \equiv \frac{2}{1 + \sqrt{1 - \bar{\lambda}_m}}, \qquad m \geq m_0.$$

*then*

$$(9.10) \qquad \underline{\omega}_m \leq \underline{\omega}_{m+1} \leq \omega_b = \frac{2}{1 + \sqrt{1 - \rho^2(B)}} \leq \bar{\omega}_{m+1} \leq \bar{\omega}_m, \qquad m \geq m_0,$$

*and*

$$(9.11) \qquad \lim_{m \to \infty} \underline{\omega}_m = \lim_{m \to \infty} \bar{\omega}_m = \omega_b.$$

*Proof.* This result is an obvious consequence of Theorem 9.1 and the fact that

$$\omega(\mu) = \frac{2}{1 + \sqrt{1 - \mu^2}}, \qquad 0 \leq \mu \leq 1,$$

is a 1-1 increasing function of $\mu$.

In the case that we choose to use the cyclic Chebyshev semi-iterative method of Sec. 5.3, then only the estimates of $\rho^2(B) = \rho(\mathcal{L}_1)$ of (9.7) are needed to deduce approximations of the parameters $\omega_m$ of (9.2).

† See Exercise 2 of this section.

As a simple numerical example of this procedure, consider the matrix

$$(9.12) \qquad A = \begin{bmatrix} 2 & -1 & 0 & 0 \\ -1 & 2 & -1 & 0 \\ 0 & -1 & 2 & -1 \\ 0 & 0 & -1 & 2 \end{bmatrix},$$

which arises (Sec. 6.1) from a discrete approximation to a two-point boundary-value problem. Its associated point Jacobi matrix,

$$(9.13) \qquad B = \frac{1}{2} \begin{bmatrix} 0 & 1 & 0 & 0 \\ 1 & 0 & 1 & 0 \\ 0 & 1 & 0 & 1 \\ 0 & 0 & 1 & 0 \end{bmatrix},$$

has real eigenvalues and satisfies all the hypotheses of Theorem 9.1 and its Corollary; its spectral radius is given by $\rho(B) = \cos(\pi/5) \doteq 0.8090$, and thus from (9.1), $\omega_b = 1.2604$. Having selected a positive vector $\mathbf{x}^{(0)}$, the subsequent vectors $\mathbf{x}^{(m)}$ can be viewed as *iterates* of the point Gauss-Seidel method applied to the solution of

$$A\mathbf{x} = \mathbf{0},$$

where $\mathbf{x}^{(0)}$ is the initial vector of the iterative process. Specifically, if we choose $\mathbf{x}^{(0)}$ to have all components unity, the results of this iterative procedure are tabulated below.

| $m$ | $\underline{\lambda}_m$ | $\bar{\lambda}_m$ | $\underline{\omega}_m$ | $\bar{\omega}_m$ |
|---|---|---|---|---|
| 0 | 0.4375 | 0.8750 | 1.1429 | 1.4776 |
| 1 | 0.6071 | 0.8333 | 1.2294 | 1.4202 |
| 2 | 0.6471 | 0.8333 | 1.2546 | 1.4202 |
| 3 | 0.6534 | 0.6750 | 1.2589 | 1.2738 |

In actual practice, this iterative process is terminated when some measure of the difference $\bar{\omega}_m - \underline{\omega}_m$ is sufficiently accurate.†

The definition of (9.5) shows that the vectors $\mathbf{x}^{(m)}$ can be expressed as powers of the matrix $\mathcal{L}_1$ applied to $\mathbf{x}^{(0)}$:

$$\mathbf{x}^{(m)} = \mathcal{L}_1^m \mathbf{x}^{(0)}, \qquad m \geq 0,$$

and for this reason the method just described for finding the largest eigenvalue (in modulus) of $\mathcal{L}_1$ is basically a combination of the Perron-Frobenius theory of non-negative matrices, and the *power method*, the latter being applicable to any matrix whose largest eigenvalue in modulus is unique. (See Bodewig (1959), p. 269.) The *rate* at which this process converges depends on the *dominance ratio*:

$$(9.14) \qquad\qquad d = \frac{|\lambda_2|}{|\lambda_1|} < 1,$$

where we have ordered the eigenvalues of $\mathcal{L}_1$ as $|\lambda_1| > |\lambda_2| \geq |\lambda_3| \geq \cdots \geq |\lambda_n|$. Convergence is slow when $d$ is close to unity.

It is clear that the numerical method described for estimating $\rho^2(B)$ can be applied equally well when one uses block or multi-line successive overrelaxation iterative methods. Once a particular successive overrelaxation variant is decided upon, only obvious minor changes need be made in the basic iteration portion of the program to carry out this procedure. We should point out that this iterated use of the min-max characterization of the spectral radius of $\mathcal{L}_1$ can often be discouragingly slow when applied to matrix problems of large order where $\rho(B)$ is very close to unity. In other words, when $\rho(B)$ is close to unity, the dominance ratio $d$ of (9.14) may also be close to unity. It is interesting to mention here that there is considerable numerical evidence showing that, for a fixed Stieltjes matrix, passing from a *point* iterative method to a *block* or multiline iterative method *decreases* the dominance ratio $d$. In the cases considered, not only was the actual iterative procedure to solve a given matrix problem accelerated in passing from point to block techniques, but so was the method for finding optimum parameters. (See Exercises 3 and 4.)

This numerical method for estimating $\rho^2(B)$, leading equivalently to estimates of $\omega_b$, can be called an *a priori* iterative method in that the iterations described can be performed in advance of the actual iterations to solve a specific matrix equation $A\mathbf{x} = \mathbf{k}$, as these calculations leading to estimates of $\omega_b$ in no way depend on the vector $\mathbf{k}$. The advantages of this procedure are twofold. First, *rigorous* upper and lower bounds for $\omega_b$ are

† For example, at the Bettis Atomic Power Laboratory, $(\bar{\omega}_m - \underline{\omega}_m)/(2 - \bar{\omega}_m) \leq 0.2$ is the criterion used. This procedure for finding bounds for $\omega_b$ is often referred to in nuclear programs as the *ω-routine*. See Bilodeau, et al (1957).

determined, and second, the basic programming of the iterative method used to solve the specific matrix problem $A\mathbf{x} = \mathbf{k}$ is easily altered to allow one to find these bounds. We should add that this procedure has been used successfully for some time for large problems arising in reactor and petroleum engineering.

This specific numerical technique, however, is not the only one that has been used to estimate $\rho^2(B)$ or $\omega_b$. Because of the practical importance of good estimates, several papers have appeared describing *non* a priori methods, where the actual vector iterates of the matrix equation $A\mathbf{x} = \mathbf{k}$ are used to determine estimates of $\omega_b$. Roughly speaking, one iterates with, say, $\omega_1 = 1$, and on the basis of numerical results, one increases $\omega_1$ to $\omega_2$, and repeats the process. Since the use of these methods appears to be more of a computing *art* than a *science*, we have omitted a detailed discussion of these methods.†

The situation regarding alternating-direction implicit methods is not theoretically different. As we have seen in the commutative case, the corresponding numerical problem rests in estimating the eigenvalue bounds $\alpha$ and $\beta$ of (9.3). In practice, estimates of the upper bound $\beta$ are made simply by using Corollary 1 of Gerschgorin's Theorem 1.5, as estimates of this upper bound $\beta$ do not have a critical effect on rates of convergence. Estimates of the lower bound $\alpha$, however, *do* have a critical effect on rates of convergence, especially when $\alpha$ is close to zero. This similarity with estimating $\rho(B)$ for successive overrelaxation variants is further strengthened when it is apparent that we can again use the Perron-Frobenius theory of non-negative matrices to estimate $\alpha$. Let $H_1$ and $V_1$ be, after suitable permutations, tridiagonal Stieltjes matrices. From Sec. 7.1, $H_1$ and $V_1$ can be expressed as the direct sum of *irreducible* tridiagonal Stieltjes matrices. Thus, without loss of generality, we assume that $H_1$ is an irreducible tridiagonal Stieltjes $n \times n$ matrix, which geometrically corresponds to the coupling between mesh points of a horizontal mesh line. Since $H_1$ is an irreducible Stieltjes matrix, then (Corollary 3 of Sec. 3.5) $H_1^{-1} > O$. If the eigenvalues of $H_1$ are $0 < \sigma_1 \leq \sigma_2 \leq \cdots \leq \sigma_n$, we have further

**Lemma 9.1.** *If $H_1$ is an irreducible $n \times n$ Stieltjes matrix with eigenvalues $0 < \sigma_1 \leq \sigma_2 \leq \cdots \leq \sigma_n$, then for all $0 \leq \lambda < \sigma_1$,*

$$(9.15) \qquad\qquad (H_1 - \lambda I)^{-1} > O.$$

*Proof.* For any $\lambda$ with $0 \leq \lambda < \sigma_1$, then $H_1 - \lambda I$ is also an irreducible Stieltjes matrix (Definition 3.4), and the result follows from Corollary 3 of Sec. 3.5.

† See, however, Forsythe and Wasow (1960), p. 368, for a complete description of these various methods, as well as the references at the end of this chapter.

The procedure for finding rigorous upper and lower bounds for $\sigma_1$, the minimum eigenvalue of $H_1$, again uses the Perron-Frobenius theory of non-negative matrices. Let $\mathbf{x}^{(m)} > 0$, and assume that $0 \leq \lambda_m < \sigma_1$. Define $\mathbf{x}^{(m+1)}$ from

$$(9.16) \qquad (H_1 - \lambda_m I)\mathbf{x}^{(m+1)} = \mathbf{x}^{(m)},$$

or equivalently,

$$(9.16') \qquad \mathbf{x}^{(m+1)} = (H_1 - \lambda_m I)^{-1}\mathbf{x}^{(m)}.$$

Since $H_1$ is tridiagonal and nonsingular, we can surely *directly* solve (by the algorithm of Sec. 6.4) for the vector $\mathbf{x}^{(m+1)}$, which from Lemma 9.1 has positive components. Applying the Perron-Frobenius min-max Theorem 2.2 to (9.16'), it is obvious that

$$(9.17) \qquad \min_{1 \leq i \leq n}\left(\frac{x_i^{(m+1)}}{x_i^{(m)}}\right) \leq \rho\{(H_1 - \lambda_m I)^{-1}\} = \frac{1}{\sigma_1 - \lambda_m} \leq \max_{1 \leq i \leq n}\left(\frac{x_i^{(m+1)}}{x_i^{(m)}}\right),$$

or equivalently,

$$(9.17') \quad \underline{\sigma}_1^{(m)} \equiv \lambda_m + \min_{1 \leq i \leq n}\left(\frac{x_i^{(m)}}{x_i^{(m+1)}}\right) \leq \sigma_1 \leq \lambda_m + \max_{1 \leq i \leq n}\left(\frac{x_i^{(m)}}{x_i^{(m+1)}}\right) \equiv \bar{\sigma}_1^{(m)}.$$

If $\underline{\sigma}_1^{(m)} = \bar{\sigma}_1^{(m)}$, this common value must be $\sigma_1$ by Theorem 2.2, and the procedure terminates. On the other hand, if these bounds are unequal, we define $\lambda_{m+1} \equiv \underline{\sigma}_1^{(m)}$ and continue this process. To start the process, one can choose $\lambda_0 = 0$. In much the same manner as in Theorem 9.1, we can prove (Exercise 6)

**Theorem 9.2.** *Let $H_1$ be an $n \times n$ irreducible Stieltjes matrix with eigenvalues $0 < \sigma_1 \leq \sigma_2 \leq \cdots \leq \sigma_n$, and let $\lambda_0 = 0$ and let $\mathbf{x}^{(0)}$ be any positive vector. Then the sequences of positive numbers $\{\underline{\sigma}_1^{(m)}\}_{m=0}^{\infty}$ and $\{\bar{\sigma}_1^{(m)}\}_{m=0}^{\infty}$ defined in (9.17') are respectively nondecreasing and nonincreasing, and have the common limit $\sigma_1$.*

To illustrate this procedure numerically, consider the matrix $A$ of (9.12), which is a tridiagonal irreducible Stieltjes matrix. The bounds $\sigma_1$ and $\sigma_4$ for the eigenvalues of $A$ are

$$\sigma_1 = 2\left(1 - \cos\left(\frac{\pi}{5}\right)\right) = 0.38196; \quad \sigma_4 = 2\left(1 + \cos\left(\frac{\pi}{5}\right)\right) = 3.61804.$$

Applying Corollary 1 of Theorem 1.5 to $A$ would give us the bound $\sigma_4 \leq 4$.

With $\lambda_0 = 0$, and $\mathbf{x}^{(0)}$ having all components unity, the bounds $\underline{\sigma}_1^{(m)}$ and $\bar{\sigma}_1^{(m)}$ for $\sigma_1$ are tabulated below.

| $m$ | $\lambda_m$ | $\underline{\sigma}_1^{(m)}$ | $\bar{\sigma}_1^{(m)}$ |
|-----|-------------|------------------------------|------------------------|
| 0 | 0 | 0.3333 | 0.5000 |
| 1 | 0.3333 | 0.3810 | 0.3846 |
| 2 | 0.3810 | 0.3820 | 0.3820 |

This numerical procedure is applied to *each* irreducible submatrix of the matrices $H_1$ and $V_1$. Geometrically then, this process is applied to *each* row and column of mesh points of the discrete problem. The final estimates of $\alpha$ and $\beta$ in (9.3) are then respectively the minimum of the lower eigenvalue bounds and the maximum of the upper eigenvalue bounds.

The procedure described here is in reality a combination of Wielandt's (1944) *inverse power method*† and the Perron-Frobenius theory of non-negative matrices. A closely related application of the Perron-Frobenius theory to eigenvalue bounds for alternating-direction implicit methods was originally made by Wachspress and Habetler (1960). Of course, improvements can be made in the upper bounds of (9.17′) by using Rayleigh quotients of the vector iterates. (See Exercise 5.)

These a priori methods for obtaining key eigenvalue estimates based on the Perron-Frobenius theory of non-negative matrices have wide applications to physical problems that possess the complications of internal interfaces and nonuniform mesh spacings, as described in Chapter 6. For the more restricted class of discrete problems with uniform mesh spacings and homogeneous composition, it may be more efficient to use isoperimetric inequalities of classical analysis, as will be described in the next section.

## EXERCISES

**1.** Let $B$ be a consistently ordered weakly cyclic matrix of index 2, with real eigenvalues, and let $\rho(B) = 1 - \delta$, where $\delta = 10^{-4}$. Then $\omega_b = 1.972$, $R_\infty(\mathcal{L}_{\omega_b}) = -\ln(\omega_b - 1) \doteq 0.0275$. Using an estimated value of $\omega = 1.9$, determine the $\rho(\mathcal{L}_\omega)$ and $R_\infty(\mathcal{L}_\omega)$ and show that

$$R_\infty(\mathcal{L}_{\omega_b})/R_\infty(\mathcal{L}_\omega) > 7.$$

† See Wilkinson (1956) for an up-to-date account of the numerical behavior of this method.

2. Let $B = L + U$ be a non-negative irreducible $n \times n$ block Jacobi matrix, which further is a consistently ordered weakly cyclic matrix of index 2. Prove that the sequences $\{\lambda_m\}_{m=0}^{\infty}$ and $\{\bar{\lambda}_m\}_{m=0}^{\infty}$ defined from (9.6) tend to the common limit $\rho^2(B)$. (*Hint*: Use Theorem 4.3.)

3. For the model problem described in Sec. 6.5, determine the dominance ratios $d_{pt}$ and $d_{line}$ of (9.14) respectively for the point and line Gauss-Seidel matrices as a function of $N = 1/h$. Show that

$$0 < d_{line} < d_{pt} \qquad \text{for all } N > 2.$$

4. For the matrix $B$ of (9.13), show that the dominance ratio $d_{pt}$ of the associated point Gauss-Seidel matrix is given by

$$d_{pt} = \left( \frac{\cos (2\pi/5)}{\cos (\pi/5)} \right)^2.$$

If the matrix $A$ of (9.12) is partitioned as follows,

$$A = \begin{bmatrix} 2 & -1 & 0 & 0 \\ -1 & 2 & -1 & 0 \\ 0 & -1 & 2 & -1 \\ 0 & 0 & -1 & 2 \end{bmatrix},$$

determine the dominance ratio $d_{bk}$ for the associated block Gauss-Seidel matrix. How do $d_{pt}$ and $d_{bk}$ compare?

5. Let $H_1$ be an irreducible $n \times n$ Stieltjes matrix with eigenvalues $0 < \sigma_1 \leq \sigma_2 \leq \cdots \leq \sigma_n$, and let $\mathbf{x}^{(m)} > 0$ and $0 \leq \lambda_m < \sigma_1$. Using the notations of (9.16) and (9.17'), show that

$$\sigma_1 \leq \lambda_m + \frac{\|\mathbf{x}^{(m)}\|^2}{(\mathbf{x}^{(m)})^* (\mathbf{x}^{(m+1)})} \leq \sigma_1^{(m)}.$$

6. Prove Theorem 9.2.

## 9.2. Application of Isoperimetric Inequalities

The goal in this section is to connect the classical theory of isoperimetric inequalities to the topic of finding bounds for the optimum relaxation factor $\omega_b$ for matrix problems somewhat more restrictive than those treated in Sec. 9.1. This interesting connection was first made by Garabedian (1956). We shall derive Garabedian's estimate in a different manner and obtain a measure for the *efficiency* of his estimate.

Let $R$ be an open, bounded, and connected region in the plane with boundary $\Gamma$, such that $\bar{R}$, the closure of $R$, is the union of a finite number of squares of side $h^*$. Specifically, we seek discrete approximations to the Poisson boundary-value problem

$$(9.18) \qquad -u_{xx} - u_{yy} = f(x, y), \qquad (x, y) \in R,$$

subject to the boundary condition

$$(9.19) \qquad u(x, y) = g(x, y), \qquad (x, y) \in \Gamma.$$

For the discussion to follow, it is useful to introduce the Helmholtz equation for $R$:

$$(9.18') \qquad v_{xx} + v_{yy} + \lambda^2 v(x, y) = 0, \qquad (x, y) \in R,$$

where

$$(9.19') \qquad v(x, y) = 0, \qquad (x, y) \in \Gamma,$$

and we denote the smallest eigenvalue $\lambda^2$ of $(9.18')$ by $\Lambda^2$. Physically, $\Lambda^2$ is the eigenvalue of the fundamental mode for the membrane problem for $R$.

For discrete approximations to (9.18) and (9.19) derived in the manner of Sec. 6.3, we use *any* uniform mesh $h = h^*/N$, where $N$ is a positive integer. Such choices for mesh spacings obviously avoid complications in approximating the boundary $\Gamma$ and provide us with a sequence of mesh spacings tending to zero. For any such mesh spacing $h$, our discrete five-point approximation to (9.18)–(9.19) is

$$\begin{cases} 4u_{i,j} - (u_{i+1,j} + u_{i-1,j} + u_{i,j+1} + u_{i,j-1}) = h^2 f_{i,j}, & (x_i, y_j) \in R_h, \\ \\ u_{i,j} = g_{i,j}, & (x_i, y_j) \in \Gamma, \end{cases}$$

where, as in Sec. 6.3, $R_h$ denotes the interior mesh points of $R$. In matrix notation, this becomes

$$(9.20) \qquad A_h \mathbf{u} = \mathbf{k},$$

but the same discrete approximation to the eigenvalue problem of $(9.18')$–$(9.19')$ similarly yields

$$(9.20') \qquad A_h \mathbf{v} = \lambda_h^2 \cdot h^2 \cdot \mathbf{v}.$$

Because of the form of the differential equation of (9.18), Theorem 6.4 tells us at least that the matrix $A_h$ is real, symmetric, and positive definite. Let us denote the smallest (positive) eigenvalue of $A_h$ by $\Lambda_h^2 \cdot h^2$. If we now form the point Jacobi matrix $B_h$ from $A_h$, then since the diagonal entries of $A_h$ are all 4,

$$(9.21) \qquad B_h \equiv I - \tfrac{1}{4} A_h.$$

Again, from Theorem 6.4, we know that $B_h$ is non-negative, convergent, and weakly cyclic of index 2. Assuming $B_h$ to be consistently ordered, we know (Theorem 4.4) that the optimum relaxation factor $\omega_b(h)$ for the point successive overrelaxation iterative method is given by

$$(9.22) \qquad \omega_b(h) = \frac{2}{1 + \sqrt{1 - \rho^2(B_h)}}.$$

However, from (9.21) and the non-negative nature of the matrix $B_h$, we can relate the spectral radius $\rho(B_h)$ to the smallest eigenvalue of $A_h$ by

$$(9.23) \qquad \rho(B_h) = 1 - \frac{\Lambda_h^2 \cdot h^2}{4}.$$

The 1-1 relationship between $\rho(B_h)$ and $\Lambda_h^2$ in (9.23) can be used in two ways. First, by means of isoperimetric inequalities, estimates of $\Lambda_h^2$ can be made, which result in estimates of $\rho(B_h)$; but, from (9.22), $\omega_b(h)$ is a 1-1 function of $\rho(B_h)$, and it follows that estimates of $\Lambda_h^2$ give rise to estimates of $\omega_b(h)$. The idea for this approach is due to Garabedian. Conversely, estimates of $\rho(B_h)$ lead to estimates of $\Lambda_h^2$. From the results of Courant, Friedrichs, and Lewy (1928), it follows that we may use the weak relationship[†]

$$(9.24) \qquad \Lambda_h^2 = \Lambda^2 + o(1), \qquad h \to 0.$$

Thus, estimates of $\rho(B_h)$ give rise to estimates of $\Lambda^2$ for small mesh spacing $h$. This latter approach was essentially used by Gerberich and Sangren (1957) to numerically estimate $\Lambda^2$ for nonrectangular regions.

In finding estimates of $\omega_b(h)$, we shall be most interested in *upper bound* estimates, since we know that overestimating $\omega_b(h)$ does not cause as serious a decrease in the rate of convergence of the successive over-relaxation method as does underestimating $\omega_b(h)$. Based on a result of Pólya (1952), Weinberger (1956) obtained the inequality

$$(9.25) \qquad \Lambda^2 \leq \frac{\Lambda_h^2}{1 - 3h^2\Lambda_h^2}$$

for $1 - 3h^2\Lambda_h^2 > 0$, i.e., for all $h$ sufficiently small. Thus

$$(9.26) \qquad \Lambda_h^2 \geq \frac{\Lambda^2}{1 + 3h^2\Lambda^2}$$

---

[†] For related results, see Forsythe (1956), we could replace $o(1)$ by $O(h^2)$ in (9.24), but such refinements are unnecessary for the development.

for all $h$ sufficiently small. From (9.26) and (9.23), we have

$$(9.27) \qquad \rho(B_h) \leq 1 - \frac{\Lambda^2 h^2}{4 + 12\Lambda^2 h^2} \equiv \overset{*}{\mu}.$$

Since $\overset{*}{\mu}$ is an upper bound for $\rho(B_h)$, then

$$(9.28) \qquad \omega_2(h) \equiv \frac{2}{1 + \sqrt{1 - (\overset{*}{\mu})^2}}$$

$$= \frac{2}{1 + \left\{ \dfrac{2\Lambda^2 h^2}{4 + 12\Lambda^2 h^2} - \left( \dfrac{\Lambda^2 h^2}{4 + 12\Lambda^2 h^2} \right)^2 \right\}^{1/2}}$$

is an upper bound for $\omega_b(h)$, and

$$(9.29) \qquad \omega_b(h) \leq \omega_2(h).$$

From (9.22) and (9.23), we derive that

$$(9.30) \qquad \frac{\sqrt{2}}{h} \left( \frac{2}{\omega_b(h)} - 1 \right) = \Lambda_h \left\{ 1 - \frac{\Lambda_h^2 h^2}{8} \right\}^{1/2}.$$

Thus, with (9.24),

$$(9.31) \qquad \lim_{h \to 0} \left\{ \frac{\sqrt{2}}{h} \left( \frac{2}{\omega_b(h)} - 1 \right) \right\} = \Lambda.$$

With Garabedian, we use the isoperimetric inequality (Pólya and Szegö (1951))

$$(9.32) \qquad \Lambda a^{1/2} \geq \tilde{\Lambda} \pi^{1/2},$$

where $a$ is the area of $\bar{R}$, and $\tilde{\Lambda} = 2.405$ is the first zero of the zero-order Bessel function of the first kind. Defining

$$(9.33) \qquad \omega_1(h) \equiv \frac{2}{1 + \tilde{\Lambda} h (\pi/2a)^{1/2}} = \frac{2}{1 + 3.015 (h^2/a)^{1/2}},$$

we have that

$$\lim_{h \to 0} \left\{ \frac{\sqrt{2}}{h} \left( \frac{2}{\omega_1(h)} - 1 \right) \right\} = \frac{\tilde{\Lambda} \pi^{1/2}}{a^{1/2}},$$

and thus, from (9.32) we have for all $h$ sufficiently small

$$(9.34) \qquad\qquad \omega_1(h) \geq \omega_b.$$

The estimate $\omega_1(h)$ of $\omega_b(h)$ is Garabedian's estimate. Returning to the upper estimate $\omega_2(h)$, we have

$$\lim_{h \to 0} \left\{ \frac{\sqrt{2}}{h} \left( \frac{2}{\omega_2(h)} - 1 \right) \right\} = \Lambda.$$

Thus, while both $\omega_1(h)$ and $\omega_2(h)$ are upper estimates of $\omega_b(h)$ for all $h$ sufficiently small, Garabedian's estimate $\omega_1(h)$ requires only the knowledge of the area of $\bar{R}$. On the other hand, if the fundamental eigenvalue $\Lambda^2$ of $R$ is known, it will appear that $\omega_2(h)$ is asymptotically the better estimate of $\omega_b(h)$. It is clear, of course, that Garabedian's estimate $\omega_1(h)$ is in general simpler to obtain.

As we know from Theorem 4.4, for all $\omega \geq \omega_b$,

$$(9.35) \qquad\qquad R(\mathcal{L}_\omega) = -\ln (\omega - 1).$$

Thus, for all $h$ sufficiently small, we conclude that

$$(9.36) \qquad\qquad R(\mathcal{L}_{\omega_b(h)}) = 2 \Lambda h + o(h),$$

$$(9.36') \qquad\qquad R(\mathcal{L}_{\omega_2(h)}) = 2 \Lambda h + o(h),$$

and

$$(9.36'') \qquad\qquad R(\mathcal{L}_{\omega_1(h)}) = 2 \left( \frac{\tilde{\Lambda} \pi^{1/2}}{a^{1/2}} \right) h + o(h).$$

Forming ratios, we have

$$(9.37) \qquad\qquad \frac{R(\mathcal{L}_{\omega_b(h)})}{R(\mathcal{L}_{\omega_2(h)})} = 1 + o(1), \qquad h \to 0,$$

and

$$(9.37') \qquad\qquad \frac{R(\mathcal{L}_{\omega_b(h)})}{R(\mathcal{L}_{\omega_1(h)})} = \frac{\Lambda a^{1/2}}{\tilde{\Lambda} \pi^{1/2}} + o(1), \qquad h \to 0.$$

The change in the rate of convergence by using $\omega_1(h)$ instead of $\omega_b(h)$ is, for small $h$, dependent on the dimensionless constant

$$(9.38) \qquad\qquad K \equiv \frac{\Lambda a^{1/2}}{\tilde{\Lambda} \pi^{1/2}},$$

which, by (9.32), is greater than or equal to unity. We tabulate the value of $K$ for various regions (Pólya and Szegö (1951)).

| Region | $K$ |
|---|---|
| circle | 1.00 |
| square | 1.04 |
| equilateral triangle | 1.12 |
| semicircle | 1.12 |
| 45°-45°-90° triangle | 1.16 |
| 30°-60°-90° triangle | 1.21 |

We remark that for any of the regions listed above, using the estimate $\omega_1(h)$ for $\omega_b(h)$ will only result in approximately a 20 per cent decrease in the rate of convergence of the successive overrelaxation method, for $h$ small. This was observed by Garabedian (1956) in examples numerically carried out by Young (1955).

### EXERCISES

1. Let $\bar{R}$ be the union of three unit squares as shown in the figure. With $h = \frac{1}{2}$, consider the problem of estimating $\omega_b$ for the point successive over-relaxation iterative method applied to the discrete approximation of (9.18)–(9.19). With the notation of this section, it is known (Forsythe and Wasow (1960), p. 334) that $\Lambda^2 \doteq 9.636$ for $R$. Determine $\omega_1(\frac{1}{2})$ and $\omega_2(\frac{1}{2})$ from (9.28) and (9.33). Finally, show directly that

$$\rho(B_{1/2}) = \frac{\cos\,(\pi/6)}{2},$$

and compute $\omega_b(\frac{1}{2})$ exactly. How good are these estimates of $\omega_b(\frac{1}{2})$?

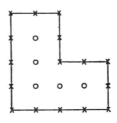

### BIBLIOGRAPHY AND DISCUSSION

**9.1.** The first application of the Perron-Frobenius theory to the problem of finding rigorous upper and lower bound estimates of $\rho(B)$ or $\omega_b$ was given by Varga (1957b). As pointed out in the text, this is but one of many methods which have been proposed. Other methods, based on results derived from the vector iterates of the nonhomogeneous matrix problem, $A\mathbf{x} = \mathbf{k}$, have been suggested by Carré (1961), Forsythe and Ortega (1960), and Kulsrud (1961). See also Forsythe and Wasow (1960), p. 368.

The first application of the Perron-Frobenius theory to the problem of estimating eigenvalue bounds for the alternating-direction implicit methods was given by Wachspress and Habetler (1960). Their procedure differs from that given in the text.

**9.2.** As stated in the text, the material for this section was inspired by Garabedian (1956). Garabedian, however, connected the successive overrelaxation iterative method with hyperbolic partial differential equations to derive his asymptotic estimate of $\omega_b$ for small mesh spacings. The fact that his estimate, for small $h$, is an *upper bound* for $\omega_b$ apparently is new, as is the measure of the efficiency of his estimate in (9.37'). It should be mentioned that his original argument contained an application to the 9-point formula approximating $u_{xx} + u_{yy} = f$.

Similar ideas for relating the rates of convergence of eigenvalues to the Helmholtz equation have been given by Parter (1961b), and Henrici (1960), and both have extended this useful notion to biharmonic problems.

# APPENDIX A

In order to gain some familiarity with the methods of derivation of finite difference equations of Chapter 6, consider the two-point boundary-value problem†

$$\text{(A.1)} \qquad -\frac{d^2y}{dx^2} + y = -x, \qquad 0 < x < 1,$$

subject to the boundary conditions

$$\text{(A.2)} \qquad y(0) = 0, \qquad y(1) = 1.$$

It is readily verified that the unique solution of this problem is given by

$$\text{(A.3)} \qquad y(x) = \frac{2 \sinh x}{\sinh 1} - x,$$

so that

$$\text{(A.4)} \qquad M_{2n} \equiv \max_{0 \le x \le 1} \left| \frac{d^{2n}y}{dx^{2n}} \right| = 2, \qquad n \ge 1.$$

First, we use a uniform mesh of $h = \frac{1}{4}$. As $y(0)$ and $y(1)$ are known from (A.2), we then seek approximations to the three values $y(ih) \equiv y_i$, $1 \le i \le 3$. The approximate Taylor's series values $z_i$ are defined in (6.7) as the solution of

$$\text{(A.5)} \qquad \begin{bmatrix} 33 & -16 & 0 \\ -16 & 33 & -16 \\ 0 & -16 & 33 \end{bmatrix} \begin{bmatrix} z_1 \\ z_2 \\ z_3 \end{bmatrix} = \begin{bmatrix} -\frac{1}{4} \\ -\frac{1}{2} \\ +\frac{61}{4} \end{bmatrix};$$

† This is also considered by L. Fox (1957, p. 68) for illustrating the method of differential corrections.

**298**

here, we have made use of the explicit representation of the matrix $A$ in (6.5). From (6.8) of Theorem 6.2 we know that

(A.6) $\qquad | y_i - z_i | \le \dfrac{M_4 h^2}{24} [ih(1 - ih)] = \dfrac{h^2}{12} [ih(1 - ih)],$

$$i \le i \le 3.$$

The solution of (A.5) is given in Table I, along with the upper bounds from (A.6) for the errors.

TABLE I

| $i$ | $y_i$: exact value | $z_i$ | $(y_i - z_i) \cdot 10^3$ | [Bounds from (A.6)] $\cdot 10^3$ |
|---|---|---|---|---|
| 1 | 0.17990479 | 0.18022950 | −0.32470812 | 0.977 |
| 2 | 0.38681887 | 0.38734835 | −0.52947467 | 1.302 |
| 3 | 0.64944842 | 0.64992647 | −0.47805076 | 0.977 |

TABLE I'

| $i$ | $y_i$: exact value | $z_i$ | $(y_i - z_i) \cdot 10^4$ | [Bounds from (A.6)] $\cdot 10^4$ |
|---|---|---|---|---|
| 1 | 0.07046741 | 0.07048938 | −0.21971591 | 0.750 |
| 2 | 0.14264090 | 0.14268364 | −0.42741807 | 1.133 |
| 3 | 0.21824367 | 0.21830475 | −0.61083032 | 1.750 |
| 4 | 0.29903319 | 0.29910891 | −0.75714761 | 2.000 |
| 5 | 0.38681887 | 0.38690415 | −0.85276142 | 2.083 |
| 6 | 0.48348014 | 0.48356844 | −0.88297274 | 2.000 |
| 7 | 0.59098524 | 0.59106841 | −0.83168838 | 1.750 |
| 8 | 0.71141095 | 0.71147906 | −0.68109581 | 1.133 |
| 9 | 0.84696337 | 0.84700451 | −0.41131190 | 0.750 |

The variational method, based on the use of continuous piecewise linear functions in the functional $F(w)$ of (6.11), gives approximate values $v_i$ defined by means of (6.16'). For the sake of comparison with (A.5), we have divided the equations of (6.16') by $h$, giving

(A.7) $\qquad \dfrac{1}{6} \begin{bmatrix} 196 & -95 & 0 \\ -95 & 196 & -95 \\ 0 & -95 & 196 \end{bmatrix} \begin{bmatrix} v_1 \\ v_2 \\ v_3 \end{bmatrix} = \begin{bmatrix} -\frac{1}{4} \\ -\frac{1}{2} \\ +\frac{181}{12} \end{bmatrix}.$

The solution of (A.7) is given in Table II. Note that there is no appreciable difference between the magnitudes of the errors $|y_i - z_i|$ and the errors $|y_i - v_i|$. It is interesting to note that all the errors $y_i - v_i$ in Table II for

TABLE II

| $i$ | $y_i$: exact value | $v_i$ | $(y_i - v_i) \cdot 10^3$ |
|---|---|---|---|
| 1 | 0.17990479 | 0.17957500 | 0.32979467 |
| 2 | 0.38681887 | 0.38628105 | 0.53782221 |
| 3 | 0.64944842 | 0.64896275 | 0.48566662 |

TABLE II′

| $i$ | $y$: exact value | $v_i$ | $(y_i - v_i) \cdot 10^4$ |
|---|---|---|---|
| 1 | 0.07046741 | 0.07044538 | 0.22023873 |
| 2 | 0.14264090 | 0.14259806 | 0.42843577 |
| 3 | 0.21824367 | 0.21818244 | 0.61228626 |
| 4 | 0.29903319 | 0.29895730 | 0.75895496 |
| 5 | 0.38681887 | 0.38673339 | 0.85480080 |
| 6 | 0.48348014 | 0.48339163 | 0.88508912 |
| 7 | 0.59098524 | 0.59090187 | 0.83368703 |
| 8 | 0.71141095 | 0.71134268 | 0.68273740 |
| 9 | 0.84696337 | 0.84692214 | 0.41230649 |

the variational method are positive. However, these values $v_i$ along with the boundary values of (A.2) serve to define the *continuous piecewise linear* function $v(x)$. If we consider instead the difference $y(x) - v(x)$ for *all* $0 \le x \le 1$, then this difference *does oscillate* in sign.

To complete the picture, we consider the method of repeated differentiations, considered in Exercise 7 of Sec. 6.2. Choosing $n = 2$ for this method gives approximate values $z_i^{(2)}$ defined as the solution of

$$(A.8) \quad \begin{bmatrix} \mu & -16 & 0 \\ -16 & \mu & -16 \\ 0 & -16 & \mu \end{bmatrix} \begin{bmatrix} z_1^{(2)} \\ z_2^{(2)} \\ z_3^{(2)} \end{bmatrix} = \begin{bmatrix} -\nu/4 \\ -\nu/2 \\ 16 - 3\nu/4 \end{bmatrix}, \quad \begin{aligned} \mu &= \tfrac{6337}{192}, \\ \nu &= \tfrac{193}{192}. \end{aligned}$$

Also from Exercise 7 of Sec. 6.2, we have that

$$(A.9) \qquad | y_i - z_i^{(2)} | \leq \frac{M_6 h^4}{6!} [ih(1 - ih)] = \frac{2h^4}{6!} [ih(1 - ih)].$$

The solution of (A.8) is given in Table III, along with the upper bounds from (A.9).

TABLE III

| $i$ | $y_i$: exact value | $z_i^{(2)}$ | $(y_i - z_i^{(2)}) \cdot 10^6$ | [Bounds from (A.9)]$\cdot 10^6$ |
|---|---|---|---|---|
| 1 | 0.17990479 | 0.17990547 | $-0.67837771$ | 2.03 |
| 2 | 0.38681887 | 0.38681998 | $-1.10651560$ | 2.71 |
| 3 | 0.64944842 | 0.64944942 | $-0.99956431$ | 2.03 |

TABLE III′

| $i$ | $y_i$: exact value | $z_i^{(2)}$ | $(y_i - z_i^{(2)}) \cdot 10^7$ | [Bounds from (A.9)]$\cdot 10^7$ |
|---|---|---|---|---|
| 1 | 0.07046741 | 0.07046741 | $-0.08538491$ | 0.250 |
| 2 | 0.14264090 | 0.14264092 | $-0.16633023$ | 0.444 |
| 3 | 0.21824367 | 0.21824369 | $-0.23825190$ | 0.583 |
| 4 | 0.29903319 | 0.29903322 | $-0.29627298$ | 0.667 |
| 5 | 0.38681887 | 0.38681891 | $-0.33506737$ | 0.694 |
| 6 | 0.48348014 | 0.48348017 | $-0.34869152$ | 0.667 |
| 7 | 0.59098524 | 0.59098527 | $-0.33039943$ | 0.583 |
| 8 | 0.71141095 | 0.71141098 | $-0.27243630$ | 0.444 |
| 9 | 0.84696337 | 0.84696339 | $-0.16580509$ | 0.250 |

Having carried through these computations and tabulations all relative to the mesh spacing $h = \frac{1}{4}$, we include also the case $h = \frac{1}{10}$. In this case, the nine values $y(ih) = y_i$ along with its approximations are similarly tabulated.

# APPENDIX B

In this appendix we consider the numerical solution of the following two-dimensional elliptic partial differential equation, typical of those encountered in reactor engineering:

$$(B.1) \qquad -(D(x, y)u_x)_x - (D(x, y)u_y)_y + \sigma(x, y)u(x, y) = S(x, y),$$

$$(x, y) \in R,$$

where $R$ is the square $0 < x, y < 2.1$ shown in Figure 30, with boundary conditions

$$(B.2) \qquad \frac{\partial u(x, y)}{\partial n} = 0, \qquad (x, y) \in \Gamma,$$

where $\Gamma$ is the boundary of $R$. The given functions $D$, $\sigma$, and $S$ are *piecewise constant*, with values given in the table. Thus, we are considering the numerical solution of a problem with *internal interfaces*, which are represented by dotted lines in the figure. This particular problem is then phys-

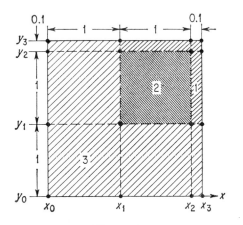

**Figure 30**

**302**

ically made up of *three* different materials. Although any piecewise constant $S(x, y)$ could equally well have been used here, we have specifically chosen $S(x, y) \equiv 0$ only to simplify the discussion of numerical results to follow.

| Region | $D(x, y)$ | $\sigma(x, y)$ | $S(x, y)$ |
|---|---|---|---|
| 1 | 1.0 | 0.02 | 0 |
| 2 | 2.0 | 0.03 | 0 |
| 3 | 3.0 | 0.05 | 0 |

It is clear that a *minimum* of 16 mesh points is necessary to describe these different regions by means of a nonuniform mesh. Using the notation of (6.31) of Sec. 6.3, we now select this minimum mesh denoted by solid circles in the figure, so that

(B.3)   $x_0 = y_0 = 0, \quad x_1 = y_1 = 1, \quad x_2 = y_2 = 2, \quad x_3 = y_3 = 2.1$

With the method of integration of Sec. 6.3 based on a *five-point* formula, we can derive the matrix equation

(B.4)   $$A\mathbf{u} = \mathbf{0},$$

which is our discrete approximation to (B.1)-(B.2). The first purpose of this appendix is to give the *actual* entries of this $16 \times 16$ matrix $A$, which can serve as a check to those derived by the reader. Numbering the mesh points in the *natural ordering*, i.e., as shown in Figure 31, the differ-

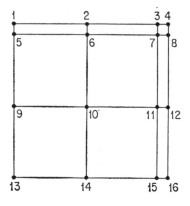

**Figure 31**

ence equations for the following mesh point are given below:

point 1: $15.15125u_1 - 0.15u_2 - 15.0u_5 = 0$;

2: $-0.15u_1 + 20.20175u_2 - 0.05u_3 - 20.0u_6 = 0$;

3: $-0.05u_2 + 6.05055u_3 - 0.5u_4 - 5.5u_7 = 0$;

4: $-0.5u_3 + 1.00005u_4 - 0.5u_8 = 0$;

(B.5) 5: $-15.0u_1 + 18.16375u_5 - 1.65u_6 - 1.5u_9 = 0$;

6: $-20.0u_2 - 1.65u_5 + 25.22175u_6 - 1.05u_7 - 2.5u_{10} = 0$;

7: $-5.5u_3 - 1.05u_6 + 13.10855u_7 - 5.5u_8 - 1.05u_{11} = 0$;

9: $-1.5u_5 + 6.025u_9 - 3.0u_{10} - 1.5u_{13} = 0$;

10: $-2.5u_6 - 3.0u_9 + 11.045u_{10} - 2.5u_{11} - 3.0u_{14} = 0$;

13: $-1.5u_9 + 3.0125u_{13} - 1.5u_{14} = 0$.

Because of the geometrical symmetry of $R$ about the line $x = y$, the remaining difference equations can be deduced from those above, e.g., $a_{15,14} = a_{5,9}$.

The second purpose of this appendix is to give numerical results concerning *numbers* of iterations for various iterative methods. Since the unique vector solution of (B.4) obviously has zero components, then the error in any vector iterate $\mathbf{u}^{(m)}$ arising from an iterative method to solve (B.4) is just the vector itself! To standardize the iterative methods considered, $\mathbf{u}^{(0)}$ was always chosen to have components all equal to $10^4$. Having selected an iterative method, we iterate until we reach the first positive integer $m$ such that *each* component $u_i^{(m)}$ is less than unity in magnitude. Tabulated below are the results of different iterative methods applied to the matrix problem of (B.4).

| Method | Number of iterations | $\omega_b$ or $\omega_\infty$ |
|---|---|---|
| point SOR | 139 | 1.9177 |
| line SOR | 116 | 1.8814 |
| 2-line cyclic Chebyshev | 31 | 1.6777 |

Certainly, from the results in Sec. 6.3, we know that the numbers in the second column must decrease, but we see in this special case that the passage from a point iterative method to a block iterative method can prove to be very beneficial.

# BIBLIOGRAPHY

The following list includes all titles referred to in the text. Numbers in italics following a reference give page numbers of this book where the reference is cited.

The journal abbreviations used here follow the usage of the *Mathematical Reviews*, vol. **17** (1956); 1423–1435.

Achieser, N. I. [1956], *Theory of Approximation*, Frederick Ungar Publishing Co., New York, translated from the Russian by Charles J. Hyman, 307 pp.; *137, 159, 223, 248.*

Aitken, A. C. [1950], "Studies in practical mathematics V. On the iterative solution of a system of linear equations," *Proc. Roy. Soc. Edinburgh* Sec. **A63**, 52–60; *159, 277.*

Alexandroff, P. and Hopf, H. [1935], *Topologie I*, Springer-Verlag, Berlin, 480 pp; *53.*

Allen, D. N. de G. [1954], *Relaxation Methods*, McGraw-Hill Book Co., New York, 257 pp; *24, 95.*

Arms, R. J. and Gates, L. D. [1956], "Iterative methods of solving linear systems, comparisons between point and block iteration," unpublished memorandum from the U. S. Naval Proving Grounds, Dahlgren, Virginia, 28 pp; *25, 96.*

Arms, R. J., Gates, L. D., and Zondek, B. [1956], "A method of block iteration," *J. Soc. Indust. Appl. Math.* **4**, 220–229; *97, 129, 208.*

Bauer, F. L. [1960], "On the definition of condition numbers and on their relation to closed methods for solving linear systems," *Information Processing, Proc. Int. Conf. Information Processing*, Unesco, Paris, 109–110; *95.*

Beckenbach, Edwin F. and Bellman, Richard [1961], *Inequalities*, Springer-Verlag, Berlin, 198 pp; *261, 280.*

Bellar, Fred J., Jr., [1961], "An iterative solution of large-scale systems of simultaneous linear equations," *J. Soc. Indust. Appl. Math.* **9**, 189–193; *159.*

Bellman, Richard, [1953], *Stability Theory of Differential Equations*, McGraw-Hill Book Co., Inc., New York, 166 pp; *279, 280.*

Bellman, Richard, [1960], *Introduction to Matrix Analysis*, McGraw-Hill Book Co., Inc., New York, 328 pp; *2, 15, 53, 54.*

Berge, Claude, [1958], *Théorie des Graphes et ses Applications*, Dunod, Paris, 275 pp; *25.*

Bernstein, Dorothy L., [1950], *Existence Theorems in Partial Differential Equations*, Princeton University Press, Princeton, 228 pp; *252.*

Bickley, W. G., Michaelson, S., and Osborne, M. R. [1961], "On finite-difference methods for the numerical solution of boundary-value problems," *Proc. Roy. Soc. London*, **A262**, 219–236; *207.*

Bilodeau, G. G., Cadwell, W. R., Dorsey, J. P., Fairey, J. M., and Varga, R. S., [1957], "PDQ—an IBM-704 code to solve the two-dimensional few-group neutron-diffusion equations," Report WAPD-TM-70, Bettis Atomic Power Laboratory, Westinghouse

Electric Corp., Pittsburgh, Pa., 66 pp; (Available from the Office of Technical Services, U.S. Dept. of Commerce, Washington, D. C.) *287*.

Birkhoff, Garrett [1957], "Extensions of Jentzsch's Theorem," *Trans. Amer. Math. Soc.* **85**, 219–227; *53*.

Birkhoff, Garrett [1961], "Positivity and criticality," Proceedings of Symposia in Applied Math. **11**, *Nuclear Reactor Theory*, Amer. Math. Soc., Providence, 116–226; *280*.

Birkhoff, Garrett and MacLane, S. [1953], *A Survey of Modern Algebra*, The MacMillan Company, New York, revised edition, 472 pp; *2, 13, 31, 35, 135*.

Birkhoff, Garrett and Varga, Richard S. [1958], "Reactor criticality and non-negative Matrices," *J. Soc. Industr. Appl. Math.* **6**, 354–377; *34, 54, 96, 259, 261, 280*.

Birkhoff, Garrett and Varga, Richard S. [1959], "Implicit alternating direction methods," *Trans. Amer. Math. Soc.* **92**, 13–24; *247, 248, 232*.

Birkhoff, Garrett, Varga, Richard S., and Young, David M., Jr. [1962], "Alternating direction implicit methods," to appear in *Advances in Computers*, Vol. 3, Academic Press, New York; *243, 247, 248*.

Blair, A., Metropolis, N., Neumann, J. v., Taub, A. H., and Tsingou, M. [1959], "A study of a numerical solution of a two-dimensional hydrodynamical problem," *Math. Tables Aids Comput.* **13**, 145–184; *159*.

Blair, P. M. [1960], M. A. Thesis, Rice University; *199, 207*.

Bodewig, E. [1959], *Matrix Calculus*, rev. ed., Interscience Publishers, Inc., New York, 452 pp; *24, 25, 94, 287*.

Bonnesen, T. and Fenchel, W. [1934], *Theorie der Konvexen Körper*, Springer-Verlag, Berlin, 164 pp; *234*.

Bonsall, F. F. [1960], "Positive operators compact in an auxiliary topology," *Pacific J. Math.* **10**, 1131–1138; *53*.

de Boor, Carl and Rice, John R. [1962], "Tchebycheff approximation by $\alpha\Pi[(x-r_j)/(x+r_j)]$ and applications to ADI iteration," to appear in *J. Soc. Indus. Appl. Math*; *248*.

Brauer, A. [1946], "Limits for the characteristic roots of a matrix," *Duke Math. J.* **13**, 387–395; *25*.

Brauer, A. [1947], "Limits for the characteristic roots of a matrix, II," *Duke Math. J.* **14**, 21–26; *22, 25*.

Brauer, A. [1957a], "The theorems of Ledermann and Ostrowski on positive matrices," *Duke Math. J.* **24**, 265–274; *54*.

Brauer, A. [1957b], "A new proof of theorems of Perron and Frobenius on non-negative matrices, I. Positive matrices," *Duke Math. J.* **24**, 367–378; *53*.

Browne, E. T. [1928], "Limits to the characteristic roots of matrices," *Amer. Math. Monthly* **46**, 252–265; *16*.

Bruce, G. H., Peaceman, D. W., Rachford, H. H., and Rice, J. D. [1953], "Calculation of unsteady-state gas flow through porous media," *Trans. Amer. Inst. Mining and Met. Engrs.* **198**, 79–91; *207*.

Callaghan, J. B., Jarvis, P. H., and Rigler, A. K. [1960], "BAFL-1—A program for the solution of thin elastic plate equations on the Philco-2000 computer," WAPD-TM-255, Bettis Atomic Power Laboratory, Pittsburgh, 26 pp; *208*.

Carré, B. A. [1961], "The determination of the optimum accelerating factor for successive over-relaxation," *Computer J.* **4**, 73–78; *297*.

Carslaw, H. S. and Jaeger, J. C. [1959], *Conduction of Heat in Solids*, 2nd ed., Clarendon Press, Oxford, 510 pp; *251, 252*.

Collatz, L. [1942a], "Einschliessungssatz für die characteristischen Zahlen von Matrizen," *Math. Z.* **48**, 221–226; *54*.

Collatz, L. [1942b], "Fehlerabschätzung für das Iterationsverfahren zur Auflösung linearer Gleichungssysteme," *Z. Angew. Math. Mech.* **22**, 357–361; *74, 95.*

Collatz, L. [1950], "Über die Konvergenzkriterien bei Iterationsverfahren für lineare Gleichungssysteme," *Math. Z.* **53**, 149–161; *96.*

Collatz, L. [1952], "Aufgaben monotoner Art," *Arch. Math.* **3**, 366–376; *87, 96.*

Collatz, L. [1955], "Über monotone Systeme linearer Ungleichungen," *J. Reine Angew. Math.* **194**, 193–194; *96.*

Collatz, L. [1960], *The Numerical Treatment of Differential Equations*, 3rd ed., Springer-Verlag, Berlin, 568 pp; *96, 161, 169, 170, 178.*

Conte, Samuel D. and Dames, Ralph T. [1958], "An alternating direction method for solving the biharmonic equation," *Math. Tables Aids Comput.* **12**, 198–205; *248.*

Conte, Samuel D. and Dames, Ralph T. [1960], "On an alternating direction method for solving the plate problem with mixed boundary conditions," *J. Assoc. Comput. Mach.* **7**, 264–273; *248.*

Cornock, A. F. [1954], "The numerical solution of Poisson's and the biharmonic equations by matrices," *Proc. Cambridge Philos. Soc.* **50**, 524–535; *197, 207.*

Courant, R., Friedrichs, K., and Lewy, H. [1928], "Über die partiellen Differenzengleichungen der mathematischen Physik," *Math. Ann.* **100**, 32–74; *206, 268, 280, 293.*

Courant, R. and Hilbert, D. [1953], *Methods of Mathematical Physics*, Vol. 1, Interscience Publishers, Inc., New York, 561 pp; *169.*

Crank, J. and Nicolson, P. [1947], "A practical method for numerical evaluation of solutions of partial differential equations of the heat-conduction type," *Proc. Cambridge Philos. Soc.* **43**, 50–67; *264, 280.*

Cuthill, Elizabeth H. and Varga, Richard S. [1959], "A method of normalized block iteration," *J. Assoc. Comput. Mach.* **6**, 236–244; *199, 207.*

Debreu, G. and Herstein, I. N. [1953], "Non-negative square matrices," *Econometrica*, **21**, 597–607; *19, 53.*

Douglas, Jim, Jr. [1956], "On the relation between stability and convergence in the numerical solution of linear parabolic and hyperbolic differential equations," *J. Soc. Indust. Appl. Math.* **4**, 20–37; *281.*

Douglas, Jim, Jr. [1957], "A note on the alternating direction implicit method for the numerical solution of heat flow problems," *Proc. Amer. Math. Soc.* **8**, 408–412; *247, 248.*

Douglas, Jim, Jr. [1959], "The effect of round-off error in the numerical solution of the heat equation," *J. Assoc. Comput. Mach.* **6**, 48–58; *207.*

Douglas, Jim, Jr. [1961a], "Alternating direction iteration for mildly nonlinear elliptic difference equations," *Numerische Math.* **3**, 92–98; *249.*

Douglas, Jim, Jr. [1961b], "A survey of numerical methods for parabolic differential equations," *Advances in Computers*, Vol. 2, edited by F. L. Alt, Academic Press, New York, 1–54; *254, 280, 281.*

Douglas, Jim, Jr. [1962], "Alternating direction methods for three space variables," *Numerische Math.* **4**, 41–63; *244, 245, 246, 248, 249.*

Douglas, Jim, Jr. and Rachford, H. H., Jr. [1956], "On the numerical solution of heat conduction problems in two or three space variables," *Trans. Amer. Math. Soc.* **82**, 421–439; *209, 240, 247, 248, 281.*

Downing, A. C., Jr. [1960], "On the convergence of steady state multiregion diffusion calculations," Report ORNL-2961, Oak Ridge National Laboratory, Oak Ridge, Tenn., 137 pp; *207.*

DuFort, E. C. and Frankel, S. P. [1953], "Stability conditions in the numerical treatment of parabolic differential equations," *Math. Tables Aids Comput.* **7**, 135–152; *281.*

Dunford, N. and Schwartz, J. T. [1958], *Linear Operators*, Part I, Interscience Pub. Inc., New York, 858 pp; *95.*

Egerváry, E. [1954], "On a lemma of Stieltjes on matrices," *Acta Sci. Math. Szeged.* **15**, 99–103; *96*.

Egerváry, E. [1960], "Über eine Methode zur numerischen Lösung der Poissonschen Differenzengleichung für beliebige Gebiete," *Acta. Math. Acad. Sci. Hungar.* **11**, 341–361; *197, 207*.

Engeli, M., Ginsburg, T., Ruthishauser, H., and Stiefel, E. [1959], *Refined Iterative Methods for Computation of the Solution and the Eigenvalues of Self-Adjoint Boundary Value Problems*, Mitteilungen aus dem Institut für angewandte Mathematik Nr. 8, Birkhäuser Verlag, Basel/Stuttgart, 107 pp; *206, 207*.

Esch, Robin E. [1960], "A necessary and sufficient condition for stability of partial difference equation problems," *J. Assoc. Comp. Mach.* **7**, 163–175; *281*.

Evans, D. J. [1960], "The solution of elliptic difference equations by stationary iterative processes," *Information Processing*, Proc. Int. Conf. Information Processing, Unesco, Paris, 79–85; *248*.

Faddeeva, V. N. [1949], "The method of lines applied to some boundary problems," *Trudy Mat. Inst. Steklov* **28**, 73–103 (Russian); *199, 279*.

Faddeeva, V. N. [1959], *Computational Methods of Linear Algebra*, translated by Curtis D. Benster, Dover Publications, Inc., New York, 252 pp; *2, 10, 24*.

Fan, K. [1958], "Topological proof for certain theorems on matrices with non-negative elements," *Monatsh. Math.* **62**, 219–237; *25; 53, 82, 87, 96*.

Fan, K. and Hoffman, A. J. [1955], "Some metric inequalities in the space of matrices," *Proc. Amer. Math. Soc.* **6**, 111–116; *25*.

Farrington, C. C., Gregory, R. T., and Taub, A. H. [1957], "On the numerical solution of Sturm-Liouville differential equations," *Math. Tables Aids Comput.* **11**, 131–150; *206*.

Fiedler, M. and Pták, V. [1956], "Über die Konvergenz des verallgemeinerten Seidelschen Verfahrens für Lösung von Systemen linearer Gleichungen," *Math. Nachr.* **5**, 31–38; *96*.

Flanders, Donald A. and Shortley, George [1950], "Numerical determination of fundamental modes," *J. Appl. Phys.* **21**, 1326–1332; *137, 159*.

Forsythe, G. E. [1953], "Solving linear algebraic equations can be interesting," *Bull. Amer. Math. Soc.* **59**, 299–329; *159*.

Forsythe, G. E. [1956], "Difference methods on a digital computer for Laplacian boundary value and eigenvalue problems," *Comm. Pure Appl. Math.* **9**, 425–434; *293*.

Forsythe, G. E. and Ortega, J. [1960], "Attempts to determine the optimum factor for successive overrelaxation," *Proc. Int. Conf. Information Processing*, Unesco, Paris, 110; *297*.

Forsythe, G. E. and Wasow, W. R. [1960], *Finite Difference Methods for Partial Differential Equations*, John Wiley and Sons, Inc., New York, London, 444 pp; *24, 130, 161, 153, 170, 206, 288, 296, 297*.

Fox, L. [1957], *The Numerical Solution of Two-Point Boundary Problems in Ordinary Differential Equations*, Oxford University Press, Fair Lawn, N. J., 371 pp; *206, 207, 298*.

Frank, Werner [1960], "Solution of linear systems by Richardson's method," *J. Assoc. Comput. Mach.* **7**, 274–286; *160*.

Frankel, S. P. [1950], "Convergence rates of iterative treatments of partial differential equations," *Math. Tables Aids Comput.* **4**, 65–75; *2, 58, 59, 95, 97, 159, 247, 281*.

Franklin, J. N. [1959a], "Conservative matrices in the numerical solution of elliptic partial differential equation," Lecture Notes (unpublished), California Institute of Technology, 21 pp; *130*.

Franklin, J. N. [1959b], "Numerical stability in digital and analog computation for diffusion problems," *J. Math. Phys.* **37**, 305–315; *279, 281*.

Friedman, B. [1957], "The iterative solution of elliptic difference equations," Report NYO-7698, Institute of Mathematical Sciences, New York University, 37 pp; *105, 130*.

Frobenius, G. [1908], "Über Matrizen aus positiven Elementen," *S.-B. Preuss. Akad. Wiss.* (Berlin), 471–476; *2*.

Frobenius, G. [1909], "Über Matrizen aus positiven Elementen," *S.-B. Preuss Akad. Wiss.*, Berlin, 514–518; *2*.

Frobenius, G. [1912], "Über Matrizen aus nicht negativen Elementen," *S.-B. Preuss. Akad. Wiss.*, Berlin, 456–477; *2, 19, 25, 30, 31, 35, 38, 40, 41, 42, 44, 54, 82, 96*.

Gantmakher, F. R. [1959], *Applications of the Theory of Matrices*, translated and revised by J. L. Brenner, Interscience Publishers, New York, 317 pp; *2, 46, 53, 206*.

Gantmakher, F. R. and Krein, M. G. [1937], "Sur les matrices oscillatoires et complètement non-négatives," *Composito Math.*, **4**, 445–476; *206*.

Garabedian, P. R. [1956], "Estimation of the relaxation factor for small mesh size," *Math. Tables Aids Comput.* **10**, 183–185; *128, 131, 281, 282, 291, 297*.

Gauss, C. F. [1823], Brief und Gerling, Werke, Vol. 9, pp. 278–281. Translated by G. E. Forsythe, *Math. Tables Aids Comput.* **5**, 1951, 255–258; *1, 24*.

Gautschi, Werner [1953a], "The asymptotic behavior of powers of matrices, I," *Duke Math. J.* **20**, 127–140; *24, 95*.

Gautschi, Werner [1953b], "The asymptotic behavior of powers of matrices, II," *Duke Math. J.* **20**, 375–379; *24, 95*.

Gautschi, Werner [1954], "Bounds of matrices with regard to an Hermitian metric," *Composito Math.* **12**, 1–16; *95*.

Geberich, C. L. and Sangren, W. C. [1957], "Codes for the classical membrane problem," *J. Assoc. Comput. Mach.* **4**, 477–486; *293*.

Geiringer, H. [1949], "On the solution of systems of linear equations by certain iterative methods," *Reissner Anniversary Volume*, J. W. Edwards, Ann Arbor, Mich., 365–393; *2, 19, 25, 57, 58, 95, 96, 160*.

Gerling, C. L. [1843], *Die Ausgleichs-Rechnungen der practischen Geometrie oder die Methode der kleinsten Quadrate mit ihren Anwendungen für geodätische Aufgaben*, Gothe, Hamburg; *96*.

Gerschgorin, S. [1930], "Fehlerabschätzung für das Differenzenverfahren zur Lösung partieller Differentialgleichungen," *Z. Angew. Math. Mech.* **10**, 373–382; *165, 206*.

Gerschgorin, S. [1931], "Über die Abrenzung der Eigenwerte einer Matrix," *Izv. Akad. Nauk SSSR Ser. Mat.* **7**, 749–754; *16, 22, 25*.

Glasstone, Samuel and Edlund, Milton C. [1052], *The Elements of Nuclear Reactor Theory*, D. Van Nostrand Co., New York, 416 pp; *251*.

Goldstine, H. H. and von Neumann, J. [1951], "Numerical inverting of matrices of high order, II," *Proc. Amer. Math. Soc.* **2**, 188–202; *95*.

Golub, Gene, H. [1959], "The use of Chebyshev matrix polynomials in the iterative solution of linear equations compared with the method of successive overrelaxation," doctoral thesis, University of Illinois, 133 pp; *144, 159*.

Golub, Gene H. and Varga, Richard S. [1961a], "Chebyshev semi-iterative methods, successive overrelaxation iterative methods, and second order Richardson iterative methods, Part I," *Numerische Math.* **3**, 147–156; *148, 159*.

Golub, Gene H. and Varga, Richard S. [1961b], "Chebyshev semi-iterative methods, successive overrelaxation iterative methods, and second order Richardson iterative methods, Part II," *Numerische Math.* **3**, 157–168; *153, 159, 160*.

Habetler, G. J. [1959], "Concerning the implicit alternating-direction method," Report KAPL-2040, Knolls Atomic Power Laboratory, Schenectady, New York, 13 pp; *248*.

Habetler, G. J. and Martino, M. A. [1961], "Existence theorems and spectral theory for the multigroup diffusion model," Proceedings of Symposia in Applied Math. **11**, *Nuclear Reactor Theory*, Amer. Math. Soc., Providence, 127–139; *280*.

Habetler, G. J. and Wachspress, E. L. [1961], "Symmetric successive over-relaxation in solving diffusion difference equations," *Math. of Comp.* **15**, 356–362; *277, 279, 282*.

Hadamard, Jacques [1952], *Lectures on Cauchy's Problem in Linear Partial Differential Equations*, Dover Publications, New York, 316 pp; *252*.

Hageman, Louis A. [1962], "Block iterative methods for two-cyclic matrix equations with special application to the numerical solution of the second-order self-adjoint elliptic partial differential equation in two dimensions," WAPD-TM-327, Bettis Atomic Power Laboratory, Pittsburgh, 135 pp; *160*.

Harary, Frank [1955], "The number of linear, directed, rosted, and connected graphs," *Trans. Amer. Math. Soc.* **78**, 445–463; *121*.

Harary, Frank [1959], "A graph theoretic method for the complete reduction of a matrix with a view toward finding its eigenvalues," *J. Math. Phys.* **38**, 104–111; *25, 55*.

Harary, Frank [1960], "On the consistency of precedence matrices," *J. Assoc. Comput. Mach.* **7**, 255–259; *25*.

Hartree, D. R. and Womersley, J. R. [1937], "A method for the numerical or mechanical solution of certain types of partial differential equations," *Proc. Roy. Soc. London*, Ser. **A161**, 353–366; *279*.

Heller, J. [1957], "Ordering properties of linear successive iteration schemes applied to multi-diagonal type linear systems," *J. Soc. Indust. Appl. Math.* **5**, 238–243; *130*.

Heller, J. [1960], "Simultaneous, successive, and alternating direction iteration schemes," *J. Soc. Indust. Appl. Math.* **8**, 150–173; *200, 208, 248*.

Henrici, P. [1960], "Estimating the best over-relaxation factor," NN-144, Ramo-Wooddridge Technical Memo, 7 pp; *297*.

Henrici, P. [1962], *Discrete Variable Methods in Ordinary Differential Equations*, John Wiley & Sons, Inc., New York, 407 pp; *162, 207*.

Herstein, I. N. [1954], "A note on primitive matrices," *Amer. Math. Monthly* **61**, 18–20; *41, 54*.

Hildebrandt, T. W. [1955], "On the reality of the eigenvalues for a one-group, $N$-region, diffusion problem," Report 55-6-35, Oak Ridge National Laboratory, Oak Ridge, Tenn., 6 pp; *206*.

Hille, Einar and Phillips, R. S. [1957], *Functional Analysis and Semi-Groups*, revised ed., Amer. Math. Soc. Colloquium Publ. **31**, Amer. Math. Soc., New York, 808 pp; *279*.

von Holdt, Richard E. [1962], "Inversion of triple-diagonal compound matrices," *J. Assoc. Comput. Mach.* **9**, 71–83.

Holladay, J. C. and Varga, R. S. [1958], "On powers of non-negative matrices," *Proc. Amer. Math. Soc.* **9**, 631–634; *42, 43, 54*.

Hotelling, H. [1943], "Some new methods in matrix calculation," *Ann. Math. Statist.* **14**, 1–33; *160*.

Householder, A. S. [1953], *Principles of Numerical Analysis*, McGraw-Hill Book Co., Inc., New York, 274 pp; *195*.

Householder, A. S. [1958], "The approximate solution of matrix problems," *J. Assoc. Comput. Mach.* **5**, 204–243; *10, 24, 25, 53, 96, 281*.

Hummel, P. M. and Seebeck, C. L. [1949], "A generalization of Taylor's Theorem," *Amer. Math. Monthly* **56**, 243–247; *269*.

Jacobi, C. G. J. [1845], "Über eine neue Auflösungsart der bei der Methode der kleinsten Quadrate vorkommenden linearen Gleichungen," *Astr. Nachr.* **22**, No. 523, 297–306; *57*.

John, Fritz [1952], "On integration of parabolic equations by difference methods," *Comm. Pure Appl. Math.* **5**, 155–211; *281*.

Juncosa, M. L. and Mulliken, T. W. [1960], "On the increase of convergence rates of relaxation procedures for elliptic partial difference equations," *J. Assoc. Comput. Mach.* **7**, 29–36; *96*.

Kahan, W. [1958], "Gauss-Seidel methods of solving large systems of linear equations," Doctoral Thesis, University of Toronto; *59, 75, 93, 95, 96, 116, 128, 130, 131, 279*.

Kantorovich, L. V. and Krylov, V. I. [1958], *Approximate Methods of Higher Analysis*, translated from the Russian by Curtis D. Benster, Interscience Publishers Inc., New York, 681 pp; *161, 170, 173, 199, 206, 279*.

Karlin, S. [1959a], "Positive operators," *J. Math. Mech.* **8**, 907–937; *53*.

Karlin, Samuel [1959b], *Mathematical Methods and Theory in Games, Programming, and Economics*, Addison-Wesley Pub. Co., Reading, Mass., 386 pp; *280*.

Karlquist, O. [1952], "Numerical solution of elliptic difference equations by matrix methods," *Tellus* **4**, 374–384; *197, 207*.

Kato, Tosio [1960], "Estimation of iterated matrices, with application to the von Neumann condition," *Numerische Math.* **2**, 22–29; *95*.

Keller, Herbert B. [1958], "On some iterative methods for solving elliptic difference equations," *Quart. Appl. Math.* **16**, 209–226; *57, 130, 208*.

Keller, Herbert B. [1960a], "Special block iterations with applications to Laplace and biharmonic difference equations," *SIAM Rev.* **2**, 277–287; *130, 208*.

Keller, Herbert B. [1960b], "The numerical solution of parabolic partial differential equations," *Mathematical Methods for Digital Computers*, edited by A. Ralston and H. S. Wilf, John Wiley and Sons, Inc., New York; 135–143; *280*.

Kemeny, J. G. and Snell, J. L. [1960], *Finite Markov Chains*, D. Van Nostrand Company, Inc., Princeton, 210 pp; *25, 50, 54*.

Kjellberg, G. [1961], "On the successive over-relaxation method for cyclic operators," *Numerische Math.* **3**, 87–91; *129, 130*.

Kolmogoroff, A. [1934], "Zur Normierbarkeit eines allgemeinen topologischen linearen Raumes," *Studia Math.* **5**, 29–33; *24*.

Kotelyanskii, D. M. [1952], "Some properties of matrices with positive elements," *Mat. Sb.* **31**, (73), 497–506; *96*.

König, D. [1950], *Theorie der endlichen und unendlichen Graphen*, Chelsea Publishing Co., New York, 258 pp; *19, 25*.

Kreiss, Heinz-Otto, [1959a], "Über Matrizen die beschrankte Halbgruppen erzeugen," *Math. Scand.* **7**, 71–80; *261, 280*.

Kreiss, Heinz-Otto [1959b], "Über die Differenzapproximation hoher Genauigkeit bei Anfangswertproblemen für partielle Differentialgleichungen," *Numerische Math.* **1**, 186–202; *281*.

Krein, M. G. and Rutman, M. A. [1948], "Linear operators leaving invariant a cone in a Banach space," *Uspehi Mat. Nauk* **3**, (23), 3–95; *53*.

Kulsrud, H. E. [1961], "A practical technique for the determination of the optimum relaxation factor of the successive over-relaxation method," *Comm. Assoc. Comput. Mach.*, **4**, 184–187; *297*.

Laasonen, P. [1949], "Über eine Methode zur Lösung der Warmeleitungsgleichung," *Acta Math.* **81**, 309–317; *280*.

Lanczos, Cornelius [1950], "An iteration method for the solution of the eigenvalue problem of linear differential and integral operators," *J. Res. Nat. Bur. of Standards* **45**, 255–282; *159*.

Lanczos, Cornelius [1952], "Solution of systems of linear equations by minimized iterations," *J. Res. Nat. Bur. of Standards* **49**, 33–53; *159*.

Lanczos, Cornelius [1956], *Applied Analysis*, Prentice Hall, Inc., Englewood Cliffs, N. J., 539 pp; *159*.

Lanczos, Cornelius [1958], "Iterative solution of large-scale linear systems," *J. Soc. Indust. Appl. Math.* **6**, 91–109; *159*.

Lax, P. D. and Richtmyer, R. D. [1956], "Survey of stability of linear finite difference equations," *Comm. Pure Appl. Math.* **9**, 267–293; *267, 281*.

Lederman, W. [1950], "Bounds for the greatest latent roots of a positive matrix," *J. London Math. Soc.* **25**, 265–268; *54*.

Lees, Milton [1959], "Approximate solution of parabolic equations," *J. Soc. Indust. Appl. Math.* **7**, 167–183; *281*.

Lees, Milton [1960], "A priori estimates for the solutions of difference approximations to parabolic partial differential equations," *Duke Math. J.* **27**, 297–311; *280*.

Lees, Milton [1961], "Alternating direction and semi-explicit difference methods for parabolic partial differential equations," *Numerische Math.* **3**, 398–412; *252, 279*.

Lees, Milton [1962], "A note on the convergence of alternating direction methods," *Math. of Comp.* **16**, 70–75; *247*.

Mac Neal, R. H. [1953], "An asymmetrical finite difference network," *Quart. Appl. Math.* **11**, 295–310; *191, 207*.

Mann, W. Robert, Bradshaw, C. L., and Cox, J. Grady [1957], "Improved approximations to differential equations by difference equations," *J. Math. Phys.* **35**, 408–415; *207*.

Marchuk, G. I. [1959], *Numerical Methods for Nuclear Reactor Calculations*, Consultants Bureau, Inc., New York, 293 pp; (This is an English translation of a work originally published in Russian as Supplement Nos. 3–4 of the *Soviet Journal of Atomic Energy*, Atomic Press, Moscow, 1958.) *206*.

Marcus, M. D. [1955], "A remark on a norm inequality for square matrices," *Proc. Amer. Math. Soc.* **6**, 117–119; *95*.

Mařik Jan and Pták Vlastimil [1960], "Norms, spectra and combinatorial properties of matrices," *Czechoslovak Math. J.* **10**, (85), 181–196; *54*.

Markoff, W. [1892], "Über Polynome, die in einem gegebenen Intervalle möglichst wenig von Null abweichem," *Math. Ann.* **77**, (1916): 213–258, (translation and condensation by J. Grossman of Russian article published in 1892); *159*.

Mehmke, R. [1892], "Über das Seidelsche Verfahren, um lineare Gleichungen bei einer sehr grossen Anzahl der Unbekannten durch sukzessive Annäherungauflösen," *Mosk. Math. Samml.* XVI, 342–345; *94*.

Mewborn, A. C. [1960], "Generalizations of some theorems on positive matrices to completely continuous linear transformations on a normed linear space," *Duke Math. J.* **27**, 273–281; *53, 54*.

Meyer, H. I. and Hollingsworth, B. J. [1957], "A method of inverting large matrices of special form," *Math Tables Aids Comput.* **11**, 94–97; *207*.

Milne, William Edmund [1953], *Numerical Solution of Differential Equations*, John Wiley and Sons, Inc., New York, 275 pp; *159*.

von Mises, R. and H. Pollaczek-Geiringer [1929], "Praktische Verfahren der Gleichungsauflösung," *Z. Angew. Math. Mech.* **9**, 58–77, 152–164; *94, 95, 96*.

Motzkin, T. S. [1949], "Approximation by curves of a unisolvent family," *Bull. Amer. Math. Soc.* **55**, 789–793; *248*.

Muskat, Morris [1937], *The Flow of Homogeneous Fluids Through Porous Media*, McGraw-Hill Book Co., Inc., New York, 763 pp; *251*.

Nekrasov, P. A. [1885], "Die Bestimmung der Unbekannten nach der Methode der kleinsten Quadrate bei einer sehr grossen Anzahl der Unbekannten," *Mat. Sb.* **12**, 189–204, (Russian); *94*.

Nekrasov, P. A. [1892], "Zum Problem der Auflösung von linearen Gleichungssysteme mit einer grossen Anzahl von Unbekannten durch sukzessive Approximationen," *Ber. Petersburger Akad. Wiss.* LXIX, N. 5: 1–18, (Russian); *94*.

von Neumann, J. [1937], "Some matrix inequalities and metrization of matrix-space," *Bull. Inst. Math. Mech. Univ. Tomsk* **1**, 286–299; *25*.

von Neumann, J. and Goldstine, H. H. [1947], "Numerical inverting of matrices of high order," *Bull. Amer. Math. Soc.* **53**, 1021–1099; *95*.

Nohel, J. A. and Timlake, W. P. [1959], "Higher order differences in the numerical solution of two-dimensional neutron diffusion equations," *Proceedings of the Second United Nations International Conference on the Peaceful Uses of Atomic Energy*, United Nations, Geneva **16**, 595–600; *207*.

O'Brien, G. G., Hyman, M. A. and Kaplan, S. [1951], "A study of the numerical solution of partial differential equations," *J. Math. Phys.* **29**, 223–251; *280, 281*.

Oldenburger, R. [1940], "Infinite powers of matrices and characteristic roots," *Duke Math. J.* **6**, 357–361; *25*.

Ostrowski, A. M. [1937], "Uber die Determinanten mit überwiegender Hauptdiagonale," *Comment. Math. Helv.* **10**, 69–96; *82, 96*.

Ostrowski, A. M. [1952], "Bounds for the greatest latent root of a positive matrix," *J. London Math. Soc.* **27**, 253–256; *54*.

Ostrowski, A. M. [1953], "On over and under relaxation in the theory of the cyclic single step iteration," *Math. Tables Aids Comput.* **7**, 152–159; *130*.

Ostrowski, A. M. [1954], "On the linear iteration procedures for symmetric matrices," *Rend. Mat. e Appl.* **14**, 140–163; *24, 75, 77, 96*.

Ostrowski, A. M. [1955], "Über Normen von Matrizen," *Math. Z.* **63**, 2–18; *24*.

Ostrowski, A. M. [1956], "Determinanten mit überwiegender Hauptdiagonale und die absolute Konvergenz von linearen Iterationsprozessen," *Comment. Math. Helv.* **30**, 175–210; *24, 94, 95, 96*.

Ostrowski, A. M. [1960], *Solution of equations and systems of equations*, Academic Press, New York, 202 pp; *24, 54*.

Ostrowski, A. M. [1961], "On the eigenvector belonging to the maximal root of a non-negative matrix," *Proc. Edinburgh Math. Soc.* **12**, 107–112; *54*.

Ostrowski, A. M. and Schneider, H. [1960], "Bounds for the maximal characteristic root of a non-negative irreducible matrix," *Duke Math. J.* **27**, 547–553; *54*.

Padé, H. [1892], "Sur la représentation approchée d'une fonction par des fractions rationnelles," thèse, Ann. de l'Éc. Nor. (3) **9**, 93 pp; *266*.

Parter, Seymour V. [1959], "On 'two-line' iterative methods for the Laplace and biharmonic difference equations," *Numerische Math.* **1**, 240–252; *208*.

Parter, Seymour V. [1961a], "The use of linear graphs in Gauss elimination," *SIAM Rev.* **3**, 119–130; *55*.

Parter, Seymour V. [1961b], "Multi-line iterative methods for elliptic difference equations and fundamental frequencies," *Numerische Math.* **3**, 305–319; *297, 205*.

Parter, Seymour V. [1961c], "Some computational results on 'two-line' iterative methods for the biharmonic difference equation," *J. Assoc. Comput. Mach.* **8**, 359–365; *208*.

Peaceman, D. W. and Rachford, H. H. Jr. [1955], "The numerical solution of parabolic and elliptic differential equations," *J. of Soc. Indust. Appl. Math.* **3**, 28–41; *209, 212, 247, 281*.

Pearcy, C. [1962], "On convergence of alternating direction procedures," *Numerische Math.* **4**, 172–176; *247*.

Perron, O. [1907], "Zur Theorie der Matrizen," *Math. Ann.* **64**, 248–263; *2, 30, 31, 48*.

Phillips, Ralph S. [1961], "Semigroup theory in the theory of partial differential equations," *Modern Mathematics for the Engineer*, second series, McGraw-Hill Book Co., Inc., New York, pp. 100–132; *279, 280*.

Polya, G. [1952], "Sur une interprétation de la méthode des différence finies qui peut fournir des bornes supérieures ou inférieures," *C.R. Acad. Sci. Paris*, **235**, 995–997; *293*.

Polya, G. and Szegö, G. [1951], *Isoperimetric Inequalities in Mathematical Physics*, Princeton University Press, Princeton, 279 pp; *294, 296*.

Potters, M. L. [1955], "A matrix method for the solution of a linear second-order difference equation in two variables," *Math. Centrum Amsterdam, Rap.*, MR-19; *207*.

Price, Harvey S. and Varga, Richard S. [1962], "Recent numerical experiments comparing successive overrelaxation iterative methods with implicit alternating direction methods," to appear as a Gulf Research and Development Co. report, Harmarville, Pa.

Pták, V. [1958], "On a combinatorial theorem and its application to non-negative matrices," *Czechoslovak Math. J.* **8**, (83), 487–495; *42, 54*.

Putnam, C. R. [1958], "On bounded matrices with non-negative elements," *Canad. J. of Math.* **10**, 587–591; *53*.

Reich, E. [1949], "On the convergence of the classical iterative method of solving linear simultaneous equations," *Ann. Math. Statist.* **20**, 448–451; *2, 75, 78, 96*.

de Rham, G. [1952], "Sur un théorème de Stieltjes relatif à certaines matrices," *Acad. Serbe Sci. Publ. Inst. Math.* **4**, 133–134; *96*.

Rice, John R. [1960], "The characterization of best nonlinear Tchebycheff approximations," *Trans. Amer. Math. Soc.* **96**, 322–340; *223, 248*.

Rice, John R. [1961], "Tschebycheff approximations by functions unisolvent of variable degree," *Trans. Amer. Math. Soc.* **99**, 298–302; *223, 248*.

Richardson, L. F. [1910], "The approximate arithmetical solution by finite differences of physical problems involving differential equations, with application to the stress in a masonry dam," *Philos. Trans. Roy. Soc. London*, Ser. **A210**, 307–357; *159, 207, 281*.

Richtmyer, R. D. [1957], *Difference Methods for Initial-Value Problems*, Interscience Publishers Inc., New York, 238 pp; *24, 280, 281*.

Riley, James, D. [1954], "Iteration procedures for the Dirichlet difference problem," *Math. Tables Aids Comput.* **8**, 125–131; *159*.

Romanovsky, V. [1936], "Recherches sur les chaîns de Markoff," *Acta Math.* **66**, 147–251; *19, 40, 49, 54*.

Rosenblatt, D. [1957], "On the graphs and symptotic forms of finite Boolean relation matrices and stochastic matrices," *Naval Res. Logist. Quart.* **4**, 151–167; *42, 54*.

Rota, Gian-Carlo [1961], "On the eigenvalues of positive operators," *Bull. Amer. Math. Soc.* **67**, 556–558; *54*.

Rutherford, D. E. [1952], "Some continuant determinants arising in physics and chemistry II," *Proc. Roy. Soc. Edinburgh*, **A63**, 232–241; *171*.

Samelson, H. [1957], "On the Perron-Frobenius theorem," *Michigan Math. J.* **4**, 57–59; *53*.

Sassenfeld, H. [1951], "Ein hinreichendes Konvergenz-kriterium und eine Fehlerabschätzung für die Iteration in Einzelschritten bei linearen Gleichungen," *Z. Angew. Math. Mech.* **31**, 92–94; *96*.

Schaefer, H. [1960], "Some spectral properties of positive linear operators," *Pacific J. Math.* **10**, 1009–1019; *53*.

Schechter, S. [1959], "Relaxation methods for linear equations," *Comm. Pure Appl. Math.* **12**, 313–335; *24*.

Schechter, S. [1960], "Quasi-tridiagonal matrices and type-insensitive difference equations," *Quart. Appl. Math.* **3**, 285–295; *197, 207*.

Schmeidler, W. [1949], *Vorträge über Determinanten und Matrizen mit Anwendungen in Physik und Technik*, Berlin, 155 pp; *96*.

Schröder, Johann, [1954], Zur Lösung von Potentialaufgaben mit Hilfe des Differenzenverfahren," *Z. Angew. Math. Mech.* **34**, 241–253; *160*.

Shaw, F. S. [1953], *An Introduction to Relaxation Methods*, Dover Publications, New York, 396 pp; *24, 95*.

Sheldon, J. [1955], "On the numerical solution of elliptic difference equations," *Math. Tables Aids Comput.* **9**, 101–112; *277*.

Sheldon, J. W. [1958], "Algebraic approximations for Laplace's equation in the neighborhood of interfaces," *Math. Tables Aids Comput.* **12**, 174–186; *207*.

Sheldon, J. W. [1959], "On the spectral norms of several iterative processes," *J. Assoc. Comput. Mach.* **6**, 494–505; *152, 155, 159, 160*.

Sheldon, J. W. [1960], "Iterative methods for the solution of elliptic partial differential equations," *Mathematical Methods for Digital Computers*, edited by A. Ralston and H. S. Wilf, John Wiley and Sons, Inc., New York, pp. 144–156; *155, 159, 160*.

Shortley, G. [1953], "Use of Tschebyscheff-polynomial operators in the numerical solution of boundary value problems," *J. Appl. Phys.* **24**, 392–396; *159*.

Shortley, G. H. and Weller, R. [1938], "The numerical solution of Laplace's equation," *J. Appl. Phys.* **9**, 334–344; *128, 131*.

Southwell, R. V. [1946], *Relaxation Methods in Theoretical Physics*, Clarendon Press, Oxford, 248 pp; *1, 24, 95, 96*.

Stark, R. H. [1956], "Rates of convergence in numerical solution of the diffusion equation," *J. Assoc. Comput. Mach.* **3**, 29–40; *206, 284*.

Stein, P. [1951], "The convergence of Seidel iterants of nearly symmetric matrices," *Math. Tables Aids Comput.* **5**, 237–240; *96*.

Stein, P. [1952], "Some general theorems on iterants," *J. Res. Nat. Bur. Standards* **48**, 82–83; *16, 25*.

Stein, P. and Rosenberg, R. L. [1948], "On the solution of linear simultaneous equations by iteration," *J. London Math. Soc.* **23**, 111–118; *2, 68, 95*.

Stiefel, E. [1956], "On solving Fredholm integral equations," *J. Soc. Indust. Appl. Math.* **4**, 63–85; *159*.

Stiefel, Eduard, L. [1958], "Kernel polynomials in linear algebra and their numerical applications," *Nat. Bur. Standards Appl. Math. Ser.* **49**, 1–22; *159*.

Stieltjes, T. J. [1887], "Sur les racines de l'equation $X_n = 0$," *Acta Math.* **9**, 385–400; *82, 96*.

Strang, Gilbert W. [1960], "Difference methods for mixed boundary-value problems," *Duke Math. J.* **27**, 221–231; *281*.

Taussky, O. [1948], "Bounds for characteristic roots of matrices," *Duke Math. J.* **15**, 1043–1044; *25*.

Taussky, O. [1949], "A recurring theorem on determinants," *Amer. Math. Monthly.* **56**, 672–676; *24, 25*.

Taussky, O. [1962], "Some topics concerning bounds for eigenvalues of finite matrices," *Survey of Numerical Analysis*, edited by John Todd, McGraw-Hill Book Company, Inc., New York 279–297; *25*.

Temple, G. [1939], "The general theory of relaxation methods applied to linear systems," *Proc. Roy. Soc. London. Ser.* **A169**, 476–500; *24*.

Tikhonov, A. N. and Samarskii, A. A. [1956], "On finite difference methods for equations with discontinuous coefficients," *Doklady Akad. Nauk SSSR* (N.S.) **108**, 393–396; *175, 181, 206*.

Todd, John [1950], "The condition of a certain matrix," *Proc. Cambridge Philos. Soc.* **46**, 116–118; *95, 171*.

Todd, John [1954], "The condition of certain matrices, II," *Arch. Math.* **5**, 249–257; *95*.

Todd, John [1956], "A direct approach to the problem of stability in the numerical solution of partial differential equations," *Comm. Pure Appl. Math.* **9**, 597–612; *280, 281*.

Todd, John [1958], "The condition of certain matrices, III," *J. Res. Nat. Bur. of Standards* **60**, 1–7; *95*.

Tornheim, L. [1950], "On *n*-parameter families of functions and associated convex functions," *Trans. Amer. Math. Soc.* **69**, 457–467; *248*.

Turing, A. M. [1948], "Rounding-off errors in matrix processes," *Quart. J. Mech. Appl. Math.* **1**, 287–308; *95*.

Ullman, J. L. [1952], "On a theorem of Frobenius," *Michigan Math. J.* **1**, 189–193; *53*.

Varga, Richard, S. [1957a], "A comparison of the successive overrelaxation method and semi-iterative methods using Chebyshev polynomials," *J. Soc. Indust. Appl. Math.* **5**, 39–46; *155, 159, 160*.

Varga, Richard S. [1957b], "Numerical solution of the two-group diffusion equation in $x - y$ geometry," *IRE Trans. of the Professional Group on Nuclear Science*, **NS-4**, 52–62; *206, 296*.

Varga, Richard, S. [1959a], "*p*-cyclic matrices: a generalization of the Young-Frankel successive overrelaxation scheme," *Pacific J. Math.* **9**, 617–628; *114, 129, 130, 248*.

Varga, Richard, S. [1959b], "Orderings of the successive overrelaxation scheme," *Pacific J. Math.* **9**, 925–939; *116, 130, 131*.

Varga, Richard, S. [1960a], "Factorization and normalized iterative methods," *Boundary Problems in Differential Equations*, edited by R. E. Langer, University of Wisconsin Press, Madison, pp. 121–142; *96, 208*.

Varga, Richard, S. [1960b], "Overrelaxation applied to implicit alternating direction methods," *Information Processing*, Proc. Int. Conf. Information Processing, Unesco, Paris, 85–90; *217, 247, 248*.

Varga, Richard, S. [1961], "On higher order stable implicit methods for solving parabolic partial differential equations," *J. Math. Physics* **40**, 220–231; *267, 279, 280, 281*.

Wachspress, E. L. [1955], "Iterative methods for solving elliptic-type differential equations with application to two-space-dimension multi-group analysis," Report KAPL-1333, Knolls Atomic Power Laboratory, Schenectady, New York, 75 pp; *155, 160*.

Wachspress, E. L. [1957], "CURE: A generalized two-space-dimension multigroup coding for the IBM-704," Report KAPL-1724, Knolls Atomic Power Laboratory, Schenectady, New York, 132 pp; *223, 247, 248*.

Wachspress, E. L. [1960], "The numerical solution of boundary value problems," *Mathematical Methods for Digital Computers*, ed. by A. Ralston and H. S. Wilf, John Wiley & Sons, Inc., New York, 121–127; *206*.

Wachspress, E. L. [1962], "Optimum alternating-direction-implicit iteration parameters for a model problem," *J. Soc. Indust. Appl. Math.* **10**, 339–350; *223, 224, 248*.

Wachspress, E. L. and Habetler, G. J. [1960], "An alternating-direction-implicit iteration technique," *J. Soc. Indust. Appl. Math.* **8**, 403–424; *240, 242, 243, 246, 248, 290, 297*.

Wachspress, E. L., Stone, P. M. and Lee, C. E. [1958], "Mathematical techniques in two-space-dimension multigroup calculations," *Proceedings of the Second United Nations International Conference on the Peaceful Uses of Atomic Energy*, United Nations, Geneva **16**, 483–488; *159*.

Wall, H. S. [1948], *Analytic Theory of Continued Fractions*, D. van Nostrand Company, Inc., Princeton, 433 pp; *266, 269*.

Wedderburn, J. [1934], *Lectures on Matrices*, Colloquium Publications Amer. Math. Soc., New York **17**, 200 pp; *64*.

Weinberger, H. F. [1956], "Upper and lower bounds for eigenvalues by finite difference methods," *Comm. Pure Appl. Math.* **9**, 613–623; *293*.

Weissinger, J. [1951], "Über das Iterationsverfahren," *Z. Angew. Math. Mech.* **31**, 245–246; *96*.

Weissinger, J. [1952], "Zur Theorie und Anwendung des Iterationsverfahren," *Math. Nachr.* **8**, 193–212; *96*.

Weissinger, J. [1953], "Verallgemeineurungen des Seidelschen Iterationsverfahren," *Z. Angew. Math. Mech.* **33**, 155–163; *96*.

Widder, David V. [1947], *Advanced Calculus*, Prentice-Hall, Inc., Englewood Cliffs, N.J., 432 pp; *184*.

Wielandt, H. [1944], "Bestimmung höheren Eigenwerte durch gebrochene Iteration," *Bericht der aerodynamischen Versuchsanstalt Göttingen*, Report 44/J/37; *290*.

Wielandt, H. [1950], "Unzerlegbare, nicht negativen Matrizen," *Math. Z.* **52**, 642–648; *19, 38, 42, 52, 53, 54*.

Wilf, Herbert, S. [1961], "Perron-Frobenius theory and the zeros of polynomials," *Proc. Amer. Math. Soc.* **12**, 247–250; *48*.

Wilkinson, J. H. [1956], "The calculations of the latent roots and vectors of matrices on the Pilot Model of the A.C.E.," *Proc. Cambridge Philos. Soc.* **50**, 536–566; *290*.

Wilkinson, J. H. [1961], "Error analysis of direct methods of matrix inversion," *J. Assoc. Comput. Mach.* **8**, 281–330; *196, 207*.

Wittmeyer, H. [1936], "Über die Lösung von linearen Gleichungssystemen durch Iteration," *Z. Angew. Math. Mech.* **16**, 301–310; *95*.

Young, David, M. [1950], "Iterative methods for solving partial difference equations of elliptic type," Doctoral Thesis, Harvard University; *2, 93, 95, 97, 106, 110, 129, 130, 131*.

Young, David, M. [1954a], "Iterative methods for solving partial difference equations of elliptic type," *Trans. Amer. Math. Soc.* **76**, 92–111; *67, 95, 97, 114, 130*.

Young, David, M. [1954b], "On Richardson's method for solving linear systems with positive definite matrices," *J. Math. Phys.* **32**, 243–255; *141, 159*.

Young, David, M. [1955], "ORDVAC solutions of the Dirichlet problem," *J. Assoc. Comput. Mach.* **2**, 137–161; *128, 284, 296*.

Young, David, M. [1956a], "On the solution of linear systems by iteration," *Proc. Sixth Sym. in Appl. Math.*, McGraw-Hill Book Co., Inc., New York, pp. 283–298; *159*.

Young, David M. [1956b], "Notes on the alternating direction implicit iterative method," NN-28, 9 pp. and "Notes on an iterative method of Douglas and Rachford," NN-32, 8 pp., Digital Computing Center, The Ramo-Wooldridge Corp; *247, 248*.

Young, David, M. [1961], "The numerical solution of elliptic and parabolic partial differential equations," *Modern Mathematics for the Engineer*, Second Series, McGraw-Hill Company, Inc., New York, pp. 373–419; *280*.

Young, David, M. and Ehrlich, Louis [1960], "Some numerical studies of iterative methods for solving elliptic difference equations," *Boundary Problems in Differential Equations*, University of Wisconsin Press, Madison, pp. 143–162; *227, 229, 245, 248*.

# INDEX